My Book

This book belongs to

Name: _____

Grade 7
Vol 1
Index

Grade 7
Vol 1
Index

Copy right © 2019 MATH-KNOTS LLC

All rights reserved, no part of this publication may be reproduced, stored in any system or transmitted in any form, or by any means, electronic, mechanical, photocopying, recording, or otherwise without the written permission of MATH-KNOTS LLC.

Cover Design by :
Gowri Vemuri

First Edition :
November , 2023

Second Edition : December , 2023

Author :
Gowri Vemuri

Edited by :
Ritvik Pothapragada

Questions: mathknots.help@gmail.com

NOTE : CCSSO or NCTM or VDOE is neither affiliated nor sponsors or

Dedication

This book is dedicated to:
My Mom, who is my best critic, guide and supporter.
To what I am today, and what I am going to become tomorrow,
is all because of your blessings, unconditional affection and support.

This book is dedicated to the
strongest women of my life,
my dearest mom
and
to all those moms in this universe.

G.V.

Grade 7
Vol 1
Index

Week	Page No
Notes	9 - 56
Week 1	57 - 88
Week 2	89 - 114
Week 3	115 - 140
Week 4	141 - 158
Week 5	159 - 188
Week 6	189 - 220
Week 7	221 - 248
Week 8	249 - 268
Week 9	269 - 286
Week 10	287 - 328
Week 11	329 - 344
Week 12	345 - 362
Week 13	363 - 374
Week 14	375 - 402
Week 15	403 - 436
Week 16	437 - 460
Week 17	461 - 478
Week 18	479 - 490
Answer Keys	491 - 526

©All rights reserved-Math-Knots LLC., VA-USA   www.math-knots.com | www.a4ace.com

Grade 7 Volume 1 Notes

Grade 7
Volume 1
Notes

Real Number System

Real number system

Irrational numbers
$\sqrt{2}, \sqrt{3}, \sqrt{5}, ...$

Rational numbers

Integers

Whole numbers

Natural numbers
1, 2, ...

0, 1, 2, 3...

..., -3, -2, -1, 0, 1, 2, 3,...

$-\infty ..... -2, \frac{-3}{2}, -1, \frac{-1}{2}, \frac{-1}{3}, 0, \frac{1}{3}, \frac{1}{2}, 1, \frac{3}{2}, 2...\infty$

Decimals notes

The word standard means regular. Numbers in standard form are whole numbers or natural numbers.

Example : The number "six hundred twenty five" in standard form is 625.

To name a decimal from its standard form, follow these steps :

1. Name the number in front of the decimal. (Do not include the word "and").

2. The word "and" is used for the decimal point.

3. The number in the decimal part is similar to the number in front of the decimal.

4. Name the last place value given (of the digit farther to the right).

Rounding Decimals to a Given Place Value:

To round a decimal number to a given place value, look at the digit to the right of the desired place value and follow the rounding rules:

"5 and above, give it a shove! 4 or below, leave it alone!"

(The number in the desired place value gets "bumped up" to the next consecutive value if the digit to the right of it is 5 or more. The number in the desired place value does not change if the digit to the right of it has a value of 5 or less.)

The purpose of rounding is to provide *an* estimate and get an approximate value. Rounding involves losing some accuracy.

Example : If 5,953 people attend a Soccer game. We can approximately say 6,000 people
watched the game.

Comparing and Ordering Decimals:

To order the given decimals, compare the digits of all decimals according to their place value.

Tip : Line the numbers up vertically according to their place value and compare from the left to the right. When comparing like place values from the left, the number with the higher digit is the larger number.

Also, added zeros to the right of a decimal number does not change the value of the decimal.

For example, 0.8 = 0.80 = 0.800 = 0.8000 = 0.80000

Naming/Reading Decimal :

Read the number to the left of the decimal. Say the word "and" for the decimal place. Read the number to the right of the decimal as you would read it if the number were on the left. End the name with the place value of the digit that is furthest to the right.

Converting Between Fractions and Decimals :

Fractions and decimals represent a certain "part of a whole", all fractions can be written as decimals and terminating/repeating decimals can be written as fractions.

Multiplying Decimals :

To multiply decimals, first ignore the decimals and simply multiply the digits. Then count the total number of spaces from the right to the decimal (or digits to the right of the decimal)
of both numbers; place the decimal that number of spaces from the right into your answer.

Dividing Decimals:

Divide the decimals as a whole numbers ignoring the decimals.

1. Count the number of digits of the dividend after the decimal.
2. Count the number of digits of the divisor.
3. Subtract the number obtained in step 1 from the number obtained instep 2.
4. Count the number of digits from the right to the number obtained in step 3.
5. Place the decimal point.

Integers notes

Integers are a group, or set of numbers that consist of "whole numbers and their opposites"

1. Natural numbers and whole numbers are subset of integers.

2. The set does not include fractions or decimals.

3. The set includes positive and negative numbers.

4. Integers include : $-\infty....., -5, -4, -3, -2, -1, 0, 1, 2, 3, 4, 5 ...... +\infty$

5. Integers greater than zero are called positive integers.

6. Integers less than zero are called negative integers.

7. Zero is neither negative nor positive.

8. Negative integers are the numbers to the left of 0.
   Example : -5, -4, -3

9. Negative numbers have a negative (-) sigh in front of the number.

10. Positive integers are the numbers to the right of 0.

11. Positive numbers do not require the + sign in front.

12. If a number has no sign, it is a positive number.
    Example : 2, 3, 4, 10, 20

13. Negative numbers are frequently used in measurements.
    Example :   To measure temperatures, depth etc
    4^0 C below zero degree celsius is represented as -4^0
    100 ft below sea level is represented as -100 ft.

14. Arrows on a number line represent the numbers continuing for ever.

15. Positive numbers are represented on the right side of zero on the number line.

16. Negative numbers are represented on the left side of zero on the number line.

17. Number are placed at equal intervals on the number line. Not necessarily one unit.

Integers can be represented on a number line as below

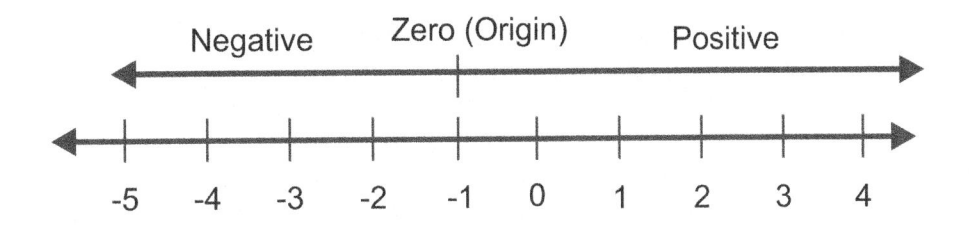

```
          Negative      Zero (Origin)      Positive
    <----|----|----|----|----|----|----|----|----|---->
        -5   -4   -3   -2   -1   0    1    2    3    4
```

Absolute value :

The number line can be used to find the absolute value. The absolute value of an integer is the distance the number is from zero on the number line.

The absolute value of 2 is 2. Using the number line, 2 is a distance of 2 to the right of zero. The absolute value of -2 is also 2. Again using the number line, the distance from -2 to zero is 2. A measure of distance is always positive.

The symbol for absolute value of any number, x, is | x |.

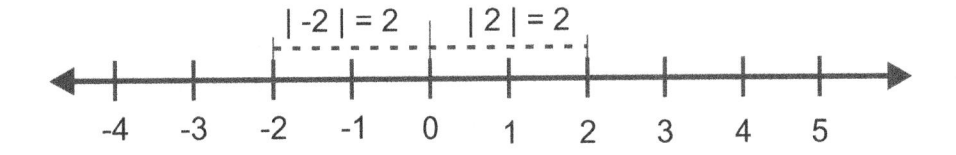

```
                  | -2 | = 2   | 2 | = 2
    <----|----|----|----|----|----|----|----|----|---->
        -4   -3   -2   -1   0    1    2    3    4    5
```

Opposite integers :

The opposite of an integer is the number that is at the same distance from zero in the opposite direction. Every integer has an opposite value, but the opposite of zero is itself.

The opposite of -4 is 4 because it is located the same distance from zero as 4 is, but in opposite direction.

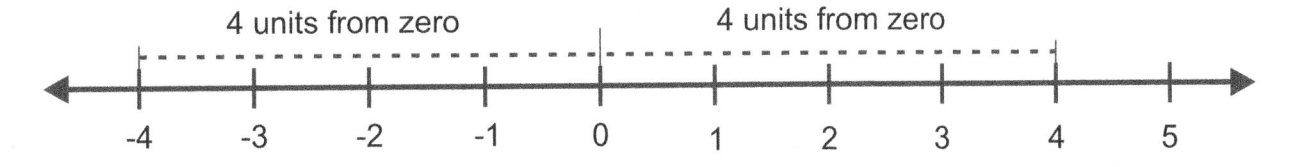

```
         4 units from zero        4 units from zero
    <----|----|----|----|----|----|----|----|----|----|---->
        -4   -3   -2   -1   0    1    2    3    4    5
```

©All rights reserved-Math-Knots LLC., VA-USA   www.math-knots.com | www.a4ace.com

Adding integers using a number line :

The number line is visual representation to understand the addition of positive and negative numbers. Start with the one value on the number line, then add the second value. If the second value (that is added) is positive, we move to the right that many spaces.

If the second value (that is added) is negative, we move to the left that many spaces.
The value where we land on the number line is the solution for the addition of two integers.

Example 1 : (-4) + (5) = 1
Start at the first number, -4, and travel 5 units to the right.

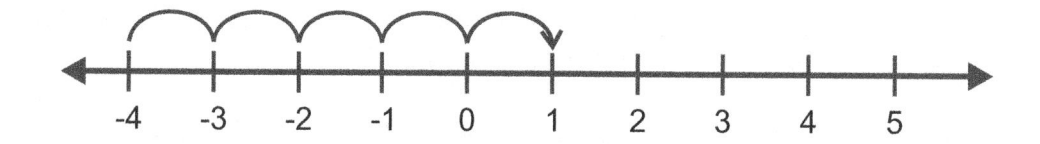

Example 2 : (5) + (-7) = -2
Start at the first number, 5, and travel 7 units to the left.

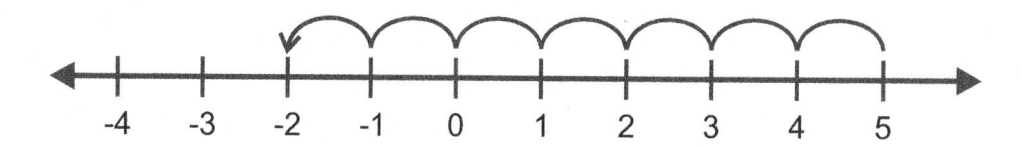

Adding integers using the rules :

Rules for adding integers :

If the signs are the same, add their absolute values, and keep the common sign.

If the signs are different, find the difference between the absolute values of the two numbers, and keep the sign of the number with the greater numerical value.

To the Tune of "Row Your Boat"

Same signs add and keep
Different signs subtract
Keep the sign of the greater digits
then you'll be exact

Subtacting integers using a number line :

A number line is helpful in understanding subtraction of positive and negative values. Start with the first value on the number line, then subtract the second value. If the second value (that is subtracted) is positive, we move to the left that many spaces.

If the second value (that is subtracted) is negative, we move to the right that many spaces. This is because subtraction a negative is the same as adding.
The value where we end on the number line is the answer.

Example 1 : (-2) + (5) = 3
Start at the first number, -2, and travel 5 units to the right.

Subtacting integers using the rules :

Every subtraction problem can be written as an additional problem. When we subtract two integers, just <u>ADD THE OPPOSITE.</u> Subtracting a positive is the same as adding a negative. Subtracting a negative is the same as adding a positive.

Multiplying Integers :

Multiplying integers is same as multiplying whole numbers, except we must keep track of the signs associated to the numbers.

To multiply signed integers, always multiply the absolute values and use these rules to determine the sign of the product value

When we multiply two integers with the same signs, the result is always a positive value.

Positive number X Positive number = Positive number

Negative number X Negative number = Positive number

When we multiply two integers with different signs, the result is always a negative value.

Positive number X Negative number = Negative number

Negative number X Positive number = Negative number

Positive X Positive :	7 X 6 = 42	negative X negative : -7 X -6 = 42
Positive X negative :	7 X -6 = -42	negative X Positive : -7 X 6 = -42

Dividing Integers :

Division of integers is similar to the division of whole numbers, except we must keep track of the signs associated.

To divide signed integers, we must always divide the absolute values and use the below rules to find the quotient value.

When we divide two integers with the same signs, the result is always a positive value.

$$\text{Positive} \div \text{Positive} = \text{Positive}$$

$$\text{Negative} \div \text{Negative} = \text{Positive}$$

When we divide two integers with opposite signs, the result is always a negative value.

$$\text{Positive} \div \text{Negative} = \text{Negative}$$

$$\text{Negative} \div \text{Positive} = \text{Negative}$$

Examples :

Positive ÷ Positive : 81 ÷ 9 = 9   Positive ÷ negative : 81 ÷ -9 = -9

negative ÷ negative : -81 ÷ -9 = 9   negative ÷ Positive : -81 ÷ 9 = -9

Golden Rules of Integers :

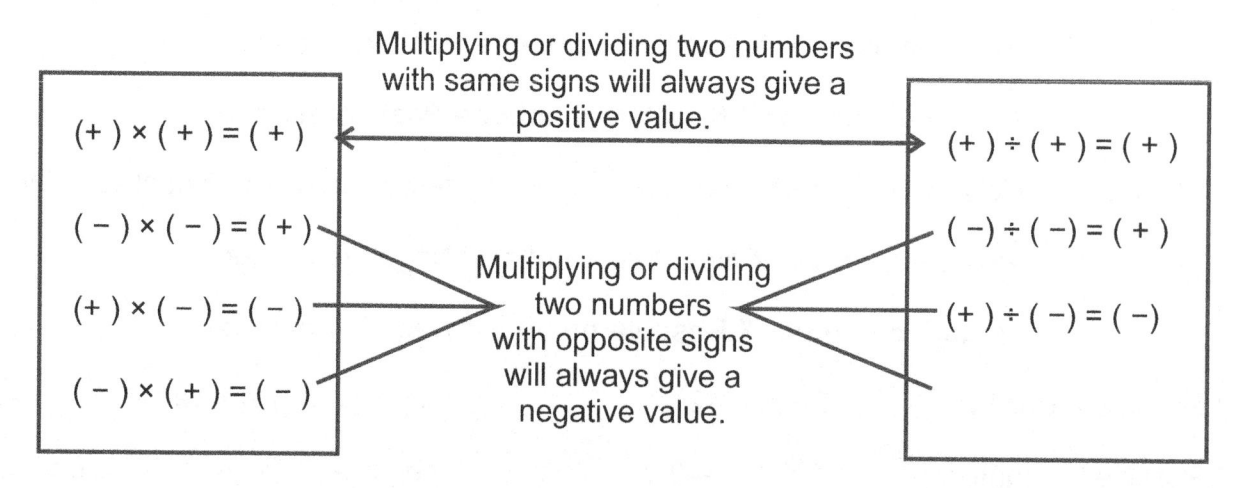

$(+) \times (+) = (+)$ $(+) \div (+) = (+)$

$(-) \times (-) = (+)$ $(-) \div (-) = (+)$

$(+) \times (-) = (-)$ $(+) \div (-) = (-)$

$(-) \times (+) = (-)$

Multiplying or dividing two numbers with same signs will always give a positive value.

Multiplying or dividing two numbers with opposite signs will always give a negative value.

Divisibility:

In general, if two natural numbers a and b are such that, when 'a' is divided by 'b', a remainder of zero is obtained, we say that 'a' is divisible by 'b'.

For example, 12 is divisible by 3 because 12 when divided by 3, the remainder is zero. Also, we say that 12 is not divisible by 5, because 12 when divided by 5, it leaves a remainder 2.

Tests of Divisibility:

We now study the methods to test the divisibility of natural numbers with 2, 3, 4, 5, 6, 8, 9 and 11 without performing actual division.

Test of Divisibility by 2:

A natural number is divisible by 2, if its units digit is divisible by 2, i.e., the units place is either 0 or 2 or 4 or 6 or 8.

Examples : The numbers 4096, 23548 and 34052 are divisible by '2' as they end with 6, 8 and 2 respectively.

Test of Divisibility by 3:

A natural is divisible by 3 if the sum of its digits is divisible by 3.

Example: Consider the number 2143251. The sum of the digits of 2143251 (2 + 1 + 4 + 3 + 2 + 5+1) is 18.

As 18 is divisible by 3, the number 2143251, is divisible by 3.

Test of Divisibility by 4:

A natural number is divisible by 4, if the number formed by its last two digits is divisible by 4.

Examples : 4096, 53216, 548 and 4000 are all divisible by 4 as the numbers formed by taking the last two digits in each case is divisible by 4.

Test of Divisibility by 5:

A natural number is divisible by 5, if its units digit is either 0 or 5.

Examples : The numbers 4095 and 235060 are divisible by 5 as they have in their units place 5 and 0 respectively.

Test of Divisibility by 6:

A number is divisible by 6, if it is divisible by both 2 and 3.

Examples : Consider the number 753618

Since its units digit is 8, so it is divisible by 2. Also its sum of digits = 7 + 5 + 3 + 6 + 1 + 8 = 30, As 30 is divisible by 3, so 753618 is divisible by 3.
Hence 753618 is divisible by 6.

Test of Divisibility by 8:

A number is divisible by 8, if the number formed by its last three digits is divisible by 8.

Examples : 15840, 5432 and 7096 are all divisible by 8 as the numbers formed by last three digits in each case is divisible by 8.

Test of Divisibility by 9:

A natural number is divisible by 9, if the sum of its digits is divisible by 9.

Examples :

(i)   Consider the number 125847.
Sum of digits = 1 + 2 + 5 + 8 + 4 + 7 = 27. As 27 is divisible by 9, the number 125847 is
divisible by 9.

(ii)   Consider the number 145862.
Sum of digits = 1 + 4 + 5 + 8 + 6 + 2 = 26. As 26 not divisible by 9, the number 145862 is not
divisible by 9.

Test of Divisibility by 11:

A number is divisible by 11, if the difference between the sum of the digits in odd places and sum of the digits in even places of the number is either 0 or a multiple of 11.

Examples :
(i)   Consider the number 9582540
Now (sum of digits in odd places) - (sum of digits in even places)
= (9 + 8 + 5 + 0) - (5 + 2 + 4)
= 11, which is divisible by 11.
Hence 958254 is divisible by 11.

(ii)   Consider the number 1453625
Now, (sum of digits at odd places) - (sum of digits at even places)
= (1 + 5 + 6 + 5) - (4 + 3 + 2)

Factors and Multiples :

Lets learn the concepts of factors and multiples.

Definition :

If 'b' divides 'a' leaving a zero remainder, then 'b' is called a factor or divisor of 'a' and 'a' is called the multiple of 'b'.
For example, 6 = 2 3
Here 2 and 3 are factors of 6 (or) 2 and 3 are divisors of 6.
And, 6 is a multiple of 3, 6 is a multiple of 2.

Examples : (i) The factors of 24 = {1, 2, 3, 4, 6, 8, 12, 24}
(ii) The factors of 256 = {1, 2, 4, 8, 16, 32, 64, 128, 256}

Observations :

One is the factor of every natural number and it is the least of the factors of any natural number.
Every natural number is the factor of itself and it is the greatest.

Some Additional Results :

Unique Prime Factorisation Theorem : "Any natural number greater than 1 can be divided into a prime number or a composite number."

For example:

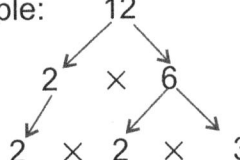

$$12 \\ 2 \times 6 \\ 2 \times 2 \times 3$$

6 is a composite number

All factors are prime numbers.

Greatest Common Divisor [GCD] (or) Greatest Common Factor [GCF]

Definition :

"The greatest common factor of two or more natural numbers is the largest factor in the set of common factors of those numbers." In other words, the GCD (or) GCF of two or more numbers is the largest number that divides each of them exactly.

Example : Find the GCF of 72 and 60.

Solution : Let the set of factors of 72 be A.
A = {1, 2, 3, 4, 6, 8, 9, 12, 18, 24, 36, 72}
Let the set of factors of 60 be B.
B = {1, 2, 3, 4, 5, 6, 10, 12, 15, 20, 30, 60}
The set of common factors for 72 and 60 is A B = {1, 2, 3, 4, 6, 12}
The greatest element in this set is 12
The GCF (or) GCD for 72 and 60 is 2.

Methods of finding GCF :

Factors Method :

When the numbers whose GCF has to be found are relatively small, this is the best suited method. Here we resolve the given numbers into their prime factors and find out the largest factor in the set of common factors to given numbers.

This method can be easily applied to any number of numbers.

Examples : (i)   Find the GCF of 24 and 36.

Solution :

Resolving given numbers into product of prime factors, we have
36 =   ×   × × 3
24 = 2   ×   ×   ×
The common factors to both the numbers are circled.
Now GCF = product common factors of given numbers = 2 × 2 × 3 = 12
GCF (24, 36) = 12

(ii)   Find the GCF of 12, 18 and 24.

Solution :

Resolving given numbers into product of prime factors;
12 = ② × 2 × ③
18 = ② × 3 × ③
24 = ② × 2 × 2 × ③

GCF = product of common factors of 12, 18 and 24
    = 2 × 3 = 6
GCF = 6

Least Common Multiple [LCM] :

Definition :

"The least common multiple of two or more natural numbers is the least of their common multiples". In other words, the LCM of two or more numbers is the least number which can be divided exactly by each of the given numbers.

Example : Find L.C.M. of 24 and 36.

Solution : Resolving 24 and 36 into product of prime factors
24 =   ×   × 2 ×
36 =   ×   × × 3

The common prime factors of 24 and 36 are 2, 2 and 3. (which are circled)

The remaining prime factors of 24 is 2. (which is not circled).

The remaining prime factors of 36 is 3. (which is not circled).

LCM = Common factors of 24 the prime factors left in 24 the prime factors left in 36
= 2 × 2 × 3 × 2 × 3
= 72

Methods of finding LCM :

Factors Method :

Here the given numbers are decomposed into product of prime factors; from which, the least common multiple is found by multiplying the terms containing factors of numbers raised to their highest powers.

Example : Find the LCM of 32 and 24.

Solution : Resolving given numbers into product of common factors, we have
$32 = 2^5$; $24 = 2^3 \times 3$
LCM = product of terms containing highest powers of factors 2, 3
$= 2^5 \; 3 = 96$

LCM of three numbers using factors method:

The method above can be extended in a similar way to three numbers. This is illustrated below:

Example : Find the LCM of 12, 48 and 36

Solution : Resolving given numbers into product of common factors, we have
$12 = 2^2 \times 3^1$; $48 = 2^4 \times 3^1$; $36 = 2^2 \times 3^2$. Then their LCM = $2^2 \times 3^2 = 16 \times 9 = 144$

Synthetic Division Method of Finding LCM:

LCM of numbers can also be found using synthetic division method. This is illustrated below:

Example : Find the LCM of 144 and 156.

Solution : Using synthetic division, we have:

```
2 | 144, 156
2 |  72,  78
3 |  36,  39
  |  12,  13
```

Examples : Find the LCM of 12, 18 and 24

Solution : Using synthetic division;

```
2 | 12, 18, 24
2 |  6,  9, 12
2 |  3,  9,  6
  |  1,  3,  2
```

LCM = 2 2 2 1 3 2 = 48

Relationship between LCM and GCF :

The LCM and GCF of two given numbers are related to the given numbers by the following relationship.
Product of the numbers = LCM × GCF

where, LCM denotes the LCM of the given numbers and GCF denotes the GCF of the given numbers.

Example : Consider two numbers, 24 and 36.
These can be resolved into product of prime factors as below :
$24 = 2^3 \times 3$
$36 = 2^2 \times 3^2$
Now LCM (24, 36) = $2^3 \times 3^2 = 72$
GCF (24, 36) = $2^2 \times 3 = 12$

Now; Product of numbers = 24 × 36 = $2^5 \times 3^3$ = 864
Product of LCM and GCF = 72 × 12 = $2^5 \times 3^3$ = 864

Clearly, Product of the numbers = Product of the LCM and GCF.

Laws of indices:

For all real numbers a and b and all rational numbers m and n, we have

(i) $a^m \times a^n = a^{m+n}$

   Examples: (1) $2^3 \times 2^6 = 2^{3+6} = 2^9$

   (2) $\left(\dfrac{5}{6}\right)^4 \times \left(\dfrac{5}{6}\right)^5 = \left(\dfrac{5}{6}\right)^{4+5} = \left(\dfrac{5}{6}\right)^9$

   (3) $5^{2/3} \times 5^{4/3} = 5^{(2/3 + 4/3)} = 5^{6/3} = 5^2$

   (4) $2^3 \times 2^4 \times 2^5 \times 2^8 = 2^{(3+4+5+8)} = 2^{20}$

   (5) $(\sqrt{7})^3 \times (\sqrt{7})^{\frac{5}{2}} = (\sqrt{7})^{3 + \frac{5}{2}} = (\sqrt{7})^{\frac{11}{2}}$

(ii) $a^m \div a^n = a^{m-n}$, $a \neq 0$

   Examples: (a) $7^8 \div 7^3 = 7^{8-3} = 7^5$

   (b) $\left(\dfrac{7}{3}\right)^9 \div \left(\dfrac{7}{3}\right)^5 = \left(\dfrac{7}{3}\right)^{9-5} = \left(\dfrac{7}{3}\right)^4$

   (c) $9^{\frac{2}{3}} \div 9^{\frac{1}{6}} = 9^{\left(\frac{2}{3} - \frac{1}{6}\right)} = 9^{\left(\frac{4-1}{6}\right)} = 9^{\frac{3}{6}} = 9^{\frac{1}{2}}$

   (d) $\left(\dfrac{5}{7}\right)^{\frac{8}{9}} \div \left(\dfrac{5}{7}\right)^{\frac{1}{3}} = \left(\dfrac{5}{7}\right)^{\left(\frac{8}{9} - \frac{1}{3}\right)} = \left(\dfrac{5}{7}\right)^{\frac{8-3}{9}} = \left(\dfrac{5}{7}\right)^{\frac{5}{9}}$

   Note: $a^n \div a^n = 1$
   or $a^{n-n} = a^0 = 1$
   $\therefore a^0 = 1$, $a \neq 0$

(iii) $(a^m)^n = a^{m \times n}$

   Examples: (a) $(5^2)^3 = 5^{2 \times 3} = 5^6$

   (b) $\left[\left(\dfrac{2}{3}\right)^4\right]^5 = \left(\dfrac{2}{3}\right)^{4 \times 5} = \left(\dfrac{2}{3}\right)^{20}$

   (c) $\left[\left(\dfrac{5}{7}\right)^{\frac{2}{3}}\right]^{\frac{9}{8}} = \left(\dfrac{5}{7}\right)^{\left(\frac{2}{3} \times \frac{9}{8}\right)} = \left(\dfrac{5}{7}\right)^{\frac{3}{4}}$

(iv) $\left(\dfrac{a}{b}\right)^n = \dfrac{a^n}{b^n}$

Example: $\left(\dfrac{4}{5}\right)^7 = \dfrac{4^7}{5^7}$

Note: Conversely we can write $\left(\dfrac{a^n}{b^n}\right) = \left(\dfrac{a}{b}\right)^n$

Example: $\dfrac{8}{27} = \dfrac{2^3}{3^3} = \left(\dfrac{2}{3}\right)^3$

(v) $(ab)^n = a^n \times b^n$

Examples: (a) $20^5 = (4 \times 5)^5 = 4^5 \times 5^5$
(b) $(42)^7 = (2 \times 3 \times 7)^7 = 2^7 \times 3^7 \times 7^7$

Note: Conversely we can write $a^n \times b^n = (ab)^n$

Examples: (a) $4^8 \times 5^8 = (4 \times 5)^8 = 20^8$
(b) $\left(\dfrac{2}{3}\right)^5 \times \left(\dfrac{9}{8}\right)^5 = \left(\dfrac{2}{3} \times \dfrac{9}{8}\right)^5 = \left(\dfrac{3}{4}\right)^5$

(vi) $a^{-n} = \dfrac{1}{a^n}$, $a \neq 0$

Example: $2^{-4} = \dfrac{1}{2^4}$, $5^{-1} = \dfrac{1}{5}$

Note: $a^{-1} = \dfrac{1}{a^1} = \dfrac{1}{a}$

(vii) $\left(\dfrac{a}{b}\right)^n = \left(\dfrac{b}{a}\right)^{n}$

Examples: (a) $\left(\dfrac{5}{9}\right)^3 = \left(\dfrac{9}{5}\right)^{-3}$

(b) $\left(\dfrac{1}{5}\right)^{-1} = \left(\dfrac{5}{1}\right)^1 = 5$

Note: $\left(\dfrac{1}{a}\right)^{-1} = \left(\dfrac{a}{1}\right)^1 = a$

(viii) If $a^m = a^n$, then m = n, where $a \neq 0$, $a \neq 1$

Examples: (a) If $5^p = 5^3 \Rightarrow p = 3$
(b) If $4^p = 256$
$4^p = 4^4 \Rightarrow p = 4$

(ix) For positive numbers a and b, if $a^n = b^n$, $n \neq 0$, then a = b (when n is odd)

Examples: (a) If $5^7 = p^7$, then clearly p = 5.
(b) If $(5)^{2n-1} = (3 \times p)^{2n-1}$, then clearly 5 = 3p or p = 5/3

(x) If $p^m \times q^n \times r^s = p^a q^b r^c$, then m = a, n = b, s = c, where p, q, r are different primes.

Examples: (a) If $40500 = 2^a \times 5^b \times 3^c$, then find $a^a \times b^b \times c^c$

2	40,500
2	20,250
5	10,125
5	2,025
5	405
3	81
3	27
3	9
	3

$\therefore 40500 = 2^2 \times 5^3 \times 3^4 = 2^a \times 5^b \times 3^c$

$\therefore$ a = 2, b = 3, c = 4, [Using the above law].

$\therefore a^a \times b^b \times c^c = 2^2 \times 3^3 \times 4^4 = 27,648$

Example 8 : (a) $20)^5 = (4 \times 5)^5 = 4^5 \times 5^5$

(b) $(42)^7 = (2 \times 3 \times 7)^7 = 2^7 \times 3^7 \times 7^7$

Note: Conversely we can write $a^n \times b^n = (ab)^n$

Example 9 : (a) $\left((5)^3\right)^2 = 5^{3 \times 2} = (5)^6 = 5 \times 5 \times 5 \times 5 \times 5 \times 5 = 15625$

(b) $(2)^3 \times (2)^5 = (2)^{3+5} = (2)^8 = 2 \times 2 \times 2 \times 2 \times 2 \times 2 \times 2 \times 2 = 256$

(c) $(7)^0 = 1$ $\boxed{\text{Any base value rise to the power zero is always equal to 1}}$

(d) $(3)^{-4} = \dfrac{1}{(3)^4} = \dfrac{1}{3 \times 3 \times 3 \times 3} = \dfrac{1}{81}$

(d) $\dfrac{(8)^7}{(8)^5} = (8)^{7-5} = (8)^2 = 64$

(e) $\dfrac{(9)^4}{(9)^7} = (9)^{4-7} = (9)^{-3} = \dfrac{1}{(9)^3} = \dfrac{1}{9 \times 9 \times 9} = \dfrac{1}{729}$

(f) $(2)^4 = 2 \times 2 \times 2 \times 2 = 16$ (g) $(-2)^4 = -2 \times -2 \times -2 \times -2 = 16$

(h) $-(2)^4 = -(2 \times 2 \times 2 \times 2) = -16$ (i) $-(2)^3 = -(2 \times 2 \times 2) = -8$

(j) $(-2)^3 = -2 \times -2 \times -2 = -8$ (k) $-(-2)^3 = -(-2 \times -2 \times -2) = -(-8) = 8$

Tip 1 : When the exponent is an even number the simplified value is always positive, when the base has a positive or negative value.

Tip 2 : When the exponent is an odd number the simplified value is always positive, when the base has a positive value.

Tip 3 : When the exponent is an odd number the simplified value is always negative, when the base has a negative value.

Grade 7
Volume 1
Notes

Fractions Notes

Fractions are part of a whole. It is also an expression representing quotient of two quantities.

Example 1 : $\frac{2}{4}$, $\frac{1}{4}$

Fraction can also be represented as ratios.

Example 2 : 1 : 2 or $\frac{1}{2}$

3 : 7 of $\frac{3}{7}$

Adding simple fractions, follow the below steps :

#1 : Fractions with like denominators can be added by adding their numerators.

Example 3 : $\frac{3}{5} + \frac{1}{5} = \frac{3+1}{5} = \frac{4}{5}$

#3 : To add fractions with unlike denominators, convert the fractions to equivalent fractions with like denominators and follow #1. To convert them into equivalent fractions you can multiply the numerator and denominator with a common factor for each of the fractions to be added separately and then add the fractions.

Method 1 :

Example 4 : $\frac{2}{5} + \frac{1}{2}$

Step 1 : $\frac{2}{5} = \frac{2 \times 2}{5 \times 2} = \frac{4}{10}$

Step 2 : $\frac{1}{2} = \frac{1 \times 5}{2 \times 5} = \frac{5}{10}$

Step 3 : $\frac{2}{5} + \frac{1}{2} = \frac{4}{10} + \frac{5}{10} = \frac{4+5}{10} = \frac{9}{10}$

$\frac{2}{5} + \frac{1}{2} = \frac{9}{10}$

©All rights reserved-Math-Knots LLC., VA-USA   www.math-knots.com | www.a4ace.com

Another method to add fractions with unlike denominators is by finding the **Least Common Multiple** (LCM) of the denominators and then follow the steps as described below in same order.

Method 2 :

Example 5 : $\frac{5}{12} + \frac{1}{8}$

Step 1 :   Find the LCM of the denominators 12 , 8

$$\begin{array}{r|l} 2 & 12, 8 \\ 2 & 6, 4 \\ 3 & 3, 2 \\ 2 & 1, 2 \\ & 1, 1 \end{array}$$

LCM = 2 X 2 X 3 X 2 = 24

Step 2 :   Converting $\frac{5}{12}$ into an equivalent fraction with the denominator as 24 (LCM).

Divide the LCM value with the number in the denominator of the fraction to obtain the common factor.

$$12 \overline{)\begin{array}{c} 2 \\ 24 \\ -24 \\ \hline 0 \end{array}}$$

Common factor obtained is 2

$\frac{5}{12} = \frac{5 \times 2}{12 \times 2} = \frac{10}{24}$; $\frac{5}{12} = \frac{10}{24}$

Step 3 :   Converting $\frac{1}{8}$ into an equivalent fraction with the denominator as 24 (LCM).

Divide the LCM value with the number in the denominator of the fraction to obtain the common factor.

$$8 \overline{)\begin{array}{c} 3 \\ 24 \\ -24 \\ \hline 0 \end{array}}$$

Common factor obtained is 3

$\frac{1}{8} = \frac{1 \times 3}{8 \times 3} = \frac{3}{24}$; $\frac{1}{8} = \frac{3}{24}$

Step 4 :   Substitute the equivalent fractions obtained in step 2 and 3.

$$\frac{5}{12} + \frac{1}{8} = \frac{10}{24} + \frac{3}{24}$$

Step 5 :   The denominators of both fractions are same (Like denominators). Add the numerators.

$$\frac{10 + 3}{24} = \frac{13}{24}$$

Note : To add more than two fractions, repeat step 2 or step 3 for each of the fractions to convert them into equivalent fractions and then proceed with step 4 and 5.

$$\frac{5}{12} + \frac{1}{8} = \frac{13}{24}$$

Subtracting simple fractions, follow the below steps :

#1 :   Fractions with like denominators can be subtracted by subtracting their numerators.

Example 6 : $\frac{3}{5} - \frac{1}{5} = \frac{3-1}{5} = \frac{2}{5}$

#2 :   To subtract fractions with unlike denominators, convert the fractions to equivalent fractions with like denominators and follow #1. To convert them into equivalent fractions you can multiply the numerator and denominator with a common factor for each of the fractions to be subtracted separately and then subtract the fractions.

Method 1 :

Example 7 : $\frac{3}{5} - \frac{1}{2}$

Step 1 : $\frac{3}{5} = \frac{3 \times 2}{5 \times 2} = \frac{6}{10}$

Step 2 : $\frac{1}{2} = \frac{1 \times 5}{2 \times 5} = \frac{5}{10}$

Step 3 : $\frac{3}{5} - \frac{1}{2} = \frac{6}{10} - \frac{5}{10} = \frac{6-5}{10} \quad \frac{1}{10}$

$$\frac{3}{5} - \frac{1}{2} = \frac{1}{10}$$

Another method to subtract fractions with unlike denominators is by finding the **Least Common Multiple** (LCM) of the denominators and then follow the steps as described below in same order.

Method 2 :

Example 8 : $\dfrac{5}{12} - \dfrac{1}{8}$

Step 1 :   Find the LCM of the denominators 12 , 8

$$\begin{array}{r|l} 2 & 12,\ 8 \\ 2 & 6,\ 4 \\ 3 & 3,\ 2 \\ 2 & 1,\ 2 \\ & 1,\ 1 \end{array}$$

LCM = 2 X 2 X 3 X 2 = 24

Step 2 :   Converting $\dfrac{5}{12}$ into an equivalent fraction with the denominator as 24 (LCM).

Divide the LCM value with the number in the denominator of the fraction to obtain the common factor.

$$12\overline{)24} \\ \underline{-24} \\ \ \ 0 \\ \text{(quotient 2)}$$

Common factor obtained is 2

$$\dfrac{5}{12} = \dfrac{5 \times 2}{12 \times 2} = \dfrac{10}{24} \ ; \ \dfrac{5}{12} = \dfrac{10}{24}$$

Step 3 :   Converting $\dfrac{1}{8}$ into an equivalent fraction with the denominator as 24 (LCM).

Divide the LCM value with the number in the denominator of the fraction to obtain the common factor.

$$8\overline{)24} \\ \underline{-24} \\ \ \ 0 \\ \text{(quotient 3)}$$

Common factor obtained is 3

$$\dfrac{1}{8} \quad \dfrac{1 \times 3}{8 \times 3} \quad \dfrac{3}{24} \quad \dfrac{1}{8} \quad \dfrac{3}{24}$$

Step 4 :   Substitute the equivalent fractions obtained in step 2 and 3.

$$\frac{5}{12} - \frac{1}{8} = \frac{10}{24} - \frac{3}{24}$$

Step 5 :   The denominators of both fractions are same (Like denominators). Subtract the numerators.

$$\frac{10-3}{24} = \frac{7}{24}$$

> **Note :** To Subtract more than two fractions, repeat step 2 or step 3 for each of the fractions to convert them into equivalent fractions and then proceed with step 4 and 5.

$$\frac{5}{12} - \frac{1}{8} = \frac{7}{24}$$

Adding Mixed numbers :

#1 :   Add the whole numbers together.

#2 :   Add the fractional parts together. (Find a common denominator if necessary) (Follow the steps as described in the previous pages)

#3 :   Write the whole number obtained from step 1 and the fraction obtained from step 2.

#4 :   If the fractional part is an improper fraction, change it to a mixed number. Add the whole part of the mixed number to the original whole numbers. Rewrite the fraction in the lowest possible value.

Example 9 : $3\frac{2}{7} + 5\frac{1}{7}$

Step 1 :   Add the whole part 3 from $3\frac{2}{7}$ to the whole part 5 from $5\frac{1}{7}$

$3 + 5 = 8$

Step 2 :   Add the fractional part $\frac{2}{7}$ from $3\frac{2}{7}$ to the fractional part $\frac{1}{7}$ from $5\frac{1}{7}$

$$\frac{2}{7} + \frac{1}{7} = \frac{2+1}{7} = \frac{3}{7}$$

$$3\frac{2}{7} + 5\frac{1}{7} = 8\frac{3}{7}$$

Example 10 : $4\frac{1}{6} + 5\frac{5}{6}$

Step 1 :   Add the whole part 4 from $4\frac{1}{6}$ to the whole part 5 from $5\frac{5}{6}$

$4 + 5 = 9$

Step 2 :   Add the fractional part $\frac{1}{6}$ from $4\frac{1}{6}$ to the fractional part $\frac{5}{6}$ from $5\frac{5}{6}$

$\frac{1}{6} + \frac{5}{6} = \frac{1+5}{6} = \frac{6}{6}$ (Numerators are added for fractions from like denominators)

Step 3 : $4\frac{1}{6} + 5\frac{5}{6} = 9\frac{6}{6} = 10$

($\frac{6}{6}$ = one whole part, add this one whole part to 9 making it equal to 10)

Example 11 : $7\frac{1}{2} + 5\frac{3}{20}$

Step 1 :   Add the whole part 7 from $7\frac{1}{2}$ to the whole part 5 from $5\frac{3}{20}$

$7 + 5 = 12$

Step 2 :   Add the fractional part $\frac{1}{2}$ from $7\frac{1}{2}$ to the fractional part $\frac{3}{20}$ from $5\frac{3}{20}$

$\frac{1}{2} + \frac{3}{20}$ (The fractions has unlike denominators)

Finding the LCM of 2 and 20

```
2 | 2 , 20
2 | 1 , 10     LCM of 2 and 20 = 2 X 2 X 5 = 20
5 | 1 , 5
    1 , 1
```

Step 3 :   Let's make $\frac{1}{2}$ as an equivalent fraction with a denominator of 20.

```
     10
 2) 20      Common factor is 10
   -20
     0
```

$$\frac{1}{2} = \frac{1 \times 10}{2 \times 10} = \frac{10}{20}$$

Step 4 : The fractional part $\frac{3}{20}$ of $5\frac{3}{20}$ has the same common denominator as 20.

We do not need to convert $\frac{3}{20}$ into another equivalent fraction.

Remember : The fractions can vary from problem to problem and students need to follow step 3 for all the fractional parts to convert them to equivalent fractions.

Step 5 : $\frac{1}{2} + \frac{3}{20} = \frac{10}{20} + \frac{3}{20} = \frac{10+3}{20} = \frac{13}{20}$

Step 6 : $7\frac{1}{2} + 5\frac{3}{20} = 8\frac{13}{20}$

Subtracting Mixed numbers :

#1 : Subtract the whole numbers together.

#2 : Subtract the fractional parts together. (Find a common denominator if necessary) (Follow the steps as described in the previous pages)

#3 : Write the whole number obtained from step 1 and the fraction obtained from step 2.

#4 : If the fractional part is an improper fraction, change it to a mixed number. Add the whole part of the mixed number to the whole number obtained in step 1. Rewrite the fraction in the lowest possible value.

Example 12 : $9\frac{2}{7} - 5\frac{1}{7}$

Step 1 : Subtract the whole part 5 from $5\frac{1}{7}$ from the whole part 9 from $3\frac{2}{7}$

$9 - 5 = 4$

Step 2 : Subtract the fractional part $\frac{1}{7}$ from $5\frac{1}{7}$ from the fractional part $\frac{2}{7}$ from $3\frac{2}{7}$

$\frac{2}{7} - \frac{1}{7} = \frac{2-1}{7} = \frac{1}{7}$

Step 3 : $9\frac{2}{7} - 5\frac{1}{7} = 4\frac{1}{7}$

Example 13 : $7\dfrac{5}{6} - 5\dfrac{1}{6}$

Step 1 : Subtract the whole part 5 from $5\dfrac{1}{6}$ from the whole part 7 from $7\dfrac{1}{6}$

7 - 5 = 2

Step 2 : Subtract the fractional part $\dfrac{1}{6}$ from $5\dfrac{1}{6}$ to the fractional part $\dfrac{5}{6}$ from $7\dfrac{5}{6}$

$\dfrac{5}{6} - \dfrac{1}{6} = \dfrac{5-1}{6} = \dfrac{4}{6}$ (Numerators are added for fractions from like denominators)

Step 3 : $7\dfrac{5}{6} - 5\dfrac{1}{6} = 2\dfrac{4}{6} = 2\dfrac{2}{3}$

($\dfrac{4}{6} = \dfrac{2}{3}$ Equivalent fractions)

Example 14 : $7\dfrac{1}{2} - 5\dfrac{3}{20}$

Step 1 : Subtract the whole part 5 from $5\dfrac{3}{20}$ from the whole part 7 from $7\dfrac{1}{2}$

7 - 5 = 2

Step 2 : Subtract the fractional part $\dfrac{3}{20}$ from $5\dfrac{3}{20}$ to the fractional part $\dfrac{1}{2}$ from $7\dfrac{1}{2}$

$\dfrac{3}{20} - \dfrac{1}{2}$ (The fractions has unlike denominators)

Finding the LCM of 2 and 20

```
2 | 2 , 20
2 | 1 , 10     LCM of 2 and 20 = 2 X 2 X 5 = 20
5 | 1 , 5
    1 , 1
```

Step 3 : Let's make $\dfrac{1}{2}$ as an equivalent fraction with a denominator of 20.

```
      10
   2) 20     Common factor is 10
    - 20
       0
```

$$\frac{1}{2} = \frac{1 \times 10}{2 \times 10} = \frac{10}{20}$$

Step 4 : The fractional part $\frac{3}{20}$ of $5\frac{3}{20}$ has the same common denominator as 20.

We do not need to convert $\frac{3}{20}$ into another equivalent fraction.

Remember : The fractions can vary from problem to problem and students need to follow step 3 for all the fractional parts to convert them to equivalent fractions.

Step 5 : $\frac{1}{2} - \frac{3}{20} = \frac{10}{20} - \frac{3}{20} = \frac{10-3}{20} = \frac{7}{20}$

Step 6 : $7\frac{1}{2} - 5\frac{3}{20} = 2\frac{7}{20}$

Example 15 : $5\frac{1}{7} - 3\frac{3}{7}$

Step 1 : Subtract the whole part 3 from $3\frac{3}{7}$ from the whole part 5 from $5\frac{1}{7}$

5 - 3 = 2

Step 2 : Subtract the fractional part $\frac{3}{7}$ from $3\frac{3}{7}$ from the fractional part $\frac{1}{7}$ from $5\frac{1}{7}$

We cannot subtract $\frac{3}{7}$ from $\frac{1}{7}$

So we need to rewrite the fraction $5\frac{1}{7}$

$5\frac{1}{7} = 4\frac{8}{7}$ (Remember in this fraction one whole part equals to seven. so when we take one whole part into the fraction form we need to add 7 to the value in the numerator which equals to 7 + 1 = 8)

Step 3 : Repeat step 1

$5\frac{1}{7} - 3\frac{3}{7} = 4\frac{8}{7} - 3\frac{3}{7}$

Subtract the whole part 3 from $3\frac{3}{7}$ from the whole part 4 from $4\frac{8}{7}$

4 - 3 = 1

Step 4 :   Subtract the fractional part $\frac{3}{7}$ from $3\frac{3}{7}$ from the fractional part $\frac{8}{7}$ from $4\frac{8}{7}$

$$\frac{8}{7} - \frac{3}{7} = \frac{8-3}{7} = \frac{5}{7}$$

Step 5 : $5\frac{1}{7} - 3\frac{3}{7} = \frac{5}{7}$

Multiplying Fractions :

#1 :   Verify if the fractions are in lowest possible values. If not convert them into lowest possible values.

#2 :   Using cross simplification method simplify the fractions, meaning a numerator can be simplifies with a denominator only and vice versa.

#3 :   Do not cross simplify numerator with a numerator value and denominator with a denominator value

#4 :   Multiply the numerator with the remaining numerator values and the denominator with the denominator values

Remember : "Top times the top over the bottom times the bottom".
All the answers must be written in simplest form.

Example 16 : $\frac{6}{15} \times \frac{3}{10}$

$$\frac{\overset{2}{\cancel{6}}}{\underset{5}{\cancel{15}}} \times \frac{3}{10} = \frac{\overset{1}{\cancel{2}}}{5} \times \frac{3}{\underset{5}{\cancel{10}}} = \frac{1}{5} \times \frac{3}{5}$$

$$= \frac{1 \times 3}{5 \times 5} = \frac{3}{25}$$

Multiplying Mixed Numbers :

#1 : To multiply mixed numbers, convert them to improper fractions.

Converting mixed number to improper fractions :
Multiply the denominator of the fraction to the whole part and then add the product to the numerator.

Example 17 : $2\dfrac{1}{3} = \dfrac{2 \times 3 + 1}{3} = \dfrac{6+1}{3} = \dfrac{7}{3}$

#2 : Verify if the fractions are in lowest possible values. If not convert them into lowest possible values.

#3 : Using cross simplification method simplify the fractions, meaning a numerator can be simplifies with a denominator only and vice versa.

#4 : Do not cross simplify numerator with a numerator value and denominator with a denominator value

#5 : Multiply the numerator with the remaining numerator values and the denominator with the denominator values

Remember : "Top times the top over the bottom times the bottom".
All the answers must be written in simplest form.
All improper fractions must be change back to mixed numbers.

Note : MULTIPLICATION CAN BE WRITTEN WITH THE SYMBOLS X OR . IN BETWEEN, .

**Grade 7
Volume 1
Notes**

Dividing Fractions :

#1 : To divide fractions, convert the division problem into a multiplication problem.
Do this by multiplying the first fraction by the reciprocal of the second fraction.
In other words convert the division to multiplication and interchange the numerator and denominator of the second fraction.

Remember : "When two fractions we divide, flip the second and multiply."

Note : Don't forget to check for "cross simplification" when multiplying.
All answers must be written in simplest form.
All improper fractions must be change back to mixed numbers.

Example 18 : $\frac{6}{15} \div \frac{3}{10}$

$$\frac{6}{15} \div \frac{3}{10} = \frac{\overset{2}{\cancel{6}}}{\underset{5}{\cancel{15}}} \times \frac{10}{3} = \frac{2}{5} \times \frac{\overset{2}{\cancel{10}}}{3} = \frac{2}{1} \times \frac{2}{3}$$

$$= \frac{2 \times 2}{1 \times 3} = \frac{4}{3}$$

Dividing Mixed Numbers :

#1 : To divide mixed fractions, first change them to improper fractions.

#2 : To divide fractions, convert the division problem into a multiplication problem.
Do this by multiplying the first fraction by the reciprocal of the second fraction.
In other words convert the division to multiplication and interchange the numerator and denominator of the second fraction.

Remember : "When two fractions we divide, flip the second and multiply."

Note : Don't forget to check for "cross simplification" when multiplying.
All answers must be written in simplest form.
All improper fractions must be change back to mixed numbers.

Example 19 : $2\frac{6}{15} \div 5\frac{3}{10}$

$$2\frac{6}{15} \div 5\frac{3}{10} = \frac{2 \times 15 + 6}{15} \div \frac{5 \times 10 + 3}{10} = \frac{30 + 6}{15} \div \frac{50 + 3}{10}$$

$$= \frac{30+6}{15} \div \frac{50+3}{10}$$

$$= \frac{36}{15} \div \frac{53}{10}$$

$$= \frac{36}{15} \times \frac{10}{53} \quad \longleftarrow \text{(Remember when we change division to multiplication, flip the second fraction)}$$

$$= \frac{\cancel{36}^{12}}{\cancel{15}_{5}} \times \frac{10}{53}$$

$$= \frac{12}{\cancel{5}_1} \times \frac{\cancel{10}^2}{53}$$

$$= \frac{12}{1} \times \frac{2}{53}$$

$$= \frac{12 \times 2}{1 \times 53}$$

$$= \frac{24}{53}$$

$$2\frac{6}{15} \div 5\frac{3}{10} = \frac{24}{53}$$

The percentage symbol is a representation of percentage. In statistics, percentages are often left in their base form of 0 - 1, where 1 represents the whole. We multiply the decimal by a factor of 100 to find the percentage.

Percent Equation form :

Percentages can be setup as proportions. The parts of a percent can be found by setting up a proportion as below.

$$\frac{is}{of} = \frac{percent}{100} \quad OR \quad \frac{part}{total} = \frac{percent}{100}$$

Before we calculate a percentage, we should understand exactly what a percentage is ? The word percentage comes from the word percent. If you split the word percent into its root words, you see "per" and "cent." Cent is an old European word with French, Latin, and Italian origins meaning "hundred". So, percent is translated directly to "per hundred." If we have 39 percent, we literally have 39 per 100. If it snowed 16 times in the last 100 days, it snowed 16 percent of the time.

Whole numbers, decimals and fractions are converted to percentages. Decimal format is easier to convert into a percentage. Fractions can be converted into decimals and then to percentages.

Converting a decimal to a percentage is as simple as multiplying it by 100. To convert 0.92 to a percent, simply multiple 0.92 by 100.

0.92 × 100 = 92%

A percentage is an expression of part of the whole. Nothing is represented by 0%, and the whole amount is 100%. Everything else is somewhere in between 0 and 100.

For example, say you have 20 muffins. If you share 12 muffins with your friends, then you have shared 12 out of the 20 muffins ($\frac{12}{20}$ × 100% = 60% shared). If 20 muffins are 100% and you share 60% of them, then 100% - 60% = 40% of the muffins are still left.

Determine the part of the whole :

Given the value for part of the whole and the whole.
OR
Given two parts that make up the whole.

It is important to differentiate what the percentage is "of."
Example : A jar contains 99 gold beads and 51 blue beads. A total of 150 beads are in the jar. In this example, 150 beads makes up a whole jar of beads, i.e. 100%.

1. **Put the two values into a fraction.** The part goes on top of the fraction (numerator), and the whole goes on the bottom (denominator). Therefore the fraction in this case is $\frac{99}{150}$ (part/whole) or $\frac{51}{150}$ (part/whole).

2. **Convert the fraction into a decimal.** Convert $\frac{99}{150}$ into decimal, divide 99 by 150

    $\frac{99}{150} = 0.66$

Converting the decimal into a percent :

Multiply the result obtained in the step above by 100% (per 100 = *per cent*).
0.66 multiplied by 100 equals 66%. Gold beads are 66% of the total beads in the jar.

Converting the percentage into a decimal : Working backward from before, divide the percentage by 100, or you can multiply by 0.01.

$66\% = \frac{66}{100} = 0.66$

Re-word the problem with your new values :

Given in the form of "**X** of **Y** is **Z**." X is the decimal form of your percent, "of" means to multiply, Y is the whole amount, and Z is the answer.

Example: 3% of $25 is 0.75.
The amount of interest accrued each day on a 3% of loan amount $25 s 0.75

Discounts :

Discounts are offered on the original price of the item on sale.

1. The discount percent need to be subtracted from the whole, meaning if a discount of 20% is offered on an item priced at $60 then sale price is 80% of $60.
   100% - 20% = 80% (whole percent - discount percent).

   80% of $60 = $\frac{80}{100}$ X 60 = $48

Like Terms :

Two or more terms are said to be alike if they have the same variable and the same degree.
Coefficients of like terms are not necessarily be same.

An expression is in its simplest form when

1. All like terms are combined.
2. All parentheses are opened and simplified.

Like Terms can combined by adding or subtracting their coefficients (pay attention to the positive and negative signs of the coefficient and apply rules of adding integers)

Example 1 : $-2x + 5x + 7 = 3x + 7$

> Note : $-2x$ and $5x$ are like terms and can be combined using rules of integers

Example 2 : $-11y + 5 + 8y - 7 = -3y - 2$

> Note : $-11y$ and $8y$ are like terms and can be combined using rules of integers. 5 and -7 are like terms and can be combined using rules of integers

Example 3 : $-12a - 5a + 8 - 3 = -17a + 5$

> Note : $-12a$ and $-5a$ are like terms and can be combined using rules of integers. 8 and -3 are like terms and can be combined using rules of integers

Example 4 : $-5b + 7 - 3b + 2a - a + 10 = -8b + a + 17$

> Note : $-5b$ and $-3b$ are like terms and can be combined using rules of integers. $2a$ and $-a$ are like terms and can be combined using rules of integers. 7 and 10 are like terms and can be combined using rules of integers.

Distributing with the negative sign :

Remember to apply the integer rules of positive and negative numbers while distributing.

Combining like terms on the opposite side of the equal sign :

When the like terms are on opposite sides, we have to combine like terms by using the inverse operation and by undoing the equation.

Example 5 :   -2x + 5 = -7x

$$-2x + 5 = -7x$$
$$+7x \qquad\ \ 7x$$
$$\overline{}$$
$$-2x +7x + 5 = -7x+ 7x$$
$$5x + 5 = 0$$

Solving equations using the distributive property :

The number in front of the parentheses needs to be multiplied with every term within the parentheses. After the distribution and opening up the parentheses, combine like terms and solve.

Distributing with the negative sign :

Remember to apply the integer rules of positive and negative numbers while distributing.

```
+ X + = +
- X - = +
- X + = -
+ X - = -
```

Example 6 :   (a) 2(5x + 7) = 2(5x) + 2(7) = 10x + 14
(b) -7(3a + 8) = (-7)(3a) + (-7)(8) = -21a + (-56) = -21a - 56
(c) 3(-5b - 2) = (3)(-5b) - (3)(2) = -15b - 6
(d) 6(-4a + 5) = (6)(-4a) + (6)(5) = -24a + 30
(e) 4(2a - 8) = (4)(2a) - (4)(8) = 8a - 32
(f) -5(a - 7) = (-5)(a) - (-5)(7) = -5a - (-35) = -5a + 35
(g) -9(-2a + 10) = (-9)(-2a) + (-9)(10) = 18a + (-90) = 18a - 90
(h) -8(-5a - 6) = (-8)(-5a) - (-8)(6) = 40a - (-48) = 40a + 40a
(i) -(a + 7) = -a - 7
(j) -(x - 5) = (-1)(x) - (-1)(5) = -x - (-5) = -x + 5
(k) -(-a - b) = (-1)(-a) - (-1)(b) = a - (-b) = a + b
(l) -(-a + 2b) = (-1)(-a) + (-1)(2b) = a + (-2b) = a - 2b

Example 7 : $2x + 3 = x + 7$
$2x + 3 = x + 7$
$\underline{-x \ -3 \ -x \ -3}$
$x + 0 = 0 + 4$
$x = 4$

(Inverse operation for addition is subtraction)

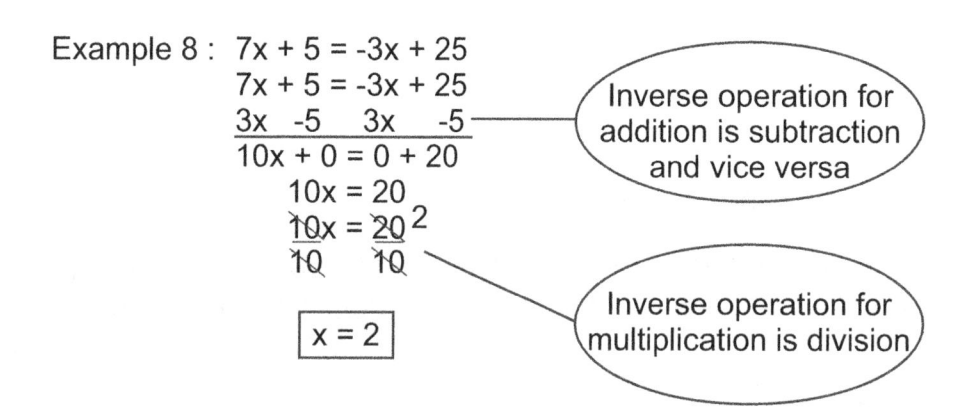

Example 8 : $7x + 5 = -3x + 25$
$7x + 5 = -3x + 25$
$\underline{3x \ -5 \ \ \ 3x \ \ -5}$
$10x + 0 = 0 + 20$
$10x = 20$
$\dfrac{10x}{10} = \dfrac{20}{10}^{2}$
$\boxed{x = 2}$

(Inverse operation for addition is subtraction and vice versa)

(Inverse operation for multiplication is division)

Example 9 : $\dfrac{2x}{5} + 5 = 15$

$\dfrac{2x}{5} + 5 = 15$
$\underline{\phantom{\dfrac{2x}{5}} \ -5 \ \ -5}$
$\dfrac{2x}{5} + 0 = 10$

$\dfrac{2x}{5} = 10$

$5 \cdot \dfrac{2x}{5} = 5 \cdot 10$

$\dfrac{2x}{2} = \dfrac{50}{2}^{25}$

$\boxed{x = 25}$

(Inverse operation for addition is subtraction and vice versa)

(Inverse operation for division is multiplication)

(Inverse operation for multiplication is division)

Inequality :

An inequality is a relation between two expressions that are not equal. As a mathematical statement an inequality states one side of the equation is less than, less than or equal to or greater than or greater than equal to the other side.

If the inequality has **less than** or **greater than** symbol,
1. The graph starts with the open circle.
2. For less than the graphing line goes toward the left.
3. For greater than the graphing line goes toward the right.

If the inequality has **less than or equal to** or **greater than or equal** to symbol,
1. The graph starts with the closed circle.
2. For less than or equal to the graphing line goes toward the left.
3. For greater than or equal to the graphing line goes toward the right.

Inequality statement	Inequality verbal expression	Inequality graph
x > -3 or -3 < x	x is greater than -3	(number line: open circle at -3, line going right; -4 -3 -2 -1 0 1 2 3)
x < 3 or 3 > x	x is less than 3	(number line: open circle at 3, line going left; -3 -2 -1 0 1 2 3 4)
x >= -1 or -1 <= x	x is greater than or equal to -1	(number line: closed circle at -1, line going right; -3 -2 -1 0 1 2 3)
x <= 1 or 1 <= x	x is less than or equal to 1	(number line: closed circle at 1, line going left; -3 -2 -1 0 1 2 3)

Basic inequalities :

Solving inequalities is same as solving for an equation except for one special rule.

Example 10 : x + 9 > 11
Step 1 (subtract 9 from both sides) : x + 9 - 9 > 11 - 9
Step 2 (combine like terms) : x > 2

Example 11 : 2x + 5 > 10
Step 1 (subtract 5 from both sides) : 2x + 5 - 5 > 10 - 5
Step 2 (combine like terms) : 2x > 5
Step 3 (divide both sides by the coefficient of x which is 2) : $\frac{\cancel{2}^1 x}{\cancel{2}_1} > \frac{5}{2}$

Step 4 (simplify both sides as needed) : $x > \frac{5}{2}$

Example 12 : 5x - 1 > 9
Step 1 (add 1 to both sides) : 5x - 1 + 1 > 9 + 1
Step 2 (combine like terms) : 5x > 10
Step 3 (divide both sides by the coefficient of x which is 5) : $\frac{\cancel{5}^1 x}{\cancel{5}_1} > \frac{10}{5}$

Step 4 (simplify both sides as needed) : x > 2

Example 13 : 2x - 8 > -11
Step 1 (add 8 to both sides) : 2x - 8 + 8 > -11 + 8
Step 2 (combine like terms) : 2x > -3
Step 3 (divide both sides by the coefficient of x which is 2) : $\frac{\cancel{2}^1 x}{\cancel{2}_1} > \frac{-3}{2}$

Step 4 (simplify both sides as needed) : $x > \frac{-3}{2}$

Example 14 : -4x + 7 > 10
Step 1 (subtract 7 from both sides) : -4x + 7 - 7 > 10 - 7
Step 2 (combine like terms) : -4x > 3
Step 3 (divide both sides by the coefficient of x which is -4) : $\frac{\cancel{-4}^1 x}{\cancel{-4}_1} < \frac{3}{-4}$

> When an inequality is multiplied or divided with negative number, the inequality changes to the opposite

Step 4 (simplify both sides as needed) : $x < \frac{3}{-4}$

$$x < \frac{-3}{4}$$

Example 15 : -5x - 1 < -11
Step 1 (add 1 to both sides) : -5x - 1 + 1 < -11 + 1
Step 2 (combine like terms) : -5x < -10
Step 3 (divide both sides by the coefficient of x which is -5) : $\frac{\cancel{-5}^1 x}{\cancel{-5}_1} > \frac{\cancel{-10}^2}{\cancel{-5}_1}$

> When an inequality is multiplied or divided with negative number, the inequality changes to the opposite

Step 4 (simplify both sides as needed) : x > 2

Example 16 : -8x + 3 < -30
Step 1 (subtract 3 from both sides) : -8x + 3 - 3 < -30 - 3
Step 2 (combine like terms) : -8x < -33
Step 3 (divide both sides by the coefficient of x which is -8) : $\frac{\cancel{-8}^1 x}{\cancel{-8}_1} > \frac{-33}{-8}$

Step 4 (simplify both sides as needed) : $x > \frac{-33}{-8}$

$$x > \frac{33}{8}$$

$$x > 4\frac{1}{8}$$

FORMULA SHEET

1. Area of a triangle

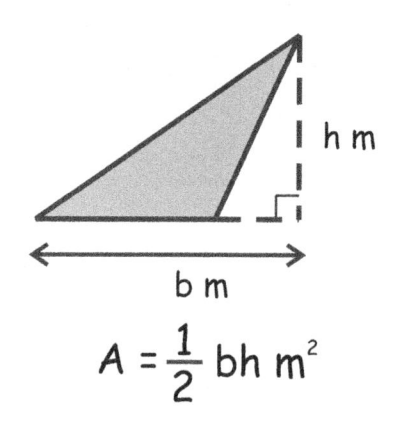

$A = \frac{1}{2} bh \; m^2$

2. Area of a parellelogram

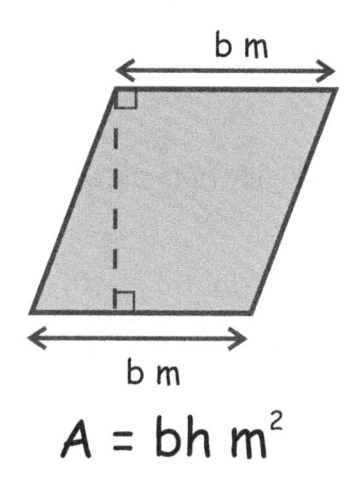

$A = bh \; m^2$

3. Volume and Surface area of a rectangular prism

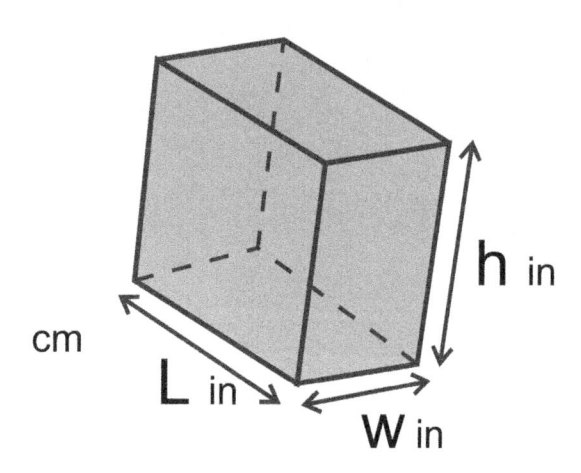

$V = lwh \; in^3$
$S.A = 2(lw + lh + wh) \; in^2$

4. Volume and Surface area of a square prism

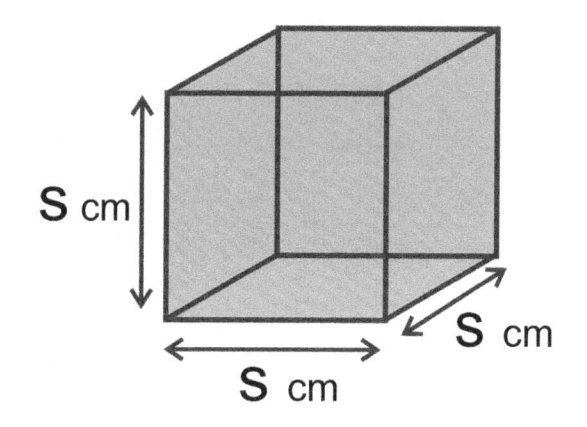

$V = s^3 \; cm^3$
$S.A = 6s^2 \; cm^2$

5. Perimeter and Area of a Square

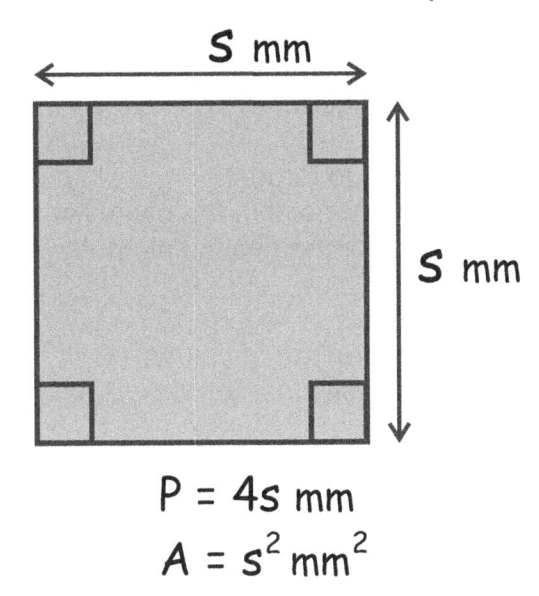

P = 4s mm
A = s^2 mm^2

6. Area of a Trapezium

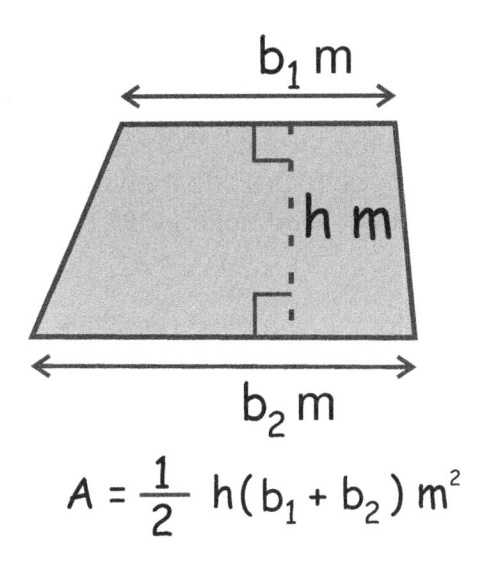

$A = \frac{1}{2} h(b_1 + b_2)$ m^2

11. Perimeter and Area of a Rectangle

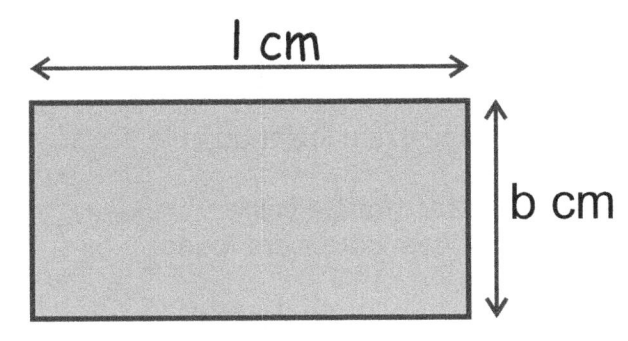

Area = l x b cm^2
Perimeter = 2(l + b) cm

Average or Mean : It is the sum of all the values divided by the number of values.

To calculate add all the values given which is called as sum. Count how many values are given
Dividing sum by count will give the average or mean value for the given data set.

Median : First arrange all the data points from smallest to largest.
If the data points are odd then median is the middle data point in the given list of values.
If the data points are even then median is the average of the two middle data points in the list.

Mode : Mode is the most repetitive data point in the given set of values.
Sometimes if two or more data points are repetitive equally then mode will have multiple values.
If no data point is repeated then mode equals to zero.

Range : First arrange all data points from smallest to largest, then subtract smallest value from
the largest which equals to range value.

Lower quartile : First arrange all the data points from smallest to largest.
If the data points are odd then median is the middle data point in the given list of values.
If the data points are even then median is the average of the two middle data points in the list.
The middle value from the smallest to the median value is the lower quartile or 1st quartile.
We follow the same process as median to find the lower quartile value.
Lower quartile value is the value under which 25% of data points are found.

Upper quartile : First arrange all the data points from smallest to largest.
If the data points are odd then median is the middle data point in the given list of values.
If the data points are even then median is the average of the two middle data points in the list.
The middle value from the greatest to the median value is the upper quartile or 3rd quartile.
We follow the same process as median to find the upper quartile value.
Upper quartile value is the value under which 75% of data points are found.

Inter quartile : It is a difference between upper quartile and lower quartile values.

Abbreviations

milligram	mg	volume	V
gram	g	total Square Area	S.A
kilogram	kg	area of base	B
milliliter	mL	ounce	oz
liter	L	pound	lb
kiloliter	kL	quart	qt
millimeter	mm	gallon	gal.
centimeter	cm	inches	in.
meter	m	foot	ft
kilometer	km	yard	yd
square centimeter	cm^2	mile	mi.
cubic centimeter	cm^3	square inch	sq in.
		square foot	sq ft
		cubic inch	cu in.
		cubic foot	cu ft

Grade 7
Volume 1
Week 1

**Grade 7
Volume 1
Week 1**

1) Name the set or sets to which the number belongs.

$$-\sqrt{64}$$

A) Q, R   B) Z, Q, R

C) I, R   D) W, Z, Q, R

2) Name the set or sets to which the number belongs.

$$-\sqrt{16}$$

A) Z, Q, R   B) W, Z, Q, R

C) I, R   D) Q, R

3) Name the set or sets to which the number belongs.

$$1$$

A) I, R   B) N, W, Z, Q, R

C) Q, R   D) W, Z, Q, R

4) Name the set or sets to which the number belongs.

$$8$$

A) Z, Q, R   B) N, W, Z, Q, R

C) I, R   D) Q, R

5) Name the set or sets to which the number belongs.

$$\frac{17}{4}$$

A) Q, R

B) N, W, Z, Q, R

C) W, Z, Q, R

D) I, R

6) Name the set or sets to which the number belongs.

$$\frac{8}{-2}$$

A) I, R   B) Q, R

C) Z, Q, R   D) W, Z, Q, R

7) Name the set or sets to which the number belongs.

$$\frac{3}{11}$$

A) W, Z, Q, R

B) N, W, Z, Q, R

C) I, R

D) Q, R

8) Name the set or sets to which the number belongs.

$$15$$

A) I, R

B) N, W, Z, Q, R

C) W, Z, Q, R

D) Q, R

9) Name the set or sets to which the number belongs.

$$\sqrt{4}$$

A) Q, R
B) Z, Q, R
C) N, W, Z, Q, R
D) W, Z, Q, R

10) Name the set or sets to which the number belongs.

−11

A) W, Z, Q, R   B) I, R
C) Q, R   D) Z, Q, R

11) Name the set or sets to which the number belongs.

$$\frac{16}{13}$$

A) N, W, Z, Q, R
B) W, Z, Q, R
C) I, R
D) Q, R

12) Name the set or sets to which the number belongs.

3

A) N, W, Z, Q, R
B) I, R
C) Q, R
D) W, Z, Q, R

13) Name the set or sets to which the number belongs.

$$\sqrt{0}$$

A) I, R   B) W, Z, Q, R
C) Z, Q, R   D) N, W, Z, Q, R

14) Name the set or sets to which the number belongs.

$$\frac{19}{10}$$

A) W, Z, Q, R
B) N, W, Z, Q, R
C) Q, R
D) Z, Q, R

15) Name the set or sets to which the number belongs.

$$\sqrt{91}$$

A) W, Z, Q, R   B) Q, R
C) Z, Q, R   D) I, R

16) Name the set or sets to which the number belongs.

$$\frac{30}{2}$$

A) I, R
B) Q, R
C) N, W, Z, Q, R
D) W, Z, Q, R

17) Name the set or sets to which the number belongs.

14

A) Z, Q, R
B) W, Z, Q, R
C) N, W, Z, Q, R
D) Q, R

Grade 7
Volume 1
Week 1

18) Name the set or sets to which the number belongs.

$$\sqrt{58}$$

A) N, W, Z, Q, R
B) W, Z, Q, R
C) Q, R
D) I, R

19) Name the set or sets to which the number belongs.

$$\sqrt{52}$$

A) Q, R
B) I, R
C) N, W, Z, Q, R
D) W, Z, Q, R

20) Name the set or sets to which the number belongs.

$$\sqrt{30}$$

A) W, Z, Q, R
B) N, W, Z, Q, R
C) Q, R
D) I, R

21) Name the set or sets to which the number belongs.

$$\sqrt{17}$$

A) I, R B) Q, R
C) N, W, Z, Q, R D) Z, Q, R

22) Name the set or sets to which the number belongs.

$$-13$$

A) Q, R
B) Z, Q, R
C) N, W, Z, Q, R
D) W, Z, Q, R

23) Name the set or sets to which the number belongs.

$$\frac{-64}{-20}$$

A) W, Z, Q, R
B) Z, Q, R
C) Q, R
D) N, W, Z, Q, R

24) Name the set or sets to which the number belongs.

$$\frac{12}{52}$$

A) N, W, Z, Q, R
B) W, Z, Q, R
C) Z, Q, R
D) Q, R

25) Name the set or sets to which the number belongs.

$$-10$$

A) N, W, Z, Q, R B) I, R
C) W, Z, Q, R D) Z, Q, R

26) Name the set or sets to which the number belongs.

$$\frac{-12}{-10}$$

A) Z, Q, R B) Q, R
C) I, R D) W, Z, Q, R

27) Name the set or sets to which the number belongs.

$$\sqrt{50}$$

A) Z, Q, R
B) N, W, Z, Q, R
C) W, Z, Q, R
D) I, R

28) Name the set or sets to which the number belongs.

$$\frac{3}{10}$$

A) W, Z, Q, R
B) Z, Q, R
C) Q, R
D) N, W, Z, Q, R

29) Name the set or sets to which the number belongs.

$$-5$$

A) N, W, Z, Q, R
B) Z, Q, R
C) I, R
D) W, Z, Q, R

30) Name the set or sets to which the number belongs.

$$\sqrt{\frac{2225}{5}}$$

A) Q, R B) I, R
C) Z, Q, R D) N, W, Z, Q, R

31) Name the set or sets to which the number belongs.

$$\frac{0}{-3}$$

A) W, Z, Q, R
B) Z, Q, R
C) Q, R
D) N, W, Z, Q, R

32) Name the set or sets to which the number belongs.

$$\frac{-28}{-2}$$

A) Z, Q, R
B) Q, R
C) N, W, Z, Q, R
D) W, Z, Q, R

33) Name the set or sets to which the number belongs.

$$\sqrt{26}$$

A) W, Z, Q, R B) I, R
C) Z, Q, R D) Q, R

34) Name the set or sets to which the number belongs.

$$\sqrt{51}$$

A) Z, Q, R B) N, W, Z, Q, R
C) I, R D) W, Z, Q, R

35) Name the set or sets to which the number belongs.

−7

A) I, R
B) N, W, Z, Q, R
C) W, Z, Q, R
D) Z, Q, R

36) Name the set or sets to which the number belongs.

−9

A) I, R
B) N, W, Z, Q, R
C) W, Z, Q, R
D) Z, Q, R

37) Name the set or sets to which the number belongs.

7

A) Z, Q, R B) Q, R
C) I, R D) N, W, Z, Q, R

38) Name the set or sets to which the number belongs.

$$\frac{0}{-5}$$

A) W, Z, Q, R B) I, R
C) Z, Q, R D) Q, R

39) Name the set or sets to which the number belongs.

$$\sqrt{86}$$

A) I, R
B) Q, R
C) N, W, Z, Q, R
D) W, Z, Q, R

40) Name the set or sets to which the number belongs.

−12

A) Z, Q, R
B) W, Z, Q, R
C) N, W, Z, Q, R
D) I, R

41) Name the set or sets to which the number belongs.

9

A) Q, R B) N, W, Z, Q, R
C) Z, Q, R D) I, R

42) Name the set or sets to which the number belongs.

0

A) Z, Q, R B) W, Z, Q, R
C) I, R D) N, W, Z, Q, R

43) Name the set or sets to which the number belongs.

−8

A) N, W, Z, Q, R B) Z, Q, R
C) Q, R D) I, R

Grade 7
Volume 1
Week 1

44) Name the set or sets to which the number belongs.

$$\frac{26}{16}$$

A) Q, R B) Z, Q, R
C) N, W, Z, Q, R D) I, R

45) Name the set or sets to which the number belongs.

$$-\frac{20}{2}$$

A) W, Z, Q, R B) I, R
C) Z, Q, R D) Q, R

46) Name the set or sets to which the number belongs.

$$\sqrt{9}$$

A) I, R
B) W, Z, Q, R
C) N, W, Z, Q, R
D) Z, Q, R

47) Name the set or sets to which the number belongs.

$$\sqrt{\frac{112}{4}}$$

A) Q, R B) Z, Q, R
C) W, Z, Q, R D) I, R

48) Name the set or sets to which the number belongs.

$$\sqrt{\frac{216}{2}}$$

A) Z, Q, R B) Q, R
C) W, Z, Q, R D) I, R

49) Name the set or sets to which the number belongs.

$$\sqrt{78}$$

A) I, R B) W, Z, Q, R
C) Q, R D) N, W, Z, Q, R

50) Name the set or sets to which the number belongs.

$$\frac{19}{9}$$

A) Q, R B) W, Z, Q, R
C) Z, Q, R D) I, R

Grade 7
Volume 1
Week 1

51) Evaluate the below expression.

$$\frac{11}{8} - 2 + \left(-\frac{6}{5}\right)$$

A) $-\dfrac{73}{40}$ B) $-\dfrac{431}{280}$

C) $-\dfrac{1}{40}$ D) $-\dfrac{631}{280}$

52) Evaluate the below expression.

$$0 - 3\frac{1}{7} + \frac{12}{7}$$

A) $-\dfrac{143}{56}$ B) $-\dfrac{75}{28}$

C) $\dfrac{1}{14}$ D) $-\dfrac{10}{7}$

53) Evaluate the below expression.

$$\left(-2\frac{5}{8}\right) - \left(-\frac{3}{7}\right) + \left(-1\frac{2}{3}\right)$$

A) $-\dfrac{2573}{840}$ B) $-\dfrac{5}{21}$

C) $-\dfrac{649}{168}$ D) $-\dfrac{817}{168}$

54) Evaluate the below expression.

$$2\frac{1}{2} + \left(-3\frac{3}{5}\right) + 3\frac{4}{7}$$

A) $\dfrac{591}{140}$ B) $\dfrac{173}{70}$

C) $\dfrac{307}{280}$ D) $\dfrac{69}{35}$

55) Evaluate the below expression.

$$\left(-\frac{4}{3}\right) + \left(-\frac{1}{8}\right) + \left(-\frac{3}{5}\right)$$

A) $\dfrac{257}{120}$ B) $-\dfrac{247}{120}$

C) $-\dfrac{391}{120}$ D) $-\dfrac{409}{840}$

56) Evaluate the below expression.

$$\frac{1}{3} - \frac{1}{2} - \frac{3}{4}$$

A) $-\dfrac{11}{12}$ B) $-\dfrac{71}{12}$

C) $-\dfrac{35}{12}$ D) $-\dfrac{29}{12}$

57) Evaluate the below expression.

$$\frac{4}{3} - 2\frac{2}{3} + \frac{1}{3}$$

A) $\dfrac{12}{5}$ B) -1

C) $-\dfrac{17}{6}$ D) $-\dfrac{5}{2}$

58) Evaluate the below expression.

$$1 + \frac{6}{5} - \left(-2\frac{6}{7}\right)$$

A) $\dfrac{1027}{210}$ B) $\dfrac{249}{70}$

C) $\dfrac{177}{35}$ D) $\dfrac{142}{35}$

©All rights reserved-Math-Knots LLC., VA-USA   www.math-knots.com | www.a4ace.com

Grade 7
Volume 1
Week 1

59) Evaluate the below expression.

$$(-2) - 2\frac{3}{4} + \left(-3\frac{1}{6}\right)$$

A) $-\dfrac{89}{12}$ B) $-\dfrac{85}{12}$

C) $-\dfrac{149}{12}$ D) $-\dfrac{95}{12}$

60) Evaluate the below expression.

$$\left(-2\frac{2}{5}\right) + \left(-3\frac{6}{7}\right) + \left(-2\frac{1}{8}\right)$$

A) $-\dfrac{2417}{280}$ B) $-\dfrac{1437}{280}$

C) $-\dfrac{1619}{280}$ D) $-\dfrac{2347}{280}$

61) Evaluate the below expression.

$$2\frac{1}{2} - \frac{1}{2} + \frac{13}{8}$$

A) $\dfrac{199}{24}$ B) $\dfrac{29}{8}$

C) $\dfrac{33}{8}$ D) 2

62) Evaluate the below expression.

$$\left(-1\frac{1}{3}\right) + 1\frac{1}{6} - \frac{7}{8}$$

A) $-\dfrac{13}{24}$ B) $-\dfrac{1}{24}$

C) $-\dfrac{25}{24}$ D) $-\dfrac{19}{6}$

63) Evaluate the below expression.

$$\left(-3\frac{7}{8}\right) + \frac{7}{8} - 1\frac{5}{8}$$

A) -4 B) $-\dfrac{37}{8}$

C) $-\dfrac{127}{24}$ D) $-\dfrac{41}{8}$

64) Evaluate the below expression.

$$2 + \left(-\frac{4}{3}\right) - \frac{5}{6}$$

A) $\dfrac{5}{2}$ B) $-\dfrac{9}{2}$

C) $-\dfrac{19}{6}$ D) $-\dfrac{1}{6}$

65) Evaluate the below expression.

$$\left(-3\frac{5}{6}\right) - \left(-2\frac{5}{6}\right) - \left(-2\frac{5}{8}\right)$$

A) $\dfrac{163}{56}$ B) $\dfrac{13}{8}$

C) $-\dfrac{25}{8}$ D) $\dfrac{55}{24}$

66) Evaluate the below expression.

$$4\frac{3}{4} - 3\frac{1}{3} - \left(-\frac{5}{4}\right)$$

A) $\dfrac{8}{3}$ B) $\dfrac{37}{15}$

C) $\dfrac{26}{3}$ D) $\dfrac{109}{15}$

67) Evaluate the below expression.

$$4\frac{7}{8} + 2\frac{2}{3} + \left(-2\frac{2}{7}\right)$$

A) $\dfrac{1163}{168}$ B) $\dfrac{43}{168}$

C) $\dfrac{341}{56}$ D) $\dfrac{883}{168}$

68) Evaluate the below expression.

$$\frac{1}{2} + 3\frac{1}{2} - \frac{1}{2}$$

A) 3 B) $\dfrac{19}{3}$

C) $\dfrac{19}{6}$ D) $\dfrac{7}{2}$

69) Evaluate the below expression.

$$1\frac{5}{8} - 1 + 4\frac{1}{4}$$

A) $\dfrac{39}{8}$ B) $\dfrac{51}{8}$

C) $\dfrac{53}{8}$ D) $\dfrac{71}{8}$

70) Evaluate the below expression.

$$4\frac{1}{6} - \frac{10}{7} + \frac{3}{4}$$

A) $\dfrac{283}{42}$ B) $\dfrac{353}{84}$

C) $\dfrac{293}{84}$ D) $\dfrac{125}{84}$

71) Evaluate the below expression.

$$\frac{7}{6} - \left(-1\frac{1}{3}\right) + 1$$

A) $\dfrac{29}{4}$ B) $\dfrac{7}{2}$

C) $\dfrac{11}{2}$ D) 2

72) Evaluate the below expression.

$$\left(-2\frac{1}{2}\right) - \frac{3}{2} + 4\frac{1}{4}$$

A) $\dfrac{1}{4}$ B) $-\dfrac{1}{4}$

C) $-\dfrac{29}{8}$ D) $\dfrac{13}{4}$

73) Evaluate the below expression.

$$\frac{9}{8} - 2 - \left(-\frac{1}{5}\right)$$

A) $-\dfrac{27}{40}$ B) $\dfrac{131}{280}$

C) $-\dfrac{549}{280}$ D) $\dfrac{411}{280}$

74) Evaluate the below expression.

$$1 - 4\frac{3}{4} - 2\frac{5}{6}$$

A) $-\dfrac{29}{12}$ B) $-\dfrac{697}{84}$

C) $-\dfrac{587}{60}$ D) $-\dfrac{79}{12}$

75) Evaluate the below expression.

$$3\frac{3}{4} + \frac{1}{2} + \left(-2\frac{1}{4}\right)$$

A) $\dfrac{7}{2}$ B) 2

C) $\dfrac{9}{2}$ D) $\dfrac{2}{5}$

76) Evaluate the below expression.

$$\left(-\frac{5}{4}\right) + \frac{1}{2} - (-2)$$

A) 2   B) $\dfrac{5}{4}$

C) $-\dfrac{11}{8}$ D) $-\dfrac{17}{8}$

77) Evaluate the below expression.

$$\left(-\frac{7}{6}\right) + (-8) - \frac{1}{6}$$

A) $-\dfrac{28}{3}$ B) $-\dfrac{131}{15}$

C) $-\dfrac{97}{12}$ D) $-\dfrac{139}{12}$

78) Evaluate the below expression.

$$\frac{9}{8} + (-6) - \left(-\frac{1}{2}\right)$$

A) -3 B) $-\dfrac{35}{8}$

C) $-\dfrac{51}{8}$ D) -7

79) Evaluate the below expression.

$$\left(-2\frac{5}{8}\right) + 1 - \left(-\frac{7}{4}\right)$$

A) $-\dfrac{47}{8}$ B) $\dfrac{29}{8}$

C) $\dfrac{1}{8}$ D) $\dfrac{15}{56}$

80) Evaluate the below expression.

$$\left(-\frac{12}{7}\right) - \left(-3\frac{1}{3}\right) - \left(-2\frac{5}{8}\right)$$

A) $\dfrac{713}{168}$ B) $\dfrac{629}{168}$

C) $\dfrac{2893}{840}$ D) $\dfrac{881}{168}$

81) Evaluate the below expression.

$$\left(-\frac{9}{8}\right) + (-2) - \frac{7}{5}$$

A) $-\dfrac{181}{40}$ B) $-\dfrac{1627}{280}$

C) $-\dfrac{141}{40}$ D) $-\dfrac{123}{20}$

82) Evaluate the below expression.

$$3\frac{2}{3} - (-1) + \left(-\frac{4}{3}\right)$$

A) $\dfrac{53}{24}$ B) $\dfrac{10}{3}$

C) $\dfrac{38}{15}$ D) $\dfrac{104}{15}$

83) Evaluate the below expression.

$$5 + \left(-3\frac{1}{6}\right) - 6$$

A) $-\dfrac{25}{6}$ B) $-\dfrac{151}{42}$

C) $-\dfrac{253}{42}$ D) $-\dfrac{20}{3}$

84) Evaluate the below expression.

$$2 + 2\frac{1}{2} - \frac{3}{4}$$

A) $\dfrac{21}{4}$ B) $\dfrac{35}{4}$

C) $\dfrac{27}{8}$ D) $\dfrac{15}{4}$

85) Evaluate the below expression.

$$\left(-3\frac{7}{8}\right) + 3\frac{2}{3} + \left(-\frac{3}{2}\right)$$

A) $-\dfrac{373}{120}$ B) $-\dfrac{41}{24}$

C) $\dfrac{43}{24}$ D) $-\dfrac{131}{24}$

86) Evaluate the below expression.

$$1\frac{1}{2} - (-6) - \left(-\frac{2}{3}\right)$$

A) $\dfrac{475}{42}$ B) $\dfrac{37}{6}$

C) $\dfrac{49}{6}$ D) $\dfrac{265}{42}$

87) Evaluate the below expression.

$$(-1) + (-2) + \left(-\frac{1}{3}\right)$$

A) $-\dfrac{17}{6}$ B) $-\dfrac{10}{3}$

C) -5 D) $-\dfrac{13}{3}$

88) Evaluate the below expression.

$$\frac{5}{3} + 3\frac{4}{7} - \left(-3\frac{1}{8}\right)$$

A) $\dfrac{755}{84}$ B) $\dfrac{1405}{168}$

C) $\dfrac{1111}{168}$ D) $\dfrac{1741}{168}$

89) Evaluate the below expression.

$$3\frac{2}{5} - 1\frac{1}{6} + 4$$

A) $\dfrac{277}{30}$ B) $\dfrac{187}{30}$

C) $\dfrac{116}{15}$ D) $\dfrac{569}{60}$

90) Evaluate the below expression.

$$\frac{4}{7} + \frac{1}{3} - 3\frac{3}{4}$$

A) $-\dfrac{239}{84}$ B) $-\dfrac{575}{84}$

C) $-\dfrac{337}{84}$ D) $-\dfrac{37}{168}$

Grade 7
Volume 1
Week 1

91) Evaluate the below expression.

$$\left(-3\frac{1}{2}\right) + \frac{1}{4} - \frac{1}{2}$$

A) $-\dfrac{1}{2}$ B) $-\dfrac{15}{4}$

C) $-\dfrac{13}{2}$ D) $-\dfrac{19}{20}$

92) Evaluate the below expression.

$$\left(-\frac{1}{7}\right) - 1 - \left(-\frac{3}{2}\right)$$

A) $\dfrac{5}{14}$ B) $-\dfrac{59}{14}$

C) $\dfrac{127}{42}$ D) $-\dfrac{10}{21}$

93) Evaluate the below expression.

$$2\frac{1}{2} + \frac{11}{6} + (-4)$$

A) $-\dfrac{13}{24}$ B) $-\dfrac{50}{21}$

C) -3 D) $\dfrac{1}{3}$

94) Evaluate the below expression.

$$\frac{3}{5} + \left(-\frac{1}{4}\right) - \left(-3\frac{1}{4}\right)$$

A) $\dfrac{18}{5}$ B) $\dfrac{147}{20}$

C) $\dfrac{61}{35}$ D) $\dfrac{14}{15}$

95) Evaluate the below expression.

$$\frac{3}{2} + \left(-2\frac{1}{2}\right) + 3\frac{3}{7}$$

A) $\dfrac{57}{35}$ B) $\dfrac{353}{56}$

C) $\dfrac{128}{21}$ D) $\dfrac{17}{7}$

96) Evaluate the below expression.

$$2\frac{2}{3} + \left(-\frac{15}{8}\right) - \left(-2\frac{1}{2}\right)$$

A) $\dfrac{23}{3}$ B) $\dfrac{61}{24}$

C) $\dfrac{841}{168}$ D) $\dfrac{79}{24}$

97) Evaluate the below expression.

$$4\frac{3}{7} - \frac{13}{7} + \frac{5}{7}$$

A) $\dfrac{108}{35}$ B) $\dfrac{94}{35}$

C) $\dfrac{19}{42}$ D) $\dfrac{23}{7}$

98) Evaluate the below expression.

$$1\frac{5}{6} - 1\frac{1}{3} - 3\frac{1}{3}$$

A) $\dfrac{1}{24}$ B) $-\dfrac{43}{12}$

C) $-\dfrac{17}{6}$ D) $\dfrac{25}{24}$

99) Evaluate the below expression.

$$\left(-1\tfrac{1}{3}\right) - 1\tfrac{1}{8} + \left(-1\tfrac{1}{7}\right)$$

A) $-\dfrac{941}{168}$ B) $-\dfrac{689}{168}$

C) $-\dfrac{295}{56}$ D) $-\dfrac{605}{168}$

100) Evaluate the below expression.

431.59 − 310.8

A) 41.006   B) 471.69

C) 180.79   D) 120.79

101) Evaluate the below expression.

388.35 + (−0.2)

A) −29.65   B) 596.15

C) 595.84   D) 388.15

102) Evaluate the below expression.

(−101.7) + 436.1

A) 183.4   B) 216.2

C) 334.4   D) −109.6

103) Evaluate the below expression.

$$\left(-3\tfrac{5}{6}\right) - \left(-\tfrac{4}{5}\right) + \left(-1\tfrac{1}{2}\right)$$

A) $-\dfrac{71}{15}$ B) $-\dfrac{68}{15}$

C) $-\dfrac{806}{105}$ D) $-\dfrac{16}{3}$

104) Evaluate the below expression.

(−413.9) + (−41.7)

A) −42.1   B) −907.6

C) −672.6   D) −455.6

105) Evaluate the below expression.

214.5 + (−445.4)

A) −415.9   B) 12.2

C) −522   D) −230.9

106) Evaluate the below expression.

(−305.5) − (−499)

A) −182.6   B) −170.2

C) 367.7   D) 193.5

Grade 7
Volume 1
Week 1

107) Evaluate the below expression.

$$333.1 - (-433.8)$$

A) 1193.5   B) 1256.5

C) 766.9   D) 1132.43

108) Evaluate the below expression.

$$271.06 + (-267.1)$$

A) 49.06   B) −189.64

C) 116.86   D) 3.96

109) Evaluate the below expression.

$$(-65.8) + 206.96$$

A) 421.06   B) 39.66

C) 292.37   D) 141.16

110) Evaluate the below expression.

$$(-162.5) + 195.9$$

A) −81.5   B) −218.7

C) −78.4   D) 33.4

111) Evaluate the below expression.

$$(-89.5) - (-313.2)$$

A) 223.7   B) 345.4

C) −103.2   D) 579.6

112) Evaluate the below expression.

$$433.9 + (-12.4)$$

A) 421.5   B) 873.5

C) 809.2   D) 172.9

113) Evaluate the below expression.

$$(-165.7) - 238.4$$

A) −825.33   B) −266.3

C) −404.1   D) −320.4

114) Evaluate the below expression.

$$(-425.3) - (-154.6)$$

A) −270.7   B) 10

C) 47.4   D) −701

115) Evaluate the below expression.

$$(-116.7) + 126$$

A) −405.8   B) −430.6

C) 9.3   D) −464.6

116) Evaluate the below expression.

$$314.9 - 235$$

A) −345.93   B) 566.5

C) −285.5   D) 79.9

117) Evaluate the below expression.

$$480.113 - 283.8$$

A) 666.113   B) 276.583

C) 196.313   D) 158.413

118) Evaluate the below expression.

$$347.9 - (-290.3)$$

A) 949.3   B) 392.15

C) 638.2   D) 348.9

119) Evaluate the below expression.

$$(-488.9) + (-271.9)$$

A) −1119.2   B) −550.2

C) −1021.3   D) −760.8

120) Evaluate the below expression.

$$(-448.9) - 286.7$$

A) −1107.3   B) −1134.6

C) −735.6   D) −1203

121) Evaluate the below expression.

$$(-381.4) - 46.53$$

A) −531.33   B) −357.44

C) −427.93   D) −564.33

122) Evaluate the below expression.

$$(-183.9) + 146.948$$

A) 64.048   B) −36.952

C) 121.248   D) 269.448

123) Evaluate the below expression.

$$269.1 + (-361.6)$$

A) 391.5 B) −119.17

C) −92.5 D) −11.07

124) Evaluate the below expression.

$$(-300.3) - (-477.9)$$

A) 89.9 B) 177.6

C) −159.8 D) 243.9

125) Evaluate the below expression.

$$281.2 - 136.2$$

A) 10.7 B) −95.6

C) −274 D) 145

126) Evaluate the below expression.

$$(-316.8) + 82.3$$

A) −234.5 B) 190.124

C) −109.5 D) 48.7

127) Evaluate the below expression.

$$(-213.6) - (-402.3)$$

A) 635.1 B) 441.6

C) −70.7 D) 188.7

128) Evaluate the below expression.

$$141 - 231$$

A) 209.2 B) −248.5

C) −90 D) −278

129) Evaluate the below expression.

$$(-34.5) - 359.18$$

A) −155.98 B) −626.88

C) −393.68 D) −162.18

130) Evaluate the below expression.

$$300.9 + (-12.955)$$

A) 385.545 B) 287.945

C) 10.475 D) 513.045

131) Evaluate the below expression.

$$436.1 - (-261.3)$$

A) 1121.3 B) 568.9

C) 697.4 D) 391.3

132) Evaluate the below expression.

$$246.2 + (-236.7)$$

A) 449.7 B) 9.5

C) −483.6 D) 141

133) Evaluate the below expression.

$$290.2 - 190.3$$

A) 99.9 B) 73.8

C) −68.2 D) 467.4

134) Evaluate the below expression.

$$395.7 - (-32.957)$$

A) 525.457 B) 428.657

C) 77.057 D) 779.557

135) Evaluate the below expression.

$$122.5 + (-499.5)$$

A) −535.5 B) −375.9

C) −377 D) −298.2

136) Evaluate the below expression.

$$407.8 + (-295)$$

A) 317.3 B) 265.3

C) 22.2 D) 112.8

137) Evaluate the below expression.

$$(-73.1) - (-393.8)$$

A) 174.6 B) 320.7

C) 360.338 D) 805.6

138) Evaluate the below expression.

$$(-230.6) + (-45.6)$$

A) −497.01 B) −276.2

C) 70 D) −538.3

139) Evaluate the below expression.

$$(-489.9) + (-197.6)$$

A) −298.3   B) −493.4

C) −687.5   D) −370.31

140) Evaluate the below expression.

$$(-66) + 256.4$$

A) 49.7   B) 190.4

C) 427.8   D) 242.7

141) Evaluate the below expression.

$$(-235.5) + 63.5$$

A) −472.5   B) −150.1

C) 105.7   D) −172

142) Evaluate the below expression.

$$387 + (-433)$$

A) −46   B) 334.2

C) −213.6   D) 138.95

143) Evaluate the below expression.

$$232.1 - 431.7$$

A) −58.1   B) −669.2

C) −540.4   D) −199.6

144) Evaluate the below expression.

$$171.1 + (-219.8)$$

A) 82.158   B) −48.7

C) −534.4   D) −466.6

145) Evaluate the below expression.

$$406.9 - 373.4$$

A) 33.5   B) −319.4

C) 413.8   D) 181.2

146) Evaluate the below expression.

$$218.1 + (-261.6)$$

A) −165.915   B) −228.89

C) −43.5   D) −478.84

Grade 7
Volume 1
Week 1

147) Evaluate the below expression.

$$(-106.7) - 4.7$$

A) −39.1   B) −574.4

C) −95.5   D) −111.4

151) Evaluate the below expression.

$$(-26.66) + (-4.1)$$

A) 388.84   B) 122.64

C) 257.64   D) −30.76

148) Evaluate the below expression.

$$309.63 - 437.7$$

A) 98.65   B) −128.07

C) 262.03   D) 246.73

152) Evaluate the below expression.

$$266.8 - 133.5$$

A) −53.9   B) −63.5

C) 505.3   D) 133.3

149) Find the product of the below.

$$2 \times -\frac{5}{6} \times \frac{25}{14}$$

A) $-\frac{31}{21}$ B) $-\frac{137}{126}$

C) $-\frac{125}{42}$ D) $-\frac{499}{210}$

153) Find the product of the below.

$$-7 \times \frac{4}{3} \times -\frac{1}{3}$$

A) $\frac{38}{65}$ B) $-\frac{38}{65}$

C) $\frac{28}{9}$ D) −6

150) Find the product of the below.

$$3\frac{6}{7} \times -1\frac{7}{9} \times \frac{4}{3}$$

A) $\frac{64}{7}$ B) $-\frac{64}{7}$

C) $-\frac{407}{56}$ D) $-\frac{619}{70}$

154) Find the product of the below.

$$6\frac{4}{5} \times -\frac{3}{2} \times \frac{5}{4}$$

A) $\frac{51}{4}$ B) $-\frac{74704}{1089}$

C) $\frac{74704}{1089}$ D) $-\frac{51}{4}$

155) Find the product of the below.

$$1\frac{7}{10} \times -\frac{4}{3} \times -\frac{9}{5}$$

A) $-\frac{43}{30}$ B) $\frac{39}{175}$

C) $\frac{102}{25}$ D) $-\frac{102}{25}$

156) Find the product of the below.

$$-8 \times 7\frac{9}{14} \times \frac{17}{12}$$

A) $-\frac{1201}{14}$ B) $-\frac{7157}{84}$

C) $-\frac{9326}{105}$ D) $-\frac{1819}{21}$

157) Find the product of the below.

$$0 \times \frac{1}{14} \times -\frac{3}{2}$$

A) $-\frac{1}{4}$ B) 0

C) $-\frac{1}{3}$ D) $\frac{13}{2}$

158) Find the product of the below.

$$-3\frac{1}{6} \times 2\frac{1}{14} \times -\frac{22}{13}$$

A) $-\frac{6061}{546}$ B) $\frac{9169}{546}$

C) $\frac{6061}{546}$ D) $\frac{29213}{2730}$

159) Find the product of the below.

$$5\frac{11}{13} \times -2 \times -\frac{4}{3}$$

A) $\frac{4981}{312}$ B) $\frac{2767}{195}$

C) $-\frac{608}{39}$ D) $\frac{608}{39}$

160) Find the product of the below.

$$-1\frac{1}{3} \times \frac{11}{7} \times \frac{4}{3}$$

A) $-\frac{32}{7}$ B) $-\frac{163}{126}$

C) $-\frac{176}{63}$ D) $\frac{11}{7}$

161) Find the product of the below.

$$9 \times -1\frac{4}{5} \times \frac{11}{12}$$

A) $-\frac{297}{20}$ B) $-\frac{83}{5}$

C) $\frac{117}{560}$ D) $\frac{297}{20}$

162) Find the product of the below.

$$-1\frac{3}{4} \times \frac{3}{5} \times \frac{1}{8}$$

A) 0 B) $\frac{47}{32}$

C) $-\frac{21}{160}$ D) $-\frac{711}{1760}$

163) Find the product of the below.

$$13\frac{7}{11} \times -\frac{11}{7} \times \frac{2}{11}$$

A) $-\dfrac{300}{77}$ B) $-\dfrac{3515}{1001}$

C) $\dfrac{575}{189}$ D) $-\dfrac{575}{189}$

164) Find the product of the below.

$$2 \times -\frac{13}{9} \times \frac{5}{9}$$

A) $\dfrac{130}{81}$ B) $-\dfrac{1147}{429}$

C) $\dfrac{1147}{429}$ D) $-\dfrac{130}{81}$

165) Find the product of the below.

$$-7 \times \frac{23}{13} \times -\frac{21}{11}$$

A) $-\dfrac{1021}{143}$ B) $\dfrac{3381}{143}$

C) $\dfrac{277}{11}$ D) $\dfrac{14188}{715}$

166) Find the product of the below.

$$\frac{11}{7} \times \frac{1}{5} \times -\frac{5}{6}$$

A) $\dfrac{293}{84}$ B) $-\dfrac{1423}{462}$

C) $\dfrac{82}{55}$ D) $-\dfrac{11}{42}$

167) Find the product of the below.

$$-10 \times 7\frac{1}{2} \times -\frac{5}{4}$$

A) $-\dfrac{375}{4}$ B) $\dfrac{1813}{20}$

C) $\dfrac{1949}{20}$ D) $\dfrac{375}{4}$

168) Find the product of the below.

$$4\frac{1}{2} \times -3\frac{7}{10} \times \frac{3}{7}$$

A) $\dfrac{999}{140}$ B) $-\dfrac{999}{140}$

C) $-\dfrac{3977}{420}$ D) $-\dfrac{849}{140}$

169) Find the product of the below.

$$-2 \times -\frac{3}{5} \times -\frac{15}{13}$$

A) $-\dfrac{244}{65}$ B) $\dfrac{461}{52}$

C) $-\dfrac{18}{13}$ D) $\dfrac{8}{13}$

170) Find the product of the below.

$$\frac{4}{5} \times -\frac{8}{5} \times \frac{3}{8}$$

A) $\dfrac{38}{25}$ B) $-\dfrac{22}{25}$

C) $-\dfrac{12}{25}$ D) $\dfrac{12}{25}$

Grade 7
Volume 1
Week 1

171) Find the product of the below.

$$-2 \times 12\frac{5}{12} \times -\frac{7}{9}$$

A) $\dfrac{3512}{135}$ B) $\dfrac{454}{27}$

C) $\dfrac{1043}{54}$ D) $-\dfrac{459}{140}$

172) Find the product of the below.

$$2\frac{12}{13} \times 4\frac{7}{12} \times -\frac{1}{2}$$

A) $-\dfrac{357}{52}$ B) $-\dfrac{5567}{1716}$

C) $-\dfrac{1045}{156}$ D) $-\dfrac{1093}{156}$

173) Find the product of the below.

$$7\frac{4}{11} \times -1\frac{1}{3} \times \frac{4}{7}$$

A) $-\dfrac{432}{77}$ B) $-\dfrac{6955}{924}$

C) $-\dfrac{6694}{1001}$ D) $\dfrac{432}{77}$

174) Find the product of the below.

$$-3\frac{5}{6} \times \frac{9}{13} \times \frac{6}{5}$$

A) $-\dfrac{1273}{910}$ B) $\dfrac{207}{65}$

C) $-\dfrac{207}{65}$ D) $-\dfrac{101}{26}$

175) Find the product of the below.

$$-3\frac{3}{7} \times \frac{7}{13} \times -\frac{15}{8}$$

A) $\dfrac{277}{65}$ B) $-\dfrac{45}{13}$

C) $\dfrac{45}{13}$ D) $\dfrac{227}{13}$

176) Find the product of the below.

$$4\frac{1}{6} \times 9 \times -\frac{13}{14}$$

A) $-\dfrac{1031}{28}$ B) $-\dfrac{975}{28}$

C) $-\dfrac{3065}{84}$ D) $\dfrac{975}{28}$

177) Find the product of the below.

$$0 \times -\frac{24}{13} \times \frac{1}{11}$$

A) $\dfrac{47}{6}$ B) $\dfrac{1}{2}$

C) 0 D) $-\dfrac{251}{143}$

178) Find the product of the below.)

$$-\frac{3}{2} \times \frac{3}{2} \times \frac{3}{2}$$

A) $-\dfrac{15}{8}$ B) $\dfrac{27}{8}$

C) $-\dfrac{27}{8}$ D) $-\dfrac{419}{72}$

**Grade 7
Volume 1
Week 1**

179) Find the product of the below.

$$5\frac{7}{9} \times -3\frac{13}{14} \times -\frac{11}{6}$$

A) $\dfrac{15919}{378}$ B) $-\dfrac{3913}{120}$

C) $\dfrac{1}{63}$ D) $\dfrac{7865}{189}$

180) Find the product of the below.

$$7\frac{1}{2} \times -\frac{22}{13} \times \frac{11}{6}$$

A) $-\dfrac{2059}{91}$ B) $\dfrac{605}{26}$

C) $-\dfrac{309}{13}$ D) $-\dfrac{605}{26}$

181) Find the product of the below.

$$2 \times 1\frac{3}{8} \times -\frac{7}{10}$$

A) $-\dfrac{73}{10}$ B) $\dfrac{649}{120}$

C) $\dfrac{107}{40}$ D) $-\dfrac{77}{40}$

182) Find the product of the below.

$$-2\frac{9}{10} \times -3\frac{4}{7} \times \frac{7}{9}$$

A) $\dfrac{653}{90}$ B) $\dfrac{1345}{234}$

C) $\dfrac{145}{18}$ D) $\dfrac{355}{72}$

183) Find the product of the below.

$$0 \times -3\frac{5}{6} \times \frac{1}{12}$$

A) $-\dfrac{43}{10}$ B) 0

C) $-\dfrac{1}{6}$ D) $-\dfrac{9}{7}$

184) Find the product of the below.

$$-\frac{4}{5} \times \frac{8}{5} \times \frac{17}{13}$$

A) $-\dfrac{544}{325}$ B) 19

C) $\dfrac{137}{65}$ D) -19

185) Find the product of the below.

$$7\frac{3}{5} \times 4\frac{2}{9} \times -\frac{23}{14}$$

A) $\dfrac{16606}{315}$ B) $-\dfrac{62959}{1260}$

C) $-\dfrac{16606}{315}$ D) $-\dfrac{14891}{315}$

186) Find the product of the below.

$$2 \times -\frac{1}{5} \times \frac{5}{3}$$

A) $-\dfrac{31}{15}$ B) $\dfrac{52}{15}$

C) $-\dfrac{2}{3}$ D) $\dfrac{2}{3}$

Grade 7
Volume 1
Week 1

187) Find the product of the below.

$$\frac{1}{10} \times \frac{7}{4} \times -\frac{1}{11}$$

A) $-\dfrac{1451}{1320}$ B) $-\dfrac{7}{440}$

C) $\dfrac{7}{440}$ D) $-\dfrac{283}{88}$

188) Find the product of the below.

$$4\frac{1}{11} \times -\frac{4}{3} \times \frac{1}{4}$$

A) $\dfrac{15}{11}$ B) $-\dfrac{15}{11}$

C) $\dfrac{39}{11}$ D) $\dfrac{95}{132}$

189) Find the product of the below.

$$-2 \times \frac{2}{5} \times \frac{1}{6}$$

A) $-\dfrac{4}{5}$ B) $\dfrac{2}{15}$

C) $-\dfrac{2}{15}$ D) $-\dfrac{32}{15}$

190) Find the product of the below.

$$-3\frac{1}{2} \times 4\frac{5}{6} \times \frac{5}{4}$$

A) $\dfrac{1015}{48}$ B) $\dfrac{31}{12}$

C) $-\dfrac{1015}{48}$ D) $-\dfrac{919}{48}$

191) Find the product of the below.

$$12 \times 1\frac{7}{10} \times -\frac{9}{5}$$

A) $\dfrac{9945}{308}$ B) $-\dfrac{918}{25}$

C) $-\dfrac{9945}{308}$ D) $\dfrac{918}{25}$

192) Find the product of the below.

$$-1\frac{2}{9} \times 2\frac{2}{7} \times -\frac{6}{11}$$

A) $\dfrac{32}{21}$ B) $\dfrac{349}{105}$

C) $\dfrac{454}{105}$ D) $\dfrac{1643}{210}$

193) Find the product of the below.

$$1\frac{8}{11} \times 6\frac{5}{8} \times -\frac{9}{7}$$

A) $\dfrac{1333}{88}$ B) $-\dfrac{1333}{88}$

C) $-\dfrac{9063}{616}$ D) $\dfrac{9063}{616}$

194) Find the product of the below.

$$5\frac{2}{5} \times -\frac{7}{9} \times \frac{8}{11}$$

A) $-\dfrac{168}{55}$ B) $\dfrac{2648}{495}$

C) $-\dfrac{7739}{715}$ D) $\dfrac{168}{55}$

©All rights reserved-Math-Knots LLC., VA-USA

www.math-knots.com | www.a4ace.com

Grade 7
Volume 1
Week 1

195) Find the product of the below.

$$3\frac{5}{7} \times -\frac{22}{13} \times \frac{19}{10}$$

A) $-\dfrac{418}{35}$ B) $-\dfrac{488}{35}$

C) $\dfrac{418}{35}$ D) $-\dfrac{381}{70}$

196) Find the product of the below.

$$\frac{9}{5} \times \frac{3}{14} \times -\frac{10}{7}$$

A) $-\dfrac{27}{49}$ B) $\dfrac{291}{539}$

C) $\dfrac{41}{70}$ D) $\dfrac{471}{98}$

197) Find the product of the below.

$$-2\frac{4}{9} \times -\frac{7}{5} \times \frac{23}{13}$$

A) $\dfrac{3542}{585}$ B) $-\dfrac{1214}{585}$

C) $\dfrac{7871}{585}$ D) $-\dfrac{3542}{585}$

198) Find the quotient of the below.

$$-3\frac{13}{14} \div -1$$

A) $\dfrac{55}{14}$ B) $\dfrac{14}{55}$

C) $-\dfrac{69}{14}$ D) $-\dfrac{55}{14}$

199) Find the product of the below.

$$-2\frac{3}{13} \times 6\frac{1}{6} \times \frac{8}{5}$$

A) $-\dfrac{4292}{195}$ B) $-\dfrac{4097}{195}$

C) $-\dfrac{2623}{130}$ D) $\dfrac{4292}{195}$

200) Find the product of the below.

$$-2 \times \frac{1}{5} \times \frac{3}{2}$$

A) $\dfrac{37}{20}$ B) $-\dfrac{3}{10}$

C) $-\dfrac{37}{20}$ D) $-\dfrac{3}{5}$

201) Find the product of the below.

$$4 \times -2 \times \frac{7}{11}$$

A) $-\dfrac{56}{11}$ B) $\dfrac{36941}{756}$

C) $-\dfrac{128}{11}$ D) $-\dfrac{36941}{756}$

202) Find the quotient of the below.

$$2\frac{9}{20} \div \frac{-12}{19}$$

A) $-\dfrac{3}{2}$ B) $-\dfrac{147}{95}$

C) $-\dfrac{931}{240}$ D) -1

©All rights reserved-Math-Knots LLC., VA-USA   www.math-knots.com | www.a4ace.com

Grade 7
Volume 1
Week 1

203) Find the quotient of the below.

$$-1\frac{1}{13} \div 10\frac{1}{2}$$

A) $-\frac{4}{39}$ B) $4\frac{13}{18}$

C) $2\frac{13}{14}$ D) $\frac{39}{4}$

204) Find the quotient of the below.

$$1\frac{8}{19} \div \frac{-5}{14}$$

A) $-\frac{1}{5}$ B) $\frac{473}{266}$

C) $\frac{378}{95}$ D) $-\frac{378}{95}$

205) Find the quotient of the below.

$$\frac{-17}{11} \div -2\frac{12}{17}$$

A) $\frac{46}{11}$ B) $-\frac{46}{11}$

C) $-\frac{289}{506}$ D) $\frac{289}{506}$

206) Find the quotient of the below.

$$1 \div \frac{-8}{5}$$

A) 15 B) $-\frac{5}{8}$

C) $\frac{5}{8}$ D) $-\frac{8}{5}$

207) Find the quotient of the below.

$$\frac{-1}{2} \div \frac{-3}{5}$$

A) $-\frac{3}{10}$ B) $\frac{5}{6}$

C) $-\frac{5}{6}$ D) $10\frac{7}{12}$

208) Find the quotient of the below.

$$0 \div 6\frac{1}{3}$$

A) $\frac{9}{13}$ B) 0

C) $\frac{4}{5}$ D) $-\frac{19}{3}$

209) Find the quotient of the below.

$$\frac{2}{3} \div \frac{-24}{13}$$

A) $\frac{13}{36}$ B) $-\frac{36}{13}$

C) $-\frac{13}{36}$ D) $\frac{98}{39}$

210) Find the quotient of the below.

$$\frac{-7}{10} \div 5\frac{2}{9}$$

A) $-\frac{63}{470}$ B) $\frac{4}{9}$

C) $2\frac{11}{20}$ D) $-\frac{470}{63}$

©All rights reserved-Math-Knots LLC., VA-USA   www.math-knots.com | www.a4ace.com

Grade 7
Volume 1
Week 1

211) Find the quotient of the below.

$$-2\frac{1}{8} \div 18$$

A) $-\frac{153}{4}$ B) $\frac{144}{17}$

C) $\frac{1}{6}$ D) $-\frac{17}{144}$

212) Find the quotient of the below.

$$2\frac{17}{20} \div -1$$

A) 17   B) $\frac{20}{57}$

C) $-\frac{57}{20}$ D) $\frac{37}{20}$

213) Find the quotient of the below.

$$\frac{-5}{8} \div 9\frac{9}{10}$$

A) $-\frac{396}{25}$ B) $\frac{25}{396}$

C) $-\frac{25}{396}$ D) $-\frac{99}{16}$

214) Find the quotient of the below.

$$-1\frac{9}{11} \div \frac{33}{17}$$

A) $-\frac{703}{187}$ B) $-\frac{340}{363}$

C) $-\frac{60}{17}$ D) $\frac{363}{340}$

215) Find the quotient of the below.

$$\frac{2}{3} \div \frac{-3}{2}$$

A) 1   B) $-2\frac{13}{15}$

C) $-\frac{5}{6}$ D) $-\frac{4}{9}$

216) Find the quotient of the below.

$$\frac{-3}{13} \div 3\frac{11}{12}$$

A) $-\frac{36}{611}$ B) $\frac{15}{11}$

C) $\frac{575}{156}$ D) $-\frac{20}{17}$

217) Find the quotient of the below.

$$5\frac{1}{18} \div -1\frac{7}{12}$$

A) $\frac{2}{5}$ B) $9\frac{9}{10}$

C) $-\frac{11}{16}$ D) $-\frac{182}{57}$

218) Find the quotient of the below.

$$\frac{-8}{11} \div 9\frac{3}{10}$$

A) $\frac{2}{5}$ B) $-\frac{1023}{80}$

C) $\frac{943}{110}$ D) $-\frac{80}{1023}$

219) Find the quotient of the below.

$$\frac{-9}{13} \div \frac{-25}{18}$$

A) $-\dfrac{487}{234}$ B) $\dfrac{25}{18}$

C) $-\dfrac{325}{162}$ D) $\dfrac{162}{325}$

220) Find the quotient of the below.

$$2 \div \frac{-31}{20}$$

A) $-\dfrac{31}{10}$ B) $\dfrac{9}{20}$

C) $-\dfrac{40}{31}$ D) $-\dfrac{31}{40}$

221) Find the quotient of the below.

$$\frac{-13}{20} \div \frac{-3}{7}$$

A) $-\dfrac{60}{91}$ B) $\dfrac{91}{60}$

C) $\dfrac{14}{15}$ D) -1

222) Find the quotient of the below.

$$\frac{-13}{7} \div \frac{-5}{8}$$

A) $-\dfrac{65}{56}$ B) $\dfrac{104}{35}$

C) $-1\dfrac{9}{16}$ D) $-\dfrac{104}{35}$

223) Find the quotient of the below.

$$-2\frac{9}{20} \div \frac{-1}{2}$$

A) $\dfrac{49}{10}$ B) $-\dfrac{49}{10}$

C) $\dfrac{49}{40}$ D) $-\dfrac{2}{5}$

224) Find the quotient of the below.

$$-1\frac{1}{6} \div 6\frac{6}{11}$$

A) $-\dfrac{432}{77}$ B) $-\dfrac{509}{66}$

C) $\dfrac{432}{77}$ D) $-\dfrac{77}{432}$

225) Find the quotient of the below.

$$14 \div \frac{-3}{2}$$

A) $\dfrac{28}{3}$ B) $\dfrac{25}{2}$

C) $-\dfrac{28}{3}$ D) $-1\dfrac{1}{16}$

226) Find the quotient of the below.

$$\frac{-13}{10} \div 9\frac{11}{12}$$

A) $\dfrac{595}{78}$ B) $-\dfrac{673}{60}$

C) $3\dfrac{11}{20}$ D) $-\dfrac{78}{595}$

Grade 7
Volume 1
Week 1

227) Find the quotient of the below.

$$0 \div \frac{10}{11}$$

A) $10\frac{9}{10}$ B) $-\frac{10}{11}$

C) $\frac{7}{19}$ D) 0

228) Find the quotient of the below.

$$\frac{-28}{15} \div \frac{-24}{13}$$

A) $\frac{91}{90}$ B) $-\frac{90}{91}$

C) $-\frac{224}{65}$ D) $\frac{29}{15}$

229) Find the quotient of the below.

$$\frac{2}{3} \div \frac{-4}{5}$$

A) $\frac{6}{5}$ B) $-\frac{5}{6}$

C) $2\frac{3}{13}$ D) $-\frac{6}{5}$

230) Find the quotient of the below.

$$\frac{6}{13} \div \frac{-17}{9}$$

A) $6\frac{9}{11}$ B) $-\frac{34}{39}$

C) $\frac{221}{54}$ D) $-\frac{54}{221}$

231) Find the quotient of the below.

$$2\frac{9}{20} \div \frac{-10}{7}$$

A) $\frac{343}{200}$ B) $\frac{5}{6}$

C) $-\frac{343}{200}$ D) $\frac{143}{140}$

232) Find the quotient of the below.

$$\frac{-27}{16} \div \frac{-20}{13}$$

A) $-\frac{135}{52}$ B) $-\frac{31}{208}$

C) $-\frac{671}{208}$ D) $\frac{351}{320}$

233) Find the quotient of the below.

$$-3\frac{10}{13} \div \frac{25}{19}$$

A) $9\frac{9}{14}$ B) $-\frac{931}{325}$

C) $8\frac{7}{12}$ D) $\frac{1225}{247}$

234) Find the quotient of the below.

$$-2 \div 2\frac{7}{9}$$

A) $\frac{18}{25}$ B) $\frac{19}{13}$

C) $-\frac{18}{25}$ D) $2\frac{1}{5}$

235) Find the quotient of the below.

$$2\frac{1}{6} \div -3\frac{7}{16}$$

A) $-\dfrac{104}{165}$ B) $-\dfrac{61}{48}$

C) $\dfrac{165}{104}$ D) $-\dfrac{715}{96}$

236) Find the quotient of the below.

$$-2\frac{5}{9} \div \frac{2}{5}$$

A) $\dfrac{9}{11}$ B) $1\dfrac{11}{16}$

C) $-\dfrac{115}{18}$ D) $\dfrac{3}{14}$

237) Find the quotient of the below.

$$1 \div \frac{-18}{11}$$

A) $-\dfrac{11}{18}$ B) $-\dfrac{7}{11}$

C) $\dfrac{15}{16}$ D) $-\dfrac{18}{11}$

238) Find the quotient of the below.

$$\frac{25}{17} \div -1\frac{1}{19}$$

A) $-\dfrac{500}{323}$ B) $3\dfrac{4}{9}$

C) $\dfrac{95}{68}$ D) $-\dfrac{95}{68}$

239) Find the quotient of the below.

$$3\frac{3}{7} \div \frac{-5}{4}$$

A) $-1\dfrac{5}{6}$ B) $-\dfrac{30}{7}$

C) $-\dfrac{96}{35}$ D) $-\dfrac{35}{96}$

240) Find the quotient of the below.

$$-2 \div \frac{4}{11}$$

A) $-\dfrac{11}{2}$ B) $-\dfrac{8}{11}$

C) $-\dfrac{14}{9}$ D) $\dfrac{11}{2}$

Grade 7
Volume 1
Week 2

Grade 7
Volume 1
Week 2

Grade 7
Volume 1
Week 2

1) Evaluate the below

218 − 391 − 459

A) −342 B) −406
C) −632 D) −365

2) Evaluate the below

(−321) − 69 − 84

A) −474 B) −526
C) −459 D) −322

3) Evaluate the below

(−66) + 349 + 356

A) 440 B) 627
C) 639 D) 979

4) Evaluate the below

150 − 231 − (−262)

A) 291 B) 181
C) 361 D) 273

5) Evaluate the below

85 + (−286) + 233

A) 32 B) −60
C) −358 D) 62

6) Evaluate the below

148 + 155 − 68

A) 235 B) 407
C) 665 D) 74

7) Evaluate the below

(−230) − (−112) + 103

A) −319 B) −15
C) 352 D) 358

8) Evaluate the below

(−50) + (−234) − (−202)

A) 72 B) −125
C) −82 D) −431

9) Evaluate the below

$(-21) + (-492) - 316$

A) −1189 B) −829
C) −467 D) −1221

10) Evaluate the below

$143 - (-228) + (-104)$

A) 579 B) −129
C) 439 D) 267

11) Evaluate the below

$(-477) + 95 - 193$

A) −223 B) −208
C) −575 D) −1066

12) Evaluate the below

$(-365) + (-301) - (-300)$

A) −558 B) −400
C) −366 D) 94

13) Evaluate the below

$(-269) - (-378) + (-200)$

A) −91 B) −63
C) −185 D) −119

14) Evaluate the below

$(-279) + 459 + 339$

A) 345 B) 519
C) 659 D) 674

15) Evaluate the below

$(-180) - 6 - (-491)$

A) 305 B) 657
C) 18 D) 239

16) Evaluate the below

$(-460) + (-425) + 358$

A) −527 B) −344
C) −908 D) −613

17) Evaluate the below

$(-138) - 144 - 339$

A) −624 B) −621
C) −1047 D) −583

18) Evaluate the below

$(-191) + 242 - (-383)$

A) 434 B) −18
C) 893 D) 468

19) Find the product of the below.

-7×16

A) −121 B) −112
C) 112 D) −120

20) Find the product of the below.

22×-2

A) −15 B) −44
C) −69 D) 44

21) Evaluate the below

$39 - 177 + 162$

A) 112 B) 24
C) 0 D) −279

22) Evaluate the below

$(-230) + 105 + (-97)$

A) 213 B) −53
C) −222 D) −678

23) Find the product of the below.

21×-7

A) −119 B) 147
C) −131 D) −147

24) Find the product of the below.

-7×-19

A) −810 B) 150
C) 133 D) 810

Grade 7
Volume 1
Week 2

25) Find the product of the below.

4 × −27

A) −108   B) 108
C) −374   D) 374

26) Find the product of the below.

−7 × 13

A) 91   B) −91
C) −81   D) 6

27) Find the product of the below.

−20 × 12

A) 240   B) −260
C) −263   D) −240

28) Find the product of the below.

−10 × −31

A) 310   B) −41
C) −310   D) −667

29) Find the product of the below.

−5 × 5

A) −38   B) 0
C) −25   D) 25

30) Find the product of the below.

13 × −28

A) −375   B) −335
C) −364   D) −336

31) Find the product of the below.

−2 × 27

A) −638   B) 638
C) −67   D) −54

32) Find the product of the below.

−23 × 2

A) −46   B) 46
C) 165   D) −21

94

33) Find the product of the below.

-24×-34

A) 800 B) −702
C) 816 D) −816

34) Find the product of the below.

20×-30

A) −622 B) −10
C) −630 D) −600

35) Find the quotient of the below.

$-361 \div -19$

A) 19 B) 30
C) 6859 D) −26

36) Find the quotient of the below.

$196 \div -7$

A) −28 B) 28
C) 203 D) −1372

37) Find the product of the below.

-28×-1

A) −28 B) 2
C) −29 D) 28

38) Find the quotient of the below.

$-48 \div 24$

A) −1152 B) −27
C) 1152 D) −2

39) Find the quotient of the below.

$-400 \div -16$

A) −0.04 B) −25
C) −384 D) 25

40) Find the quotient of the below.

$30 \div -30$

A) −1 B) −900
C) 18 D) 9

Grade 7
Volume 1
Week 2

41) Find the quotient of the below.

$$-171 \div -9$$

A) 19 B) −180
C) −1539 D) 1539

42) Find the quotient of the below.

$$-84 \div -6$$

A) −1 B) −78
C) 14 D) −14

43) Find the quotient of the below.

$$-196 \div 14$$

A) 14 B) 2744
C) −14 D) −210

44) Find the quotient of the below.

$$441 \div -21$$

A) 21 B) 420
C) 26 D) −21

45) Find the quotient of the below.

$$0 \div -24$$

A) 27 B) 0
C) −24 D) −22

46) Find the quotient of the below.

$$-528 \div 24$$

A) −9 B) 22
C) −4 D) −22

47) Find the quotient of the below.

$$-126 \div -6$$

A) 21 B) −132
C) −23 D) −120

48) Find the quotient of the below.

$$-216 \div -8$$

A) 15 B) 27
C) 21 D) −1728

Grade 7
Volume 1
Week 2

49) Find the quotient of the below.

$$-228 \div -19$$

A) 4332   B) 12

C) −247   D) −5

53) Find the quotient of the below.

$$-120 \div -15$$

A) −135   B) 20

C) −0.125   D) 8

50) Simplify the below and convert the answer to positive exponents.

$$10^0 \cdot 10^6$$

A) 10^4 B) $\dfrac{1}{10^6}$

C) 10^6 D) 10^2

54) Simplify the below and convert the answer to positive exponents.

$$6^7 \cdot 6^{-2} \cdot 6^3$$

A) 6^8 B) $\dfrac{1}{6^4}$

C) 6 D) 6^{10}

51) Simplify the below and convert the answer to positive exponents.

$$16^4 \cdot 16^6$$

A) 16^{10} B) 16^{11}

C) $\dfrac{1}{16^5}$ D) 16^5

55) Simplify the below and convert the answer to positive exponents.

$$16 \cdot 16^7$$

A) $\dfrac{1}{16^{10}}$ B) $\dfrac{1}{16^5}$

C) 16^5 D) 16^8

52) Simplify the below and convert the answer to positive exponents.

$$5^2 \cdot 5^{-8}$$

A) $\dfrac{1}{5^5}$ B) $\dfrac{1}{5^6}$

C) 5^4 D) $\dfrac{1}{5^2}$

56) Simplify the below and convert the answer to positive exponents.

$$8^8 \cdot 8^{-1}$$

A) $\dfrac{1}{8^4}$ B) 8^7

C) 8^{18} D) 1

©All rights reserved-Math-Knots LLC., VA-USA   www.math-knots.com | www.a4ace.com

Grade 7
Volume 1
Week 2

57) Simplify the below and convert the answer to positive exponents.

$$6^{-7} \cdot 6^6 \cdot 6^2$$

A) 6^{10} B) 6

C) 6^{15} D) 6^6

58) Simplify the below and convert the answer to positive exponents.

$$3^3 \cdot 3^4$$

A) 3^{13} B) 3^7

C) $\dfrac{1}{3^7}$ D) 3^3

59) Simplify the below and convert the answer to positive exponents.

$$11^7 \cdot 11^5$$

A) 11 B) 11^7

C) 11^{12} D) 11^{11}

60) Simplify the below and convert the answer to positive exponents.

$$10^{-6} \cdot 10^{-8}$$

A) $\dfrac{1}{10^{14}}$ B) 10^{15}

C) 10^7 D) 10^2

61) Simplify the below and convert the answer to positive exponents.

$$9^8 \cdot 9^{-2}$$

A) 9^6 B) 9^2

C) 9^8 D) $\dfrac{1}{9^4}$

62) Simplify the below and convert the answer to positive exponents.

$$9^5 \cdot 9^4$$

A) 9^9 B) 9

C) 9^7 D) 9^2

63) Simplify the below and convert the answer to positive exponents.

$$12^2 \cdot 12^{-5}$$

A) 12^{16} B) $\dfrac{1}{12^8}$

C) $\dfrac{1}{12^3}$ D) 12^{13}

64) Simplify the below and convert the answer to positive exponents.

$$11^8 \cdot 11^8$$

A) 11^5 B) 11^{10}

C) 11^7 D) 11^{16}

65) Simplify the below and convert the answer to positive exponents.

$$16^8 \cdot 16^3 \cdot 16^{-2}$$

A) 16^{11} B) 16^6
C) $\dfrac{1}{16^4}$ D) 16^9

66) Simplify the below and convert the answer to positive exponents.

$$15^{-8} \cdot 15^{-2}$$

A) $\dfrac{1}{15^{10}}$ B) 15^6
C) 15^2 D) 15^{10}

67) Simplify the below and convert the answer to positive exponents.

$$16^7 \cdot 16^{-3} \cdot 16^0$$

A) $\dfrac{1}{16^7}$ B) 16^4
C) 16^8 D) $\dfrac{1}{16^5}$

68) Simplify the below and convert the answer to positive exponents.

$$\dfrac{10^9}{10^{-7}}$$

A) 10^{16} B) 10^2
C) $\dfrac{1}{10^3}$ D) $\dfrac{1}{10^4}$

69) Simplify the below and convert the answer to positive exponents.

$$4^6 \cdot 4^{-8} \cdot 4^5$$

A) $\dfrac{1}{4}$ B) 4
C) 4^{13} D) 4^3

70) Simplify the below and convert the answer to positive exponents.

$$11^5 \cdot 11^{-5}$$

A) 11^{12} B) 11^4
C) 11^{11} D) 1

71) Simplify the below and convert the answer to positive exponents.

$$14^{-7} \cdot 14^7$$

A) 1 B) 14^8
C) 14 D) 14^2

72) Simplify the below and convert the answer to positive exponents.

$$\dfrac{9^{-7}}{9^{-1}}$$

A) 9^7 B) 9^{16}
C) $\dfrac{1}{9^6}$ D) 9^{12}

73) Simplify the below and convert the answer to positive exponents.

$$\frac{4^{-5}}{4^{-5}}$$

A) 1 B) 4^2
C) $\frac{1}{4}$ D) 4

74) Simplify the below and convert the answer to positive exponents.

$$\frac{5^{-8}}{5^0}$$

A) 5^5 B) $\frac{1}{5^4}$
C) $\frac{1}{5^{12}}$ D) $\frac{1}{5^8}$

75) Simplify the below and convert the answer to positive exponents.

$$\frac{9}{9^{-5}}$$

A) 9^6 B) $\frac{1}{9^2}$
C) $\frac{1}{9^{16}}$ D) 9^5

76) Simplify the below and convert the answer to positive exponents.

$$\frac{11^2}{11^3}$$

A) $\frac{1}{11^3}$ B) 11^{12}
C) $\frac{1}{11}$ D) 11^5

77) Simplify the below and convert the answer to positive exponents.

$$\frac{8^8}{8^3}$$

A) $\frac{1}{8^{15}}$ B) 8^5
C) $\frac{1}{8^3}$ D) $\frac{1}{8^{14}}$

78) Simplify the below and convert the answer to positive exponents.

$$\frac{13^0}{13^5}$$

A) $\frac{1}{13^9}$ B) $\frac{1}{13^5}$
C) $\frac{1}{13^3}$ D) 13^2

79) Simplify the below and convert the answer to positive exponents.

$$\frac{10^{-7}}{10^{-8}}$$

A) $\frac{1}{10^8}$ B) 10^7
C) 10 D) $\frac{1}{10^9}$

80) Simplify the below and convert the answer to positive exponents.

$$\frac{9^{-8}}{9^5}$$

A) $\frac{1}{9^{13}}$ B) 1
C) $\frac{1}{9^{11}}$ D) 9^6

Grade 7
Volume 1
Week 2

81) Simplify the below and convert the answer to positive exponents.

$$\frac{11^{-4}}{11^7}$$

A) $\frac{1}{11^{11}}$ B) $\frac{1}{11^{13}}$

C) 11^3 D) 11^4

82) Simplify the below and convert the answer to positive exponents.

$$\frac{18^0}{18}$$

A) $\frac{1}{18}$ B) $\frac{1}{18^5}$

C) $\frac{1}{18^{13}}$ D) $\frac{1}{18^8}$

83) Simplify the below and convert the answer to positive exponents.

$$\frac{12}{12^{-6}}$$

A) 12^8 B) 12^9

C) 12^7 D) 12^{10}

84) Simplify the below and convert the answer to positive exponents.

$$\frac{10^9}{10^6}$$

A) $\frac{1}{10^8}$ B) 10^2

C) 10^7 D) 10^3

85) Simplify the below and convert the answer to positive exponents.

$$\frac{4^2}{4^4}$$

A) $\frac{1}{4^2}$ B) $\frac{1}{4^5}$

C) 4^{11} D) $\frac{1}{4}$

86) Simplify the below and convert the answer to positive exponents.

$$\frac{14^{-1}}{14^{-9}}$$

A) $\frac{1}{14^4}$ B) 14^8

C) $\frac{1}{14^2}$ D) $\frac{1}{14^5}$

87) Simplify the below and convert the answer to positive exponents.

$$\frac{14^4}{14^5}$$

A) $\frac{1}{14^5}$ B) $\frac{1}{14^2}$

C) 1 D) $\frac{1}{14}$

88) Simplify the below and convert the answer to positive exponents.

$$\frac{13^{-3}}{13^4}$$

A) 13^2 B) $\frac{1}{13^2}$

C) $\frac{1}{13}$ D) $\frac{1}{13^7}$

89) Simplify the below and convert the answer to positive exponents.

$$\frac{18^0}{18^2}$$

A) $\frac{1}{18^7}$ B) 18^2

C) $\frac{1}{18^2}$ D) $\frac{1}{18^3}$

90) Simplify the below and convert the answer to positive exponents.

$$\left(8^3\right)^{-1}$$

A) $\frac{1}{8^3}$ B) $\frac{1}{8^8}$

C) 8^8 D) $\frac{1}{8^4}$

91) Simplify the below and convert the answer to positive exponents.

$$2^{-3}$$

A) $\frac{1}{2^3}$ B) $\frac{1}{2^{12}}$

C) $\frac{1}{2^9}$ D) $\frac{1}{2^8}$

92) Simplify the below and convert the answer to positive exponents.

$$\left(3^3\right)^3$$

A) $\frac{1}{3^6}$ B) 3^9

C) 3^2 D) 3^8

93) Simplify the below and convert the answer to positive exponents.

$$\frac{12^7}{12^0}$$

A) 12^7 B) $\frac{1}{12}$

C) 12^2 D) 12^6

94) Simplify the below and convert the answer to positive exponents.

$$\left(4^4\right)^2$$

A) 4^{12} B) 4^8

C) 1 D) $\frac{1}{4^8}$

95) Simplify the below and convert the answer to positive exponents.

$$\left(4^2\right)^{-4}$$

A) $\frac{1}{4^4}$ B) $\frac{1}{4^8}$

C) 1 D) $\frac{1}{4^2}$

96) Simplify the below and convert the answer to positive exponents.

$$7^0$$

A) 7^4 B) 1

C) $\frac{1}{7^3}$ D) 7^2

97) Simplify the below and convert the answer to positive exponents.

$$8^{-2}$$

A) 8 B) 8^{12}

C) $\dfrac{1}{8^2}$ D) 8^6

98) Simplify the below and convert the answer to positive exponents.

$$(6^{-1})^0$$

A) $\dfrac{1}{6^9}$ B) 6^2

C) 1 D) $\dfrac{1}{6^4}$

99) Simplify the below and convert the answer to positive exponents.

$$(8^2)^3$$

A) 8^6 B) $\dfrac{1}{8^8}$

C) $\dfrac{1}{8^{12}}$ D) 1

100) Simplify the below and convert the answer to positive exponents.

$$(3^4)^3$$

A) 3^4 B) 3^3

C) 3^{12} D) 3^2

101) Simplify the below and convert the answer to positive exponents.

$$(2^{-1})^{-1}$$

A) $\dfrac{1}{2^3}$ B) 2

C) $\dfrac{1}{2^{16}}$ D) $\dfrac{1}{2^2}$

102) Simplify the below and convert the answer to positive exponents.

$$(6^{-3})^2$$

A) $\dfrac{1}{6^6}$ B) 6^3

C) 1 D) 6^9

103) Simplify the below and convert the answer to positive exponents.

$$(8^3)^{-3}$$

A) 8^6 B) 8^{12}

C) 1 D) $\dfrac{1}{8^9}$

104) Simplify the below and convert the answer to positive exponents.

$$(2^{-1})^4$$

A) $\dfrac{1}{2^4}$ B) 2^4

C) $\dfrac{1}{2^2}$ D) $\dfrac{1}{2^6}$

105) Simplify the below and convert the answer to positive exponents.

$$(8^4)^4$$

A) 8^8 B) $\dfrac{1}{8^{16}}$

C) 8^6 D) 8^{16}

106) Simplify the below and convert the answer to positive exponents.

$$8^3$$

A) $\dfrac{1}{8^8}$ B) 8^{12}

C) 8^3 D) 1

107) Simplify the below and convert the answer to positive exponents.

$$5^2$$

A) 5^9 B) 5^2

C) $\dfrac{1}{5^4}$ D) 5^{12}

108) Simplify the below and convert the answer to positive exponents.

$$\dfrac{11 \cdot 11^2}{11^{-3}}$$

A) $\dfrac{1}{11^3}$ B) $\dfrac{1}{11^8}$

C) 11   D) 11^6

109) Simplify the below and convert the answer to positive exponents.

$$(2^0)^0$$

A) 2^8 B) 1

C) 2^6 D) $\dfrac{1}{2^{12}}$

110) Simplify the below and convert the answer to positive exponents.

$$(4^2)^4$$

A) $\dfrac{1}{4^9}$ B) 4^{12}

C) $\dfrac{1}{4^{12}}$ D) 4^8

111) Simplify the below and convert the answer to positive exponents.

$$2^3$$

A) $\dfrac{1}{2^4}$ B) 2^3

C) $\dfrac{1}{2^{12}}$ D) 1

112) Simplify the below and convert the answer to positive exponents.

$$\dfrac{5^{-8}}{5^0 \cdot 5^7}$$

A) 5^3 B) $\dfrac{1}{5^{15}}$

C) 5^7 D) $\dfrac{1}{5^9}$

113) Simplify the below and convert the answer to positive exponents.

$$\frac{12^6 \cdot 12^2}{12^{-3}}$$

A) 12^7 B) 12^6

C) $\frac{1}{12}$ D) 12^{11}

114) Simplify the below and convert the answer to positive exponents.

$$\frac{16 \cdot 16^0}{16^8}$$

A) 16^{11} B) 16

C) $\frac{1}{16^7}$ D) 16^8

115) Simplify the below and convert the answer to positive exponents.

$$\frac{10 \cdot 10^7}{10^{-5}}$$

A) 10^{13} B) 10^8

C) $\frac{1}{10^{12}}$ D) 10^{10}

116) Simplify the below and convert the answer to positive exponents.

$$\frac{2^7 \cdot 2^3}{2^8}$$

A) $\frac{1}{2^8}$ B) 2^2

C) 2^7 D) $\frac{1}{2^{22}}$

117) Simplify the below and convert the answer to positive exponents.

$$\frac{256}{16^7 \cdot 16^{-3}}$$

A) 16^{11} B) $\frac{1}{16^2}$

C) $\frac{1}{16^4}$ D) $\frac{1}{16^{15}}$

118) Simplify the below and convert the answer to positive exponents.

$$\frac{7^2}{7^5 \cdot 7^{-2}}$$

A) $\frac{1}{7}$ B) $\frac{1}{7^6}$

C) $\frac{1}{7^{11}}$ D) $\frac{1}{7^8}$

119) Simplify the below and convert the answer to positive exponents.

$$\frac{10^{-8}}{10 \cdot 10^4}$$

A) 10^{15} B) $\frac{1}{10^{13}}$

C) 10^4 D) 10

120) Simplify the below and convert the answer to positive exponents.

$$\frac{3^{-4} \cdot 3^{-7}}{3^{-2}}$$

A) $\frac{1}{3^{16}}$ B) $\frac{1}{3^9}$

C) $\frac{1}{3^2}$ D) $\frac{1}{3^8}$

121) Simplify the below and convert the answer to positive exponents.

$$(2^0)^{-4} \cdot 2^3$$

A) 2^{17} B) 2^{12}

C) 2^3 D) 1

122) Simplify the below and convert the answer to positive exponents.

$$4^3 \cdot (4^{-4})^2$$

A) 1 B) $\dfrac{1}{4^{13}}$

C) $\dfrac{1}{4^5}$ D) 4^{30}

123) Simplify the below and convert the answer to positive exponents.

$$(2^4 \cdot 2^4)^2$$

A) $\dfrac{1}{2^9}$ B) 2^{16}

C) 1 D) $\dfrac{1}{2^4}$

124) Simplify the below and convert the answer to positive exponents.

$$(4 \cdot 4^4)^0$$

A) 4^{10} B) $\dfrac{1}{4^4}$

C) $\dfrac{1}{4^6}$ D) 1

125) Simplify the below and convert the answer to positive exponents.

$$(2^{-2} \cdot 2^3)^2$$

A) $\dfrac{1}{2^4}$ B) 2^6

C) 2^2 D) $\dfrac{1}{2^{16}}$

126) Simplify the below and convert the answer to positive exponents.

$$2^4 \cdot 2^2$$

A) $\dfrac{1}{2^{18}}$ B) 2^6

C) $\dfrac{1}{2^{11}}$ D) 2^{15}

127) Simplify the below and convert the answer to positive exponents.

$$4^4 \cdot (4^{-2})^2$$

A) 1 B) $\dfrac{1}{4^7}$

C) $\dfrac{1}{4^{15}}$ D) $\dfrac{1}{4^6}$

128) Simplify the below and convert the answer to positive exponents.

$$(3 \cdot 3^2)^0$$

A) 3^{11} B) 1

C) $\dfrac{1}{3^8}$ D) $\dfrac{1}{3^2}$

129) Simplify the below and convert the answer to positive exponents.

$$(2^{-3})^2 \cdot 2^{-2}$$

A) 2^{13} B) $\dfrac{1}{2^8}$

C) 2^9 D) 2^3

130) Simplify the below and convert the answer to positive exponents.

$$\dfrac{4^2}{(4^3)^2}$$

A) $\dfrac{1}{4^4}$ B) $\dfrac{1}{4^2}$

C) 4^4 D) $\dfrac{1}{4^{15}}$

131) Simplify the below and convert the answer to positive exponents.

$$\dfrac{2^2}{2^3}$$

A) 2^2 B) $\dfrac{1}{2^{14}}$

C) 2^4 D) $\dfrac{1}{2}$

132) Simplify the below and convert the answer to positive exponents.

$$(3^0 \cdot 3^4)^2$$

A) 3^{15} B) 3^8

C) 3^{18} D) $\dfrac{1}{3^6}$

133) Simplify the below and convert the answer to positive exponents.

$$\dfrac{(2^3)^{-2}}{2^2}$$

A) $\dfrac{1}{2^8}$ B) $\dfrac{1}{2^{10}}$

C) $\dfrac{1}{2^4}$ D) $\dfrac{1}{2^3}$

134) Simplify the below and convert the answer to positive exponents.

$$\left(\dfrac{(2^3)^{-2}}{2^2}\right)^{-1}$$

A) 2^8 B) 2^{20}

C) 2^3 D) $\dfrac{1}{2^4}$

135) Simplify the below and convert the answer to positive exponents.

$$\frac{(2^2)^2}{2^{-1}}$$

A) 2^3 B) $\frac{1}{2^4}$

C) 2^{14} D) 2^5

136) Simplify the below and convert the answer to positive exponents.

$$\left(\frac{4^2}{4^{-1}}\right)^2$$

A) 4^7 B) 4^{17}

C) $\frac{1}{4^5}$ D) 4^6

137) Simplify the below and convert the answer to positive exponents.

$$\frac{(3^{-4})^{-2}}{3^{-2}}$$

A) 3^{10} B) 3^{15}

C) $\frac{1}{3^5}$ D) 1

138) Simplify the below and convert the answer to positive exponents.

$$\left(\frac{2^3}{(2^2)^3}\right)^4$$

A) 2^7 B) 2^{28}

C) $\frac{1}{2^{12}}$ D) 2^{16}

139) Simplify the below and convert the answer to positive exponents.

$$\frac{3^0}{(3^0)^2}$$

A) 1 B) 3^{18}

C) 3^4 D) $\frac{1}{3^4}$

140) Simplify the below and convert the answer to positive exponents.

$$\left(\frac{4}{4^4}\right)^2$$

A) 4^3 B) 4^6

C) $\frac{1}{4^6}$ D) $\frac{1}{4^8}$

141) Evaluate

$$\left(\frac{-1}{8}\right)^{(-1)}$$

142) Write 5 × 5 × 5 × 5 × 5 × 5 in form of exponents

143) Evaluate : 4^4

144) Evaluate : 6^{-3}

145) Evaluate

$$\left(\frac{3}{-5}\right)^3$$

146) Express $\frac{-81}{256}$ in exponential form.

147) Evaluate

$$\left(\frac{2}{9}\right)^9 \times \left(\frac{2}{9}\right)^5$$

148) Evaluate

$$\left(\frac{5}{11}\right)^{20} \div \left(\frac{5}{11}\right)^{21}$$

149) Evaluate

$$\left\{\left(\frac{1}{5}\right)^2\right\}^3$$

150) Express $\dfrac{-1}{-216}$ in exponential form.

151) Evaluate

$$\left(\frac{-17}{23}\right)^{-1}$$

152) By what number should we multiply 2^{-6} so that the product is 4?

153) By what number should -17^5 be divided so that the quotient is 17^{-2}

154) Find the reciprocal of $\left(\dfrac{5}{7}\right)^4$

155) Evaluate:

$$\left(\frac{-1}{2}\right)^5 \times 2^4 \times \left(\frac{3}{4}\right)^0$$

156) Evaluate:

$$\left(2^{-2} \times 2^{-3}\right)^4$$

157) Evaluate

$$3 \times (-1)^{99}$$

158) Simplify the following:

$$\left(\frac{5}{7}\right)^4 \times \left(\frac{3}{7}\right)^{-2} \times \left(\frac{3}{5}\right)^3$$

159) Evaluate

$$(\sqrt{3})^3 \times (3\sqrt{3})^4 \times \sqrt{3}$$

160) The product of two numbers is 1. If one of them is $9^{-\left(\frac{4}{5}\right)}$, what is the other number?

161) By what number should we divide 81^2 to get $\frac{1}{3}$?

162) Express 243 in exponential form.

163) Find the value of :

$$(2^3)^2 + \left(\frac{3}{4}\right)^0 + 2^5 \times \left(\frac{1}{2}\right)^3$$

164) Find the value of $(-2)^3 \times 2^{-3}$

165) Find the value of the following:

$$3^2 \times 3^{-4} \times 2^3 \times 2^{-2} \times 3$$

166) Express the following as the power of 2:

$$4^2 \times 8^{-2} \times 4^3$$

167) Simplify:

$$5^7 \times 5^{-13}$$

168) Simplify:

$$(-3)^{12} \times (-3)^{13}$$

169) Simplify the following:

$$8^5 \times 2^3 \times 4^{-4}$$

170) Evaluate

$$\left(\frac{5}{6}\right)^3$$

171) Find the value of

$$\left(\frac{5}{3}\right)^4 \times \left(\frac{2}{5}\right)^3$$

172) Evaluate

$$\left(\frac{4}{5}\right)^0 \times \left(\frac{5}{7}\right)^2 \times 7^3$$

Grade 7
Volume 1
Week 2

173) Evaluate

$$(3^{-3} \times 3^{-2})^2$$

174) Express $\dfrac{169}{196}$ in exponential form.

175) Express $\dfrac{243}{3125}$ in exponential form.

176) Evaluate

$$\left[\left(\dfrac{2}{5}\right)^3\right]^2$$

177) Evaluate

$$\left(\dfrac{2}{7}\right)^8 \div \left(\dfrac{2}{7}\right)^6$$

178) Find the reciprocal of $\left(\dfrac{7}{5}\right)^5$

179) Simplify:

$$\left(\dfrac{5}{3}\right)^3 \times \left(\dfrac{2}{3}\right)^0 \times 3^2$$

180) Simplify:

$$\left(\dfrac{8}{5}\right)^{-3} \times \left(\dfrac{6}{5}\right)^0 \times (5)^{-2} \times \left(\dfrac{1}{8}\right)^{-1}$$

181) By what number should we multiply 5^{-7} so that the product is 5^{-1}

Grade 7
Volume 1
Week 3

Grade 7
Volume 1
Week 3

1) State if the below pair of ratios form a proportion.

$$\frac{25}{21} \text{ and } \frac{5}{3}$$

A) No B) Yes

2) State if the below pair of ratios form a proportion.

$$\frac{6}{5} \text{ and } \frac{30}{25}$$

A) No B) Yes

3) State if the below pair of ratios form a proportion.

$$\frac{6}{3} \text{ and } \frac{24}{9}$$

A) Yes B) No

4) State if the below pair of ratios form a proportion.

$$\frac{18}{10} \text{ and } \frac{6}{5}$$

A) No B) Yes

5) State if the below pair of ratios form a proportion.

$$\frac{20}{16} \text{ and } \frac{5}{4}$$

A) Yes B) No

6) State if the below pair of ratios form a proportion.

$$\frac{6}{3} \text{ and } \frac{36}{24}$$

A) Yes B) No

7) State if the below pair of ratios form a proportion.

$$\frac{6}{4} \text{ and } \frac{24}{20}$$

A) No B) Yes

8) State if the below pair of ratios form a proportion.

$$\frac{4}{3} \text{ and } \frac{16}{9}$$

A) Yes B) No

9) State if the below pair of ratios form a proportion.

$$\frac{35}{15} \text{ and } \frac{5}{3}$$

A) No   B) Yes

10) State if the below pair of ratios form a proportion.

$$\frac{4}{5} \text{ and } \frac{16}{15}$$

A) No   B) Yes

11) State if the below pair of ratios form a proportion.

$$\frac{3}{4} \text{ and } \frac{15}{12}$$

A) Yes   B) No

12) State if the below pair of ratios form a proportion.

$$\frac{12}{10} \text{ and } \frac{6}{5}$$

A) No   B) Yes

13) State if the below pair of ratios form a proportion.

$$\frac{18}{36} \text{ and } \frac{3}{6}$$

A) Yes   B) No

14) State if the below pair of ratios form a proportion.

$$\frac{4}{3} \text{ and } \frac{8}{9}$$

A) No   B) Yes

15) State if the below pair of ratios form a proportion.

$$\frac{30}{24} \text{ and } \frac{5}{4}$$

A) Yes   B) No

16) State if the below pair of ratios form a proportion.

$$\frac{6}{4} \text{ and } \frac{36}{24}$$

A) No   B) Yes

17) State if the below pair of ratios form a proportion.

$$\frac{4}{3} \text{ and } \frac{20}{9}$$

A) No B) Yes

18) State if the below pair of ratios form a proportion.

$$\frac{12}{15} \text{ and } \frac{4}{5}$$

A) Yes B) No

19) State if the below pair of ratios form a proportion.

$$\frac{3.4}{2.9} \text{ and } \frac{6.8}{5.8}$$

A) Yes B) No

20) State if the below pair of ratios form a proportion.

$$\frac{3.9}{2.2} \text{ and } \frac{15.6}{8.8}$$

A) Yes B) No

21) State if the below pair of ratios form a proportion.

$$\frac{30}{12} \text{ and } \frac{5}{3}$$

A) Yes B) No

22) State if the below pair of ratios form a proportion.

$$\frac{24}{40} \text{ and } \frac{4}{5}$$

A) No B) Yes

23) State if the below pair of ratios form a proportion.

$$\frac{3.3}{3.7} \text{ and } \frac{13.2}{14.8}$$

A) Yes B) No

24) State if the below pair of ratios form a proportion.

$$\frac{2.8}{2.4} \text{ and } \frac{11.2}{9.6}$$

A) No B) Yes

25) State if the below pair of ratios form a proportion.

$$\frac{11.6}{10.4} \text{ and } \frac{2.9}{2.6}$$

A) No　　　B) Yes

26) State if the below pair of ratios form a proportion.

$$\frac{8.4}{9.6} \text{ and } \frac{2.8}{2.4}$$

A) No　　　B) Yes

27) State if the below pair of ratios form a proportion.

$$\frac{3.1}{2.5} \text{ and } \frac{6.2}{7.5}$$

A) No　　　B) Yes

28) State if the below pair of ratios form a proportion.

$$\frac{6.8}{14.8} \text{ and } \frac{3.4}{3.7}$$

A) Yes　　　B) No

29) State if the below pair of ratios form a proportion.

$$\frac{10.8}{7.8} \text{ and } \frac{3.6}{2.6}$$

A) No　　　B) Yes

30) State if the below pair of ratios form a proportion.

$$\frac{10.5}{11.4} \text{ and } \frac{3.5}{3.8}$$

A) No　　　B) Yes

31) State if the below pair of ratios form a proportion.

$$\frac{16.5}{12} \text{ and } \frac{3.3}{4}$$

A) No　　　B) Yes

32) State if the below pair of ratios form a proportion.

$$\frac{5.6}{5.4} \text{ and } \frac{2.8}{2.7}$$

A) Yes　　　B) No

Grade 7
Volume 1
Week 3

33) State if the below pair of ratios form a proportion.

$$\frac{16}{15.6} \text{ and } \frac{4}{3.9}$$

A) Yes B) No

34) State if the below pair of ratios form a proportion.

$$\frac{2.8}{2.4} \text{ and } \frac{14}{9.6}$$

A) No B) Yes

35) Solve the below proportion.

$$\frac{20\,r}{2} = \frac{14}{15}$$

A) 8 B) 7

C) $\frac{7}{75}$ D) 1

36) Solve the below proportion.

$$\frac{x}{15} = \frac{16}{5}$$

A) 13 B) $\frac{19}{10}$

C) 2 D) 48

37) State if the below pair of ratios form a proportion.

$$\frac{3.3}{3.8} \text{ and } \frac{13.2}{7.6}$$

A) Yes B) No

38) State if the below pair of ratios form a proportion.

$$\frac{3.8}{3.5} \text{ and } \frac{7.6}{10.5}$$

A) No B) Yes

39) Solve the below proportion.

$$\frac{6}{x} = \frac{9}{7}$$

A) $\frac{3}{5}$ B) 15

C) $\frac{14}{3}$ D) 7

40) Solve the below proportion.

$$\frac{20}{9} = \frac{14}{v}$$

A) $\frac{1}{5}$ B) $\frac{63}{10}$

C) $6\frac{5}{6}$ D) 2

Grade 7
Volume 1
Week 3

41) Solve the below proportion.

$$\frac{5}{a} = \frac{9}{16}$$

A) $7\frac{11}{15}$ B) 2

C) $\frac{80}{9}$ D) $\frac{3}{2}$

42) Solve the below proportion.

$$\frac{20}{n} = \frac{9}{12}$$

A) $\frac{80}{3}$ B) $\frac{13}{7}$

C) 11 D) 18

43) Solve the below proportion.

$$\frac{11}{p} = \frac{19}{12}$$

A) $\frac{19}{17}$ B) $6\frac{6}{7}$

C) $1\frac{14}{15}$ D) $\frac{132}{19}$

44) Solve the below proportion.

$$\frac{a}{8} = \frac{12}{10}$$

A) 6 B) 1

C) 16 D) $\frac{48}{5}$

45) Solve the below proportion.

$$\frac{11}{n} = \frac{20}{9}$$

A) 2 B) $\frac{13}{10}$

C) 4 D) $\frac{99}{20}$

46) Solve the below proportion.

$$\frac{15}{5} = \frac{3}{x}$$

A) 10 B) 1

C) $2\frac{9}{20}$ D) 17

47) Solve the below proportion.

$$\frac{6}{n} = \frac{17}{8}$$

A) $\frac{48}{17}$ B) $4\frac{5}{7}$

C) 2 D) 11

48) Solve the below proportion.

$$\frac{12}{13} = \frac{3}{n}$$

A) 14 B) 20

C) $\frac{8}{11}$ D) $\frac{13}{4}$

49) Solve the below proportion.

$$\frac{6}{11} = \frac{10}{b}$$

A) 12   B) $\frac{55}{3}$

C) 11   D) 7

50) Solve the below proportion.

$$\frac{6}{5} = \frac{10}{m}$$

A) 20   B) $10\frac{14}{17}$

C) 8   D) $\frac{25}{3}$

51) Solve the below proportion.

$$\frac{x}{13} = \frac{19}{15}$$

A) 3   B) $\frac{247}{15}$

C) 14   D) 2

52) Solve the below proportion.

$$\frac{18}{19} = \frac{a}{16}$$

A) $\frac{2}{19}$ B) $10\frac{8}{9}$

C) $\frac{288}{19}$ D) $\frac{18}{17}$

53) Solve the below proportion.

$$\frac{x}{10} = \frac{16}{14}$$

A) $8\frac{1}{10}$ B) $\frac{80}{7}$

C) $1\frac{5}{17}$ D) $\frac{1}{10}$

54) Solve the below proportion.

$$\frac{b}{9} = \frac{16}{6}$$

A) $6\frac{7}{8}$ B) $3\frac{9}{16}$

C) $\frac{3}{2}$ D) 24

55) Solve the below proportion.

$$\frac{v}{12} = \frac{7}{14}$$

A) 5   B) $6\frac{2}{3}$

C) 6   D) $\frac{8}{11}$

56) Solve the below proportion.

$$\frac{3}{14} = \frac{16}{n}$$

A) $\frac{19}{12}$ B) 11

C) 10   D) $\frac{224}{3}$

57) Solve the below proportion.

$$\frac{2}{15} = \frac{18n}{10}$$

A) $5\frac{5}{6}$ B) 5

C) $\frac{2}{27}$ D) 10

58) Solve the below proportion.

$$\frac{13}{18} = \frac{n}{20}$$

A) $\frac{130}{9}$ B) $\frac{2}{3}$

C) 10 D) $\frac{8}{5}$

59) Solve the below proportion.

$$\frac{17}{n} = \frac{12}{2}$$

A) $\frac{17}{6}$ B) 1

C) 6 D) $7\frac{3}{7}$

60) Solve the below proportion.

$$\frac{19}{5} = \frac{x}{16}$$

A) $\frac{304}{5}$ B) 16

C) 17 D) $\frac{7}{4}$

61) Solve the below proportion.

$$\frac{7}{2} = \frac{18}{n}$$

A) $\frac{36}{7}$ B) 1

C) 11 D) $\frac{13}{9}$

62) Solve the below proportion.

$$\frac{9}{6} = \frac{n}{9}$$

A) $\frac{27}{2}$ B) $\frac{1}{4}$

C) 4 D) $4\frac{7}{10}$

63) Solve the below proportion.

$$\frac{15}{11} = \frac{19}{x}$$

A) $\frac{209}{15}$ B) $\frac{26}{17}$

C) $\frac{2}{5}$ D) $2\frac{19}{20}$

64) Solve the below proportion.

$$\frac{20}{12} = \frac{m}{17}$$

A) $\frac{85}{3}$ B) $7\frac{15}{19}$

C) $3\frac{9}{10}$ D) 2

65) Solve the below proportion.

$$\frac{x}{6} = \frac{4}{16}$$

A) $12\frac{1}{4}$ B) 12

C) $4\frac{17}{20}$ D) $\frac{3}{2}$

66) Solve the below proportion.

$$\frac{18}{14} = \frac{x}{2}$$

A) $8\frac{14}{15}$ B) $1\frac{11}{12}$

C) $\frac{18}{7}$ D) $9\frac{1}{2}$

67) Solve the below proportion.

$$\frac{r}{19} = \frac{16}{20}$$

A) $\frac{76}{5}$ B) 9

C) 10 D) $\frac{19}{16}$

68) Solve the below proportion.

$$\frac{17}{11} = \frac{n}{5}$$

A) $\frac{1}{4}$ B) 16

C) 17 D) $\frac{85}{11}$

69) Solve the below proportion.

$$\frac{4}{2n} = \frac{9}{20}$$

A) $\frac{40}{9}$ B) $\frac{25}{18}$

C) $8\frac{3}{4}$ D) 4

70) Solve the below proportion.

$$\frac{8}{7} = \frac{11}{x}$$

A) $9\frac{3}{7}$ B) $\frac{77}{8}$

C) $\frac{26}{17}$ D) 9

71) Solve the below proportion.

$$\frac{n}{16} = \frac{2}{3}$$

A) 17 B) $10\frac{5}{6}$

C) $7\frac{3}{8}$ D) $\frac{32}{3}$

72) Solve the below proportion.

$$\frac{k}{4} = \frac{9}{5}$$

A) $\frac{36}{5}$ B) $\frac{6}{11}$

C) 2 D) $\frac{2}{3}$

73) Solve the below proportion.

$$\frac{18}{19} = \frac{x}{15}$$

A) $\frac{2}{5}$ B) $7\frac{6}{11}$

C) $\frac{270}{19}$ D) $7\frac{7}{20}$

74) Solve the below proportion.

$$\frac{19}{r} = \frac{5}{9}$$

A) $2\frac{11}{20}$ B) $9\frac{7}{10}$

C) 14   D) $\frac{171}{5}$

75) Solve the below proportion.

$$\frac{x}{13} = \frac{3}{11}$$

A) 1   B) 6.03

C) 6.7   D) 3.55

76) Solve the below proportion.

$$\frac{6}{3} = \frac{13}{x}$$

A) 13.9   B) 4

C) 6.5   D) 5.7

77) Solve the below proportion.

$$\frac{3}{2} = \frac{x}{14}$$

A) 21   B) $\frac{10}{17}$

C) $\frac{11}{19}$ D) $\frac{3}{8}$

78) Solve the below proportion.

$$\frac{7}{15} = \frac{5}{r}$$

A) $5\frac{5}{12}$ B) $\frac{75}{7}$

C) $3\frac{10}{17}$ D) $5\frac{3}{19}$

79) Solve the below proportion.

$$\frac{10}{x} = \frac{13}{8}$$

A) 13   B) 6.15

C) 10.6   D) 9

80) Solve the below proportion.

$$\frac{x}{11} = \frac{2}{7}$$

A) 3.14   B) 3.9

C) 12.1   D) 13

Grade 7
Volume 1
Week 3

81) Solve the below proportion.

$$\frac{3}{2} = \frac{k}{13}$$

A) 19.5   B) 10.4
C) 5.5   D) 9.5

82) Solve the below proportion.

$$\frac{9}{x} = \frac{2}{4}$$

A) 18   B) 2.821
C) 12.967   D) 10.579

83) Solve the below proportion.

$$\frac{13}{n} = \frac{12}{4}$$

A) 2   B) 3.6
C) 4.33   D) 9

84) Solve the below proportion.

$$\frac{10}{5} = \frac{x}{14}$$

A) 13   B) 11.7
C) 28   D) 3.7

85) Solve the below proportion.

$$\frac{6}{5} = \frac{n}{4}$$

A) 3.9   B) 1.7
C) 4.8   D) 4

86) Solve the below proportion.

$$\frac{n}{7} = \frac{3}{11}$$

A) 7.6   B) 1.91
C) 8   D) 9.1

87) Solve the below proportion.

$$\frac{x}{2} = \frac{2}{4}$$

A) 6.1   B) 1
C) 3.8   D) 9

88) Solve the below proportion.

$$\frac{8}{n} = \frac{5}{2}$$

A) 11.37   B) 11
C) 3.2   D) 13.62

89) Solve the below proportion.

$$\frac{r}{7} = \frac{8}{5}$$

A) 11.2   B) 3

C) 6   D) 4.1

90) Solve the below proportion.

$$\frac{2}{k} = \frac{13}{3}$$

A) 7.8   B) 13

C) 0.46   D) 1.1

91) Solve the below proportion.

$$\frac{4}{14} = \frac{11}{x}$$

A) 4.6   B) 14

C) 9.96   D) 38.5

92) Solve the below proportion.

$$\frac{2}{a} = \frac{14}{3}$$

A) 1   B) 3.4

C) 0.43   D) 10.204

93) Solve the below proportion.

$$\frac{5}{11} = \frac{4}{n}$$

A) 8.8   B) 12.3

C) 13   D) 2.54

94) Solve the below proportion.

$$\frac{12}{13} = \frac{k}{5}$$

A) 10   B) 2

C) 12   D) 4.62

95) Solve the below proportion.

$$\frac{3}{14} = \frac{n}{10}$$

A) 2.14   B) 7.2

C) 3.8   D) 12.1

96) Solve the below proportion.

$$\frac{14}{2} = \frac{p}{8}$$

A) 56   B) 9

C) 2.8   D) 1

Grade 7
Volume 1
Week 3

97) Solve the below proportion.

$$\frac{4.9}{4.58} = \frac{r}{7.5}$$

A) 1.9 B) 10
C) 8.02 D) 6.9

98) Solve the below proportion.

$$\frac{n}{8.8} = \frac{2.2}{2.8}$$

A) 3 B) 6.91
C) 2 D) 3.1

99) Solve the below proportion.

$$\frac{8.8}{7.8} = \frac{x}{8.81}$$

A) 4.6 B) 4.35
C) 4.2 D) 9.94

100) Solve the below proportion.

$$\frac{3.8}{x} = \frac{4.2}{6.8}$$

A) 9 B) 1.6
C) 7.8 D) 6.15

101) Solve the below proportion.

$$\frac{x}{5.1} = \frac{6.3}{2.6}$$

A) 9.5 B) 1.9
C) 3.7 D) 12.36

102) Solve the below proportion.

$$\frac{7.5}{8.6} = \frac{n}{8.7}$$

A) 4 B) 7
C) 2 D) 7.59

103) Solve the below proportion.

$$\frac{n}{4.1} = \frac{5.4}{2.76}$$

A) 8.02 B) 6
C) 6.1 D) 1

104) Solve the below proportion.

$$\frac{7.5}{3.9} = \frac{m}{8.7}$$

A) 16.73 B) 1
C) 7 D) 6

©All rights reserved-Math-Knots LLC., VA-USA   www.math-knots.com | www.a4ace.com

Grade 7
Volume 1
Week 3

105) Solve the below proportion.

$$\frac{6.9}{9.8} = \frac{2.5}{n}$$

A) 3.55 B) 5.6

C) 5 D) 1.5

106) Dan bought one package of grapes for $3. How many packages can Ming buy if she has $60? (Round to the nearest number)

A) 20 B) 180

C) 17 D) 18

107) 24 jars of peanuts cost $108. How many jars of peanuts can you buy for $36? (Round to the nearest number)

A) 8 B) 162

C) 10 D) 9

108) One tub of blueberries costs $8. How many tubs can you buy for $56? (Round to the nearest number)

A) 8 B) 5

C) 448 D) 7

109) Solve the below proportion.

$$\frac{4.1}{7.4} = \frac{5.5}{n}$$

A) 9.2 B) 7.675

C) 9.93 D) 4.7

110) Jessica bought two pens for $4. How many pens can Stephanie buy if she has $48? (Round to the nearest number)

A) 24 B) 23

C) 96 D) 21

111) Eight tubs of Almonds cost $64. How many tubs of Almonds can you buy for $128? (Round to the nearest number)

A) 21 B) 18

C) 16 D) 1024

112) Kate bought 36 onions for $72. How many bags can Kate buy if she has $18? (Round to the nearest number)

A) 11 B) 36

C) 10 D) 9

113) Aria bought 78 kiwi fruit for $39. How many kiwi can Kim buy if she has $3? (Round to the nearest number)

A) 2   B) 5
C) 6   D) 3

114) Jade bought 18 cantaloupes for $45. How many cantaloupes can Gary buy if he has $15?
(Round to the nearest number)

A) 6   B) 38
C) 8   D) 5

115) Tom bought one box of tomatoes for $3. How many boxes of tomatoes can Tim buy if he has $63?
(Round to the nearest number)

A) 189   B) 23
C) 20   D) 21

116) If you can buy nine seedless watermelons for $17, then how many can you buy with $85? (Round to the nearest number)

A) 44   B) 45
C) 41   D) 161

117) Eight packages of lolly pops cost $24. How many packages of lolly pops can you buy for $12?
(Round to the nearest number)

A) 4   B) 6
C) 36   D) 7

118) Ted bought two bulbs of yellow roses for $4. How many bulbs can Rosy buy if she has $16?
(Round to the nearest number)

A) 9   B) 8
C) 7   D) 32

119) Peter bought 40 bags of potatoes for $75. How many bags can Ron buy if he has $15?
(Round to the nearest number)

A) 28   B) 8
C) 9   D) 7

120) If you can buy six bunches of spinach for $15, then how many can you buy with $45?
(Round to the nearest number)

A) 15   B) 113
C) 13   D) 18

**Grade 7
Volume 1
Week 3**

121) Two mangoes cost $4. How many mangoes can you buy for $88?
(Round to the nearest number)

A) 38   B) 41
C) 44   D) 176

122) A rectangle is 2 in tall and 4 in wide. If it is enlarged to a width of 16 in, then how tall will it be?
(Round to the nearest number)

A) 4 in   B) 1 in
C) 9 in   D) 8 in

123) Jack reduced the size of a photo to a width of 5 in. What is the new height if it was originally 12 in tall and 20 in wide?
(Round to the nearest number)

A) 48 in   B) 3 in
C) 6 in   D) 7 in

124) Violet enlarged the size of a rectangle to a height of 6 in. What is the new width if it was originally 7 in wide and 3 in tall?
(Round to the nearest number)

A) 3 in   B) 4 in
C) 14 in   D) 12 in

125) A painting is 2 cm tall and 3 cm wide. If it is enlarged to a height of 28 cm, then how wide will it be?
(Round to the nearest number)

A) 54 cm   B) 42 cm
C) 57 cm   D) 47 cm

126) Rik reduced the size of a triangle to a height of 7 cm. What is the new width if it was originally 28 cm tall and 8 cm wide? (Round to the nearest number)

A) 1 cm   B) 2 cm
C) 28 cm   D) 32 cm

127) A photo is 7 in tall and 1 in wide. If it is enlarged to a height of 49 in, then how wide will it be?
(Round to the nearest number)

A) 7 in   B) 343 in
C) 5 in   D) 3 in

128) A photo is 4 in wide and 6 in tall. If it is enlarged to a height of 24 in, then how wide will it be?
(Round to the nearest number)

A) 12 in   B) 11 in
C) 14 in   D) 16 in

129) A photo is 1 in tall and 7 in wide. If it is enlarged to a width of 14 in, then how tall will it be? (Round to the nearest number)

A) 28 in B) 2 in

C) 1 in D) 7 in

130) Chris reduced the size of a photo to a width of 7 in. What is the new height if it was originally 49 in wide and 28 in tall? (Round to the nearest number)

A) 4 in B) 196 in

C) 2 in D) 1 in

131) Scott enlarged the size of a rectangle to a width of 12 cm. What is the new height if it was originally 2 cm wide and 5 cm tall? (Round to the nearest number)

A) 30 cm B) 2 cm

C) 72 cm D) 1 cm

132) Maria enlarged the size of a rectangle to a width of 9 cm. What is the new height if it was originally 3 cm wide and 6 cm tall? (Round to the nearest number)

A) 2 cm B) 17 cm

C) 14 cm D) 18 cm

133) Bella enlarged the size of a frame to a height of 30 cm. What is the new width if it was originally 6 cm tall and 3 cm wide? (Round to the nearest number)

A) 11 cm B) 15 cm

C) 150 cm D) 13 cm

134) A painting is 2 cm tall and 3 cm wide. If it is enlarged to a height of 18 cm, then how wide will it be? (Round to the nearest number)

A) 23 cm B) 162 cm

C) 27 cm D) 21 cm

135) Ian enlarged the size of a rectangle to a height of 10 cm. What is the new width if it was originally 5 cm tall and 6 cm wide? (Round to the nearest number)

A) 14 cm B) 3 cm

C) 12 cm D) 5 cm

136) Ivanka enlarged the size of a photo to a width of 30 cm. What is the new height if it was originally 6 cm wide and 2 cm tall? (Round to the nearest number)

A) 12 cm B) 10 cm

C) 6 cm D) 11 cm

137) The currency in South Africa is the Soles. The exchange rate is approximately $1 = 7 Soles. At this rate, how many dollars would you get if you exchanged 105 Soles? (Round to the nearest number)

   A) $16   B) $17

   C) $735   D) $15

138) The currency in Albania is the Bales. The exchange rate is approximately $1 to 96 Bales. At this rate, how many Bales would you get if you exchanged $2? (Round to the nearest number)

   A) 219 Bales   B) 192 Bales

   C) 172 Bales   D) 193 Bales

139) The currency in Algoa islands is the Onga. The exchange rate is approximately 71 Onga for every $1. At this rate, how many Onga would you get if you exchanged $3? (Round to the nearest number)

   A) 209 Onga   B) 213 Onga

   C) 182 Onga   D) 192 Onga

140) The currency in West bank is the Tola. The exchange rate is approximately $1 for 3 Tola. At this rate, how many Tola would you get if you exchanged $46? (Round to the nearest number)

   A) 152 Tola   B) 15 Tola

   C) 138 Tola   D) 170 Tola

141) The currency in Nuwait is the Dola. The exchange rate is approximately 1 Dola = $3. At this rate, how many Dola would you get if you exchanged $405? (Round to the nearest number)

   A) 1215 Dola   B) 135 Dola

   C) 151 Dola   D) 169 Dola

142) The currency in Zara island is the Zors. The exchange rate is approximately $1 = 7 Zors. At this rate, how many dollars would you get if you exchanged 98 Zors? (Round to the nearest number)

   A) $686   B) $16

   C) $14   D) $13

143) Mary was planning a trip to Thailand. Before going, she did some research and learned that the exchange rate is 33 Baht for every $1. How many Baht would shegget if she exchanged $3? (Round to the nearest number)

   A) 99 Baht   B) 96 Baht

   C) 106 Baht   D) 87 Baht

144) The money used in Malaysia is called the Sias. The exchange rate is $1 = 7 Sias. Find how many Rand you would receive if you exchanged $21. (Round to the nearest number)

   A) 160 Sias   B) 3 Sias

   C) 147 Sias   D) 138 Sias

145) The money used in Angola is called the Wanza. The exchange rate is $1 for every 80 Wanza. Find how many dollars you would receive if you exchanged 240 Wanza. (Round to the nearest number)

A) $2   B) $1

C) $3   D) $19200

146) The currency in North Shore is the Terns. The exchange rate is approximately 3 Terns for $1. At this rate, how many dollars would you get if you exchanged 150 Terns? (Round to the nearest number)

A) $450   B) $60

C) $57   D) $50

147) Teacher took a trip to Algeria. Upon leaving he decided to convert all of his Jinars back into dollars. How many dollars did he receive if he exchanged 142 Jinars at a rate of $1 for 71 Jinars? (Round to the nearest number)

A) $2   B) $3

C) $4   D) $10082

148) The currency in Saudi Arabia is the Riyal. The exchange rate is approximately $1 for every 4 Riyals. At this rate, how many Riyals would you get if you exchanged $42? (Round to the nearest number)

A) 143 Riyals   B) 11 Riyals

C) 136 Riyals   D) 168 Riyals

149) Gabriella took a trip to Ala island. Upon leaving she decided to convert all of her Lala back into dollars. How many dollars did she receive if she exchanged 195 Lala at a rate of 3 Lala to $1? (Round to the nearest number)

A) $64   B) $585

C) $70   D) $65

150) The money used in Takoma park is called the Akom. The exchange rate is $1 to 3 Akom. Find how many dollars you would receive if you exchanged 162 Akom. (Round to the nearest number)

A) $54   B) $486

C) $46   D) $52

151) The currency in Sweden is the Knor. The exchange rate is approximately $1 for every 7 Knor. At this rate, how many Knor would you get if you exchanged $26? (Round to the nearest number)

A) 4 Knor   B) 182 Knor

C) 218 Knor   D) 200 Knor

152) Ron took a trip to Poland. Upon leaving he decided to convert all of his Olas back into dollars. How many dollars did he receive if he exchanged 20.1 Olas at a rate of 1 Olas for every $0.80? (Round to the nearest number)

A) $17.69   B) $25.12

C) $18.75   D) $16.08

153) If you can buy nine plums for $7.20, then how many can you buy with $2.40? (Round to the nearest number)

A) 5 B) 2
C) 3 D) 7

154) The currency in Croatia is the Nuan. The exchange rate is approximately $3.80 to 5.6 Nuan. At this rate, how many dollars would you get if you exchanged 23.6 Nuan? (Round to the nearest number)

A) $34.78 B) $13.54
C) $14.25 D) $16.01

155) If you can buy one bag of potatoes for $1.50, then how many can you buy with $15? (Round to the nearest number)

A) 9 B) 23
C) 11 D) 10

156) Heather bought one mango for $0.90. How many mangos can Helen buy if she has $24.30? (Round to the nearest number)

A) 24 B) 27
C) 25 D) 22

157) If you can buy two packages of raspberries for $5.98, then how many can you buy with $23.92? (Round to the nearest number)

A) 9 B) 8
C) 7 D) 72

158) A rectangle is 19.8 cm wide and 14.4 cm tall. If it is reduced to a width of 1.1 cm, then how tall will it be? (Round to the nearest number)

A) 2.7 cm B) 3.2 cm
C) 259.2 cm D) 0.8 cm

159) Lola took a trip to Mexico. Upon leaving she decided to convert all of her Pesos back into dollars. How many dollars did she receive if she exchanged 26.4 Pesos at a rate of 11.1 Pesos for $4.80? (Round to the nearest number)

A) $11.42 B) $10.73
C) $61.05 D) $10.19

160) Ben was planning a trip to China. Before going, he did some research and learned that the exchange rate is 7.7 Yuan for every $1. How many Yuan would he get if he exchanged $3.30? (Round to the nearest number)

A) 0.4 Yuan B) 25.4 Yuan
C) 27.9 Yuan D) 24.8 Yuan

161) The money used in Argentina is called the Tina. The exchange rate is $1 to 3.1 Tinas. Find how many Tinas you would receive if you exchanged $8.60.
(Round to the nearest number)

A) 2.8 Tinas   B) 25.1 Tinas

C) 28.9 Tinas   D) 26.7 Tinas

162) Nora was planning a trip to Rotata. Before going, she did some research and learned that the exchange rate is $1.50 = 5.6 Rots. How many Rots would he get if he exchanged $6.60?
(Round to the nearest number)

A) 21.6 Rots   B) 20.5 Rots

C) 1.8 Rots   D) 24.6 Rots

163) The money used in Trenton is called the Tonton. The exchange rate is $3.60 for 1 Tonton. Find how many Tontons you would receive if you exchanged $107.30.
(Round to the nearest number)

A) 29.8 Tontons   B) 28.3 Tontons

C) 25.3 Tontons   D) 386.3 Tontons

164) A rectangle is 17.1 cm tall and 13.5 cm wide. If it is reduced to a height of 1.9 cm, then how wide will it be?
(Round to the nearest number)

A) 1.3 cm   B) 0.2 cm

C) 121.5 cm D) 1.5 cm

165) Vivian took a trip to Bolivia. Upon leaving he decided to convert all of his Vios back into dollars. How many dollars did he receive if he exchanged 28.8 Vios at a rate of $9.90 = 8 Vios?
(Round to the nearest number)

A) $23.27   B) $35.75

C) $35.64   D) $40.63

166) The money used in the eastern Caribbean islands is called the Aribs. The exchange rate is 1 Arib to $9.40. Find how many dollars you would receive if you exchanged 25.4 Aribs
(Round to the nearest number)

A) $238.76   B) $2.70

C) $195.40   D) $210.11

167) A photo is 0.4 cm tall and 2.4 cm wide. If it is enlarged to a width of 4.8 cm, then how tall will it be?
(Round to the nearest number)

A) 2.4 cm   B) 9.6 cm

C) 0.2 cm   D) 0.8 cm

168) Jade enlarged the size of a frame to a height of 16.8 cm. What is the new width if it was originally 4.2 cm tall and 1.2 cm wide?
(Round to the nearest number)

A) 4.8 cm   B) 0.3 cm

C) 4.2 cm   D) 5.4 cm

169) A painting is 3 in wide and 0.5 in tall. If it is enlarged to a width of 30 in, then how tall will it be?
(Round to the nearest number)

A) 300 in   B) 3 in
C) 0.1 in   D) 5 in

170) Diana took a trip to Sahara. Upon leaving she decided to convert all of her Sahas back into dollars. How many dollars did she receive if she exchanged 27.9 Sahas at a rate of 1 Saha to $4?
(Round to the nearest number)

A) $117.18   B) $107.81
C) $111.60   D) $6.97

171) If you can buy 20 bunches of cilantro for $31.80, then how many can you buy with $7.95? (Round to the nearest number)

A) 3   B) 6
C) 13   D) 5

172) If you can buy five bunches of seedless green grapes for $12.50, then how many can you buy with $25?
(Round to the nearest number)

A) 63   B) 10
C) 9   D) 11

173) 24 bananas cost $7.20. How many bananas can you buy for $0.60?
(Round to the nearest number)

A) 6   B) 2
C) 1   D) 4

174) The currency in Ronga is the Anga. The exchange rate is approximately $6.60 for every 1 Anga. At this rate, how many dollars would you get if you exchanged 21.5 Anga?
(Round to the nearest number)

A) $3.26   B) $124.87
C) $111.13   D) $141.90

175) Bella enlarged the size of a triangle to a width of 11.6 in. What is the new height if it was originally 2.9 in wide and 2.7 in tall?
(Round to the nearest number)

A) 11 in   B) 2.9 in
C) 10.8 in   D) 11.7 in

176) The money used in Everest city is called the Esto. The exchange rate is 1 Esto to $8. Find how many Estos you would receive if you exchanged $167.80.
(Round to the nearest number)

A) 17.4 Dinars   B) 1342.4 Dinars
C) 18.5 Dinars   D) 21 Dinars

Grade 7
Volume 1
Week 3

177) A frame is 13.8 cm wide and 10.2 cm tall. If it is reduced to a width of 4.6 cm, then how tall will it be?
(Round to the nearest number)

  A) 3.4 cm   B) 30.6 cm

  C) 1.5 cm   D) 3.2 cm

178) If you can buy ten cantaloupes for $24.90, then how many can you buy with $12.45?
(Round to the nearest number)

  A) 8   B) 31

  C) 9   D) 5

179) Oliver was planning a trip to China. Before going, he did some research and learned that the exchange rate is $4 = 7.7 Yuan. How many Yuan would he get if he exchanged $11.90?
(Round to the nearest number)

  A) 21.5 Yuan   B) 6.2 Yuan

  C) 23.2 Yuan   D) 22.9 Yuan

180) The currency in Kuwait is the Kuwa. The exchange rate is approximately 1 Kuwa for every $2.80. At this rate, how many dollars would you get if you exchanged 25.1 Kuwas?
(Round to the nearest number)

  A) $70.28   B) $63.95

  C) $8.96   D) $58.19

181) Bella took a trip to North shore islands. Upon leaving she decided to convert all of her Shores back into dollars. How many dollars did he receive if she exchanged 24.2 Shores at a rate of $5 for every 1 Shore?
(Round to the nearest number)

  A) $107.69   B) $121

  C) $4.84   D) $95.84

Grade 7
Volume 1
Week 3

Grade 7 Volume 1 Week 4

Grade 7
Volume 1
Week 4

**Grade 7
Volume 1
Week 4**

1) Find the missing side from the given pair of similar figures

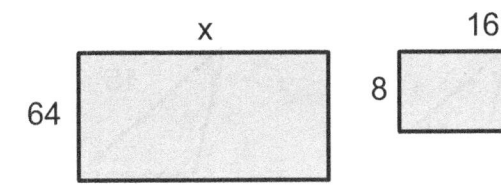

A) 16   B) 64

C) 8   D) 128

2) Find the missing side from the given pair of similar figures

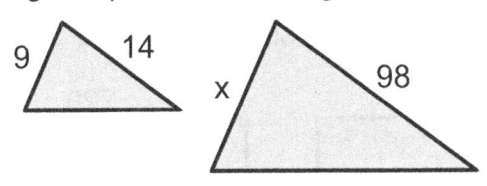

A) 7   B) 98

C) 441   D) 63

3) Find the missing side from the given pair of similar figures

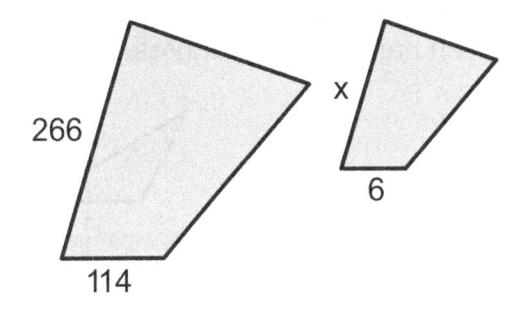

A) 19   B) 209

C) 266   D) 14

4) Find the missing side from the given pair of similar figures

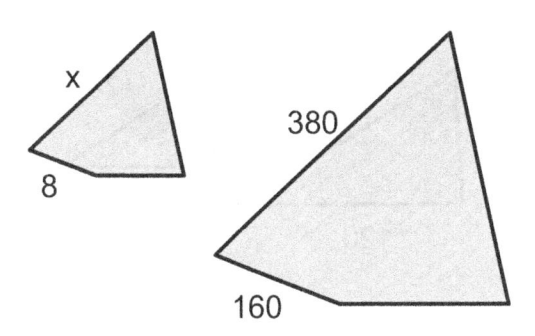

A) 19   B) 6400

C) 0.9   D) 7600

5) Find the missing side from the given pair of similar figures

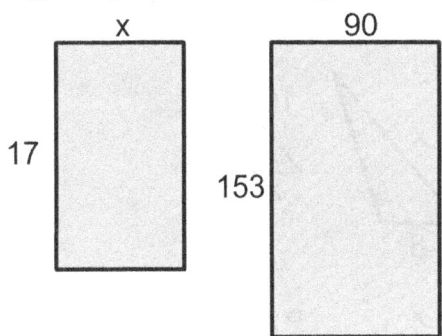

A) 90   B) 10

C) 9   D) 153

6) Find the missing side from the given pair of similar figures

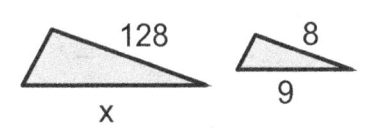

A) 2304 B) 3

C) 9   D) 144

7) Find the missing side from the given pair of similar figures

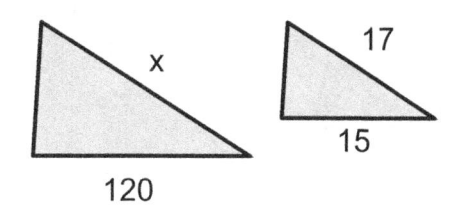

A) 136   B) 72

C) 960   D) 17

8) Find the missing side from the given pair of similar figures

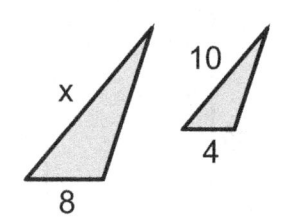

A) 4   B) 20

C) 40   D) 8

9) Find the missing side from the given pair of similar figures

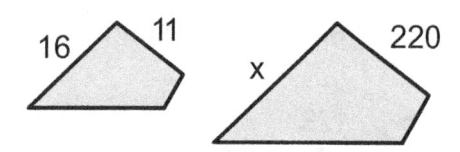

A) 100   B) 11

C) 20   D) 320

10) Find the missing side from the given pair of similar figures

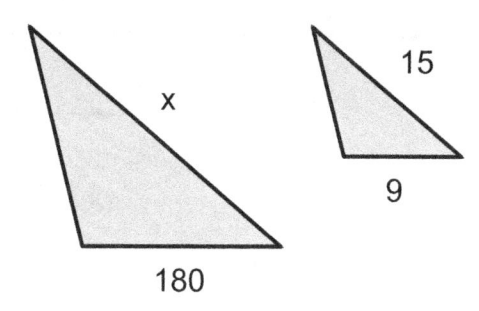

A) 6000   B) 10

C) 300   D) 200

11) Find the missing side from the given pair of similar figures

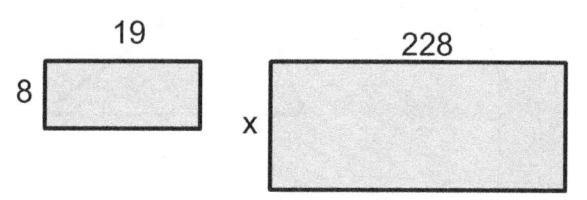

A) 96   B) 228

C) 19   D) 8

12) Find the missing side from the given pair of similar figures

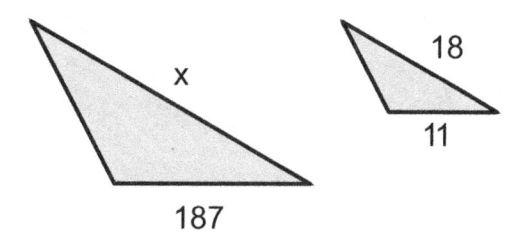

A) 170   B) 3179

C) 18   D) 306

Grade 7
Volume 1
Week 4

13) Find the missing side from the given pair of similar figures

3   45
[ ] x   [ ]
         75

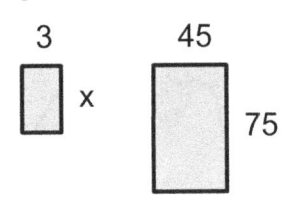

A) 1125   B) 5

C) 0.3   D) 15

14) Find the missing side from the given pair of similar figures

154   14
          x
77

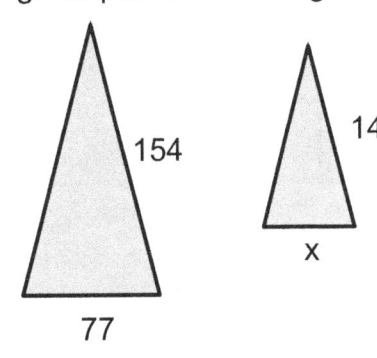

A) 154   B) 7

C) 1694   D) 14

15) Find the missing side from the given pair of similar figures

x   154
17
         187

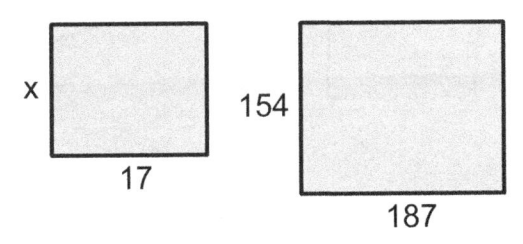

A) 14   B) 11

C) 2057   D) 187

16) Find the missing side from the given pair of similar figures

x   60
8   80

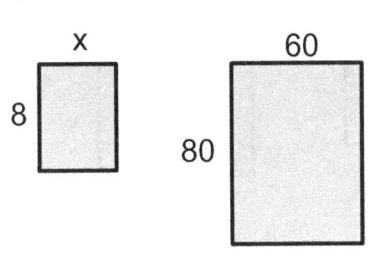

A) 6   B) 10

C) 60   D) 8

17) Find the missing side from the given pair of similar figures

x   17
     154   187

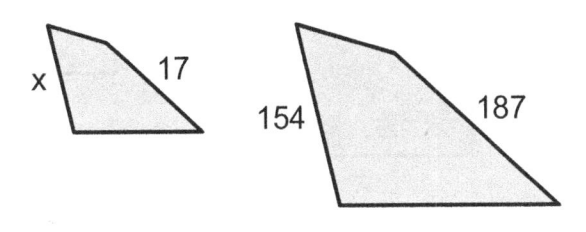

A) 154   B) 14

C) 11   D) 8

18) Find the missing side from the given pair of similar figures

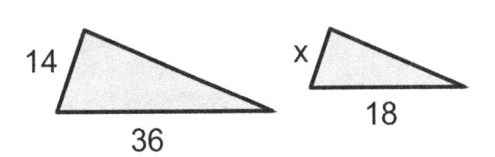

14   x
     36   18

A) 3.5   B) 31

C) 28   D) 7

19) Find the missing side from the given pair of similar figures

8
7
96
x

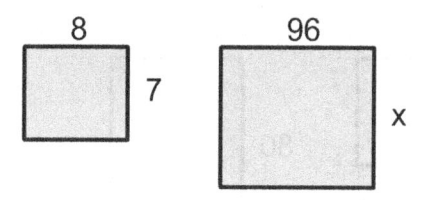

A) 96   B) 46
C) 84   D) 1008

20) Find the missing side from the given pair of similar figures

x
117
12
13

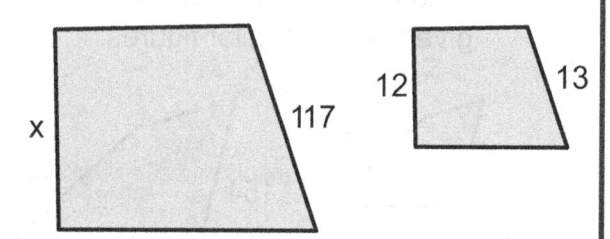

A) 12   B) 108
C) 144   D) 1053

21) Find the missing side from the given pair of similar figures

65
5
x
15

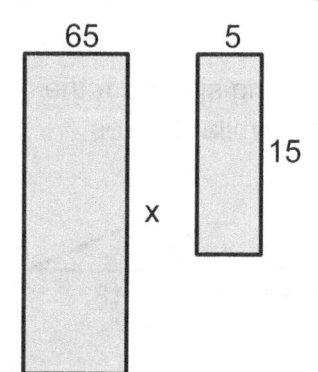

A) 1.1   B) 845
C) 5   D) 195

22) Find the missing side from the given pair of similar figures

8
16
x
192

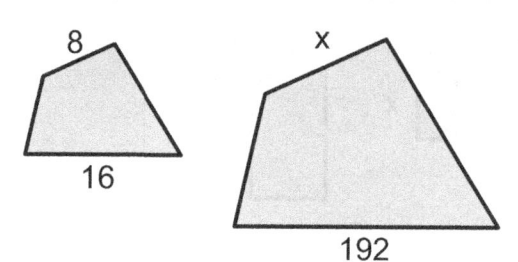

A) 156   B) 96
C) 1152   D) 192

23) Find the missing side from the given pair of similar figures

14
16
272
x

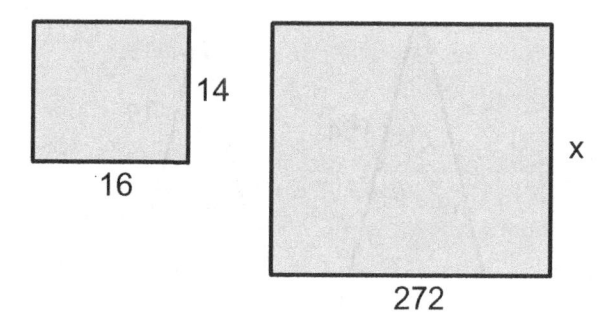

A) 4624   B) 4046
C) 238   D) 17

24) Find the missing side from the given pair of similar figures

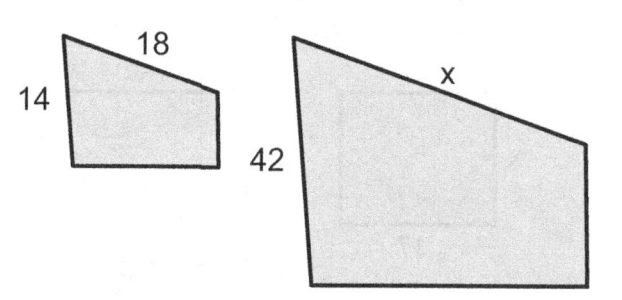

18
14
x
42

A) 3   B) 48
C) 18   D) 54

Grade 7
Volume 1
Week 4

25) Find the missing side from the given pair of similar figures

132
168
14
x

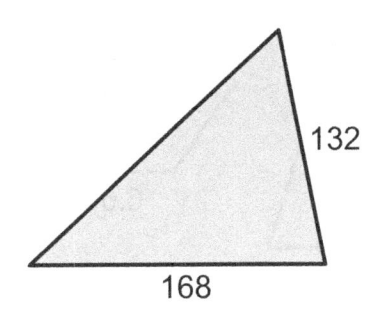

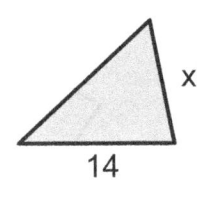

A) 168   B) 2016

C) 192   D) 11

26) Find the missing side from the given pair of similar figures

x
24
15
4

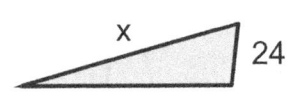

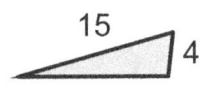

A) 2.5   B) 144

C) 4   D) 90

27) Find the missing side from the given pair of similar figures

33
6
2
x

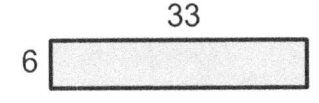

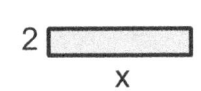

A) 11   B) 99

C) 6   D) 2

28) Find the missing side from the given pair of similar figures

234
x
18
14

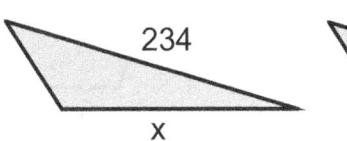

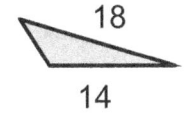

A) 182   B) 23

C) 78   D) 217

29) Find the missing side from the given pair of similar figures

11
x
209
190

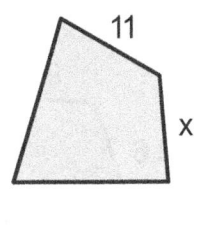

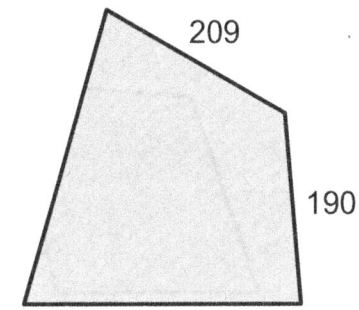

A) 11   B) 304

C) 10   D) 5776

30) Find the missing side from the given pair of similar figures

162
171
x
19

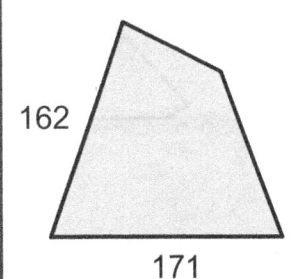

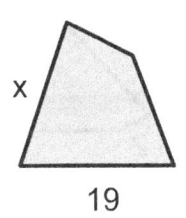

A) 1458   B) 9

C) 18   D) 1134

©All rights reserved-Math-Knots LLC., VA-USA   147   www.math-knots.com | www.a4ace.com

31) Find the missing side from the given pair of similar figures

46.9
6.7
5.2
x

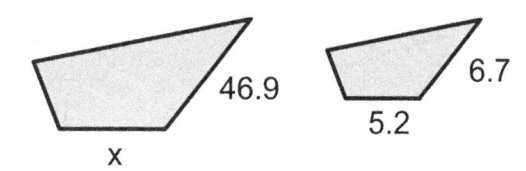

A) 46.9 B) 36.4
C) 5.2   D) 16

32) Find the missing side from the given pair of similar figures

x
9.5
21.2
10.6

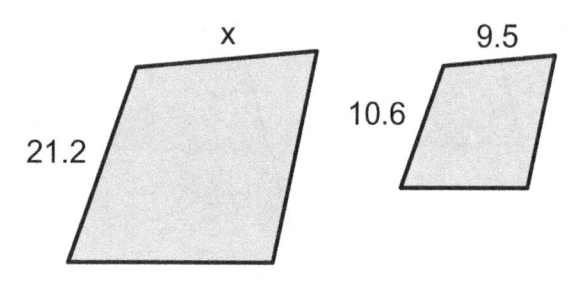

A) 22.2 B) 38
C) 19   D) 21.2

33) Find the missing side from the given pair of similar figures

64.2
x
13.2
2.2

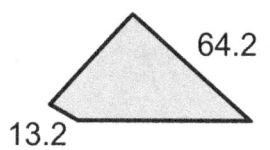

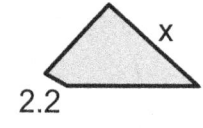

A) 11.8 B) 295.2
C) 2.2   D) 10.7

34) Find the missing side from the given pair of similar figures

x
8.9
6.6
19.8

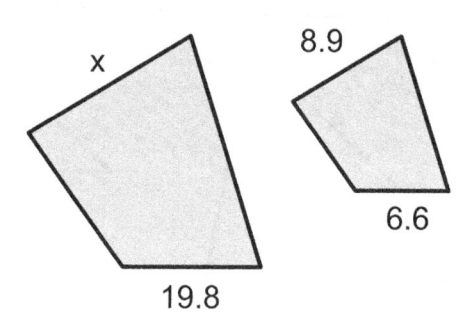

A) 22.7 B) 3
C) 26.7 D) 8.9

35) Find the missing side from the given pair of similar figures

43.2
x
9.7
38.8

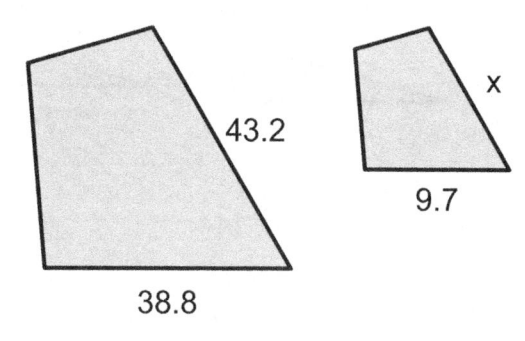

A) 5.1   B) 38.8
C) 10.8 D) 4

36) Find the missing side from the given pair of similar figures

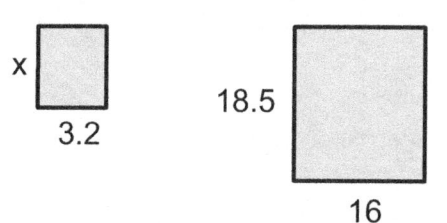

x
3.2
18.5
16

A) 3.7   B) 18.5
C) 16   D) 92.5

Grade 7
Volume 1
Week 4

37) Find the missing side from the given pair of similar figures

2.9, x; 34.8, 104.4

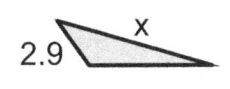

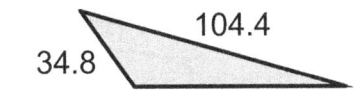

A) 104.4   B) 34.8

C) 8.7   D) 1252.8

38) Find the missing side from the given pair of similar figures

11.5, 8.8; 103.5, x

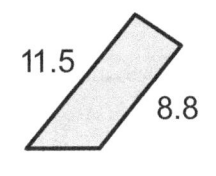

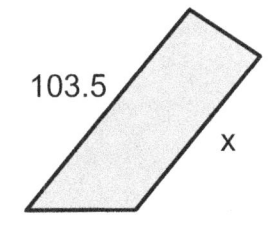

A) 11.5   B) 103.5

C) 396.9   D) 79.2

39) Find the missing side from the given pair of similar figures

71.4, x; 10.2, 10

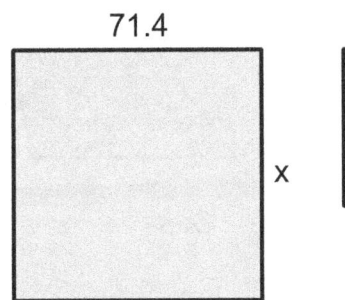

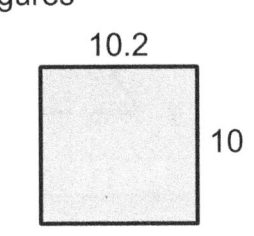

A) 1.4   B) 70

C) 490   D) 10

40) Find the missing side from the given pair of similar figures

25.2, 14.8; 6.3, x

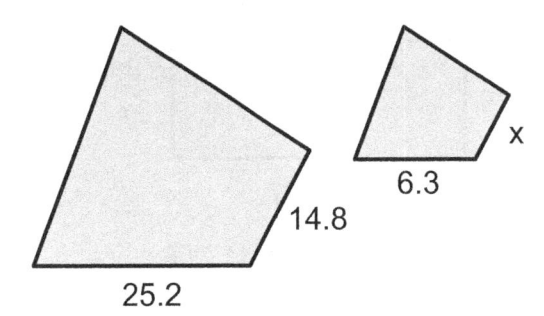

A) 14.8   B) 28.4

C) 3.7   D) 4

41) Find the missing side from the given pair of similar figures

110, x; 11, 5.4

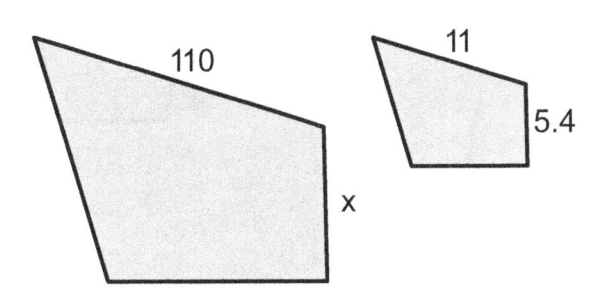

A) 54   B) 10

C) 56   D) 110

42) Find the missing side from the given pair of similar figures

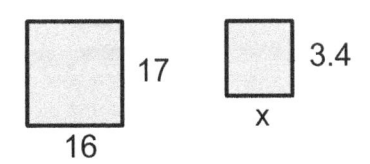

16, 17; x, 3.4

A) 16   B) 0.6

C) 17   D) 3.2

Grade 7
Volume 1
Week 4

43) Find the missing side from the given pair of similar figures

x, 106.7 ; 3.8, 9.7

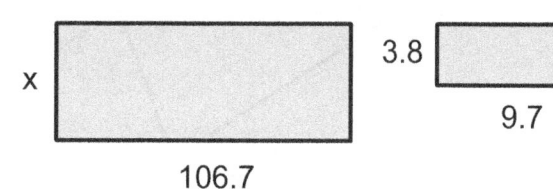

A) 72 B) 1173.6

C) 41.8 D) 5

44) Find the missing side from the given pair of similar figures

x, 52 ; 10.6, 6.5

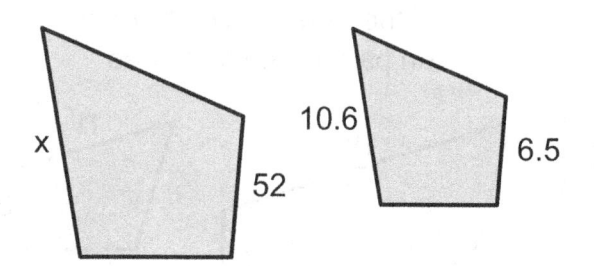

A) 82.4 B) 8

C) 84.8 D) 678.4

45) Find the missing side from the given pair of similar figures

x, 7.2 ; 60, 72

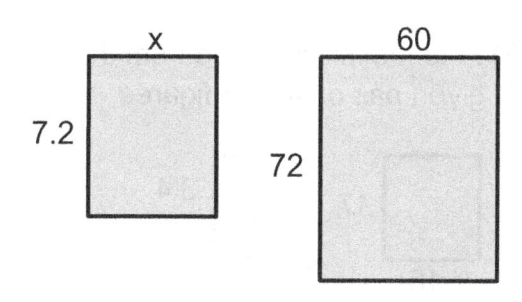

A) 60 B) 10

C) 6 D) 72

46) Find the missing side from the given pair of similar figures

11, 5.3 ; 99, x

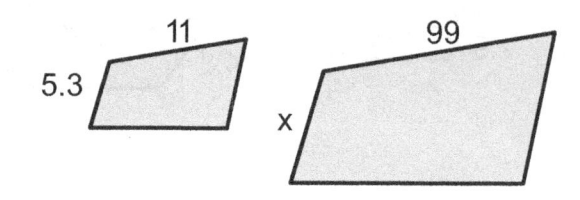

A) 47.6 B) 9

C) 98.1 D) 47.7

47) Find the missing side from the given pair of similar figures

5, 9.3 ; 55, x

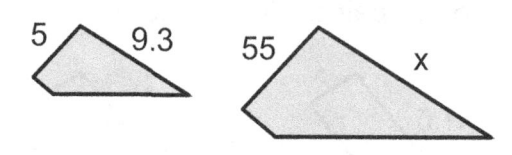

A) 11 B) 605

C) 108.9 D) 102.3

48) Find the missing side from the given pair of similar figures

4.2, x ; 46.2, 116.6

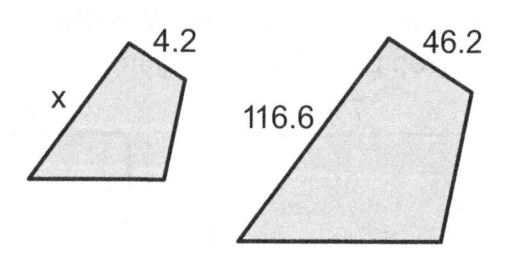

A) 11 B) 10.6

C) 116.6 D) 46.2

©All rights reserved-Math-Knots LLC., VA-USA   150   www.math-knots.com | www.a4ace.com

Grade 7
Volume 1
Week 4

49) Find the missing side from the given pair of similar figures

46.9
64.4
x
9.2

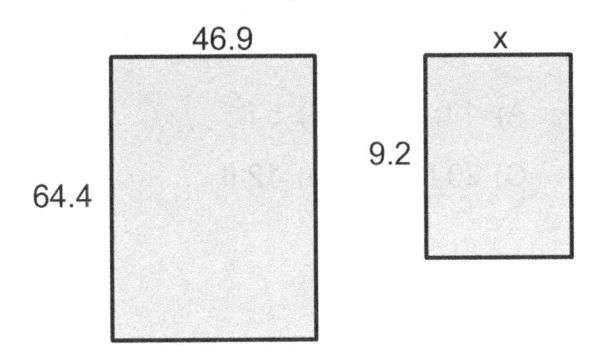

A) 64.3   B) 6.7
C) 9.2   D) 450.8

50) A model tower is 6 in wide. If it was built with a scale of 1 in : 2 ft, then how wide is the real tower?
(Round to the nearest whole number)

A) 8 ft   B) 1 ft
C) 12 ft   D) 3 ft

51) A model monument is 7 in tall. If it was built with a scale of 1 in : 2 ft, then how tall is the real monument?
(Round to the nearest whole number)

A) 9 ft   B) 2 ft
C) 14 ft   D) 4 ft

52) Tom's house is 16 ft wide. A model of it was built with a scale of 1 in : 2 ft. How wide is the model?
(Round to the nearest whole number)

A) 8 in   B) 2 in
C) 14 in   D) 32 in

53) Find the missing side from the given pair of similar figures

x
95
11
9.5

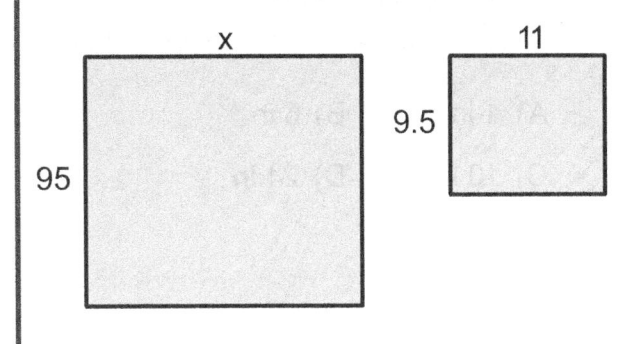

A) 110   B) 9.5
C) 1100   D) 95

54) A model Rocket is 2 in tall. If it was built with a scale of 1 in : 9 ft, then how tall is the real Rocket?
(Round to the nearest whole number)

A) 11 ft   B) 18 ft
C) 2 ft   D) 0 ft

55) A particular house is 10 ft tall. A model of it was built with a scale of 1 in : 2 ft. How tall is the model?
(Round to the nearest whole number)

A) 5 in   B) 1 in
C) 20 in   D) 8 in

56) A plane is 12 ft wide. A model of it was built with a scale of 1 in : 2 ft. How wide is the model?
(Round to the nearest whole number)

A) 2 in   B) 6 in
C) 24 in   D) 10 in

©All rights reserved-Math-Knots LLC., VA-USA
www.math-knots.com | www.a4ace.com

57) A monument is 12 ft tall. A model of it was built with a scale of 1 in : 2 ft. How tall is the model?
(Round to the nearest whole number)

A) 1 in   B) 6 in
C) 10 in   D) 24 in

58) A model house is 6 in wide. If it was built with a scale of 1 in : 3 ft, then how wide is the real house?
(Round to the nearest whole number)

A) 9 ft   B) 2 ft
C) 3 ft   D) 18 ft

59) A model train has a scale of 1 in : 3 ft. If the real train is 15 ft tall, then how tall is the model train?
(Round to the nearest whole number)

A) 5 in   B) 12 in
C) 2 in   D) 45 in

60) A model car has a scale of 1 in : 3 ft. If the real car is 12 ft long, then how long is the model car?
(Round to the nearest whole number)

A) 4 in   B) 9 in
C) 36 in   D) 1 in

61) A model SUV is 10 in long. If it was built with a scale of 1 in : 2 ft, then how long is the real SUV?
(Round to the nearest whole number)

A) 1 ft   B) 5 ft
C) 20 ft   D) 12 ft

62) An apartment building is 18 yd tall. A model of it was built with a scale of 1 in : 2 yd. How tall is the model?
(Round to the nearest whole number)

A) 9 in   B) 36 in
C) 1 in   D) 16 in

63) A satellite is 12 m wide. A model of it was built with a scale of 1 cm : 3 m. How wide is the model?
(Round to the nearest whole number)

A) 36 cm   B) 4 cm
C) 2 cm   D) 9 cm

64) A model motorcycle is 4 in long. If it was built with a scale of 1 in : 3 ft, then how long is the real motorcycle?
(Round to the nearest whole number)

A) 12 ft   B) 1 ft
C) 2 ft   D) 7 ft

Grade 7
Volume 1
Week 4

65) A model satellite is 8 cm wide. If it was built with a scale of 1 cm : 2 m, then how wide is the real satellite?
(Round to the nearest whole number)

A) 1 m   B) 4 m
C) 16 m   D) 10 m

66) A house is 12 ft tall. A model of it was built with a scale of 1 in : 2 ft. How tall is the model?
(Round to the nearest whole number)

A) 24 in   B) 2 in
C) 6 in   D) 10 in

67) A model train has a scale of 1 in : 3 ft. If the model train is 5 in tall, then how tall is the real train ?
(Round to the nearest whole number)

A) 8 ft   B) 1 ft
C) 15 ft   D) 2 ft

68) If a 10.8 ft tall statue casts a 31.3 ft long shadow, then how long is the shadow that a 1.7 ft tall lawn ornament casts?
(Round to the nearest tenth)

A) 3.7 ft   B) 4.9 ft
C) 90.8 ft   D) 0.6 ft

69) A model rocket has a scale of 1 cm : 2 m. If the real rocket is 18 m tall, then how tall is the model rocket?
(Round to the nearest whole number)

A) 36 cm   B) 9 cm
C) 16 cm   D) 1 cm

70) A car is 20 ft long. A model of it was built with a scale of 1 in : 5 ft. How long is the model?
(Round to the nearest whole number)

A) 15 in   B) 2 in
C) 100 in   D) 4 in

71) A model train has a scale of 1 in : 4 ft. If the real train is 12 ft tall, then how tall is the model train?
(Round to the nearest whole number)

A) 48 in   B) 3 in
C) 1 in   D) 8 in

72) A telephone booth that is 7.6 ft tall casts a shadow that is 15.2 ft long. Find the length of the shadow that a 2.9 ft lawn ornament casts.
(Round to the nearest tenth)

A) 30.4 ft   B) 5.8 ft
C) 3.8 ft   D) 1.5 ft

©All rights reserved-Math-Knots LLC., VA-USA   www.math-knots.com | www.a4ace.com

Grade 7
Volume 1
Week 4

73) A flagpole that is 6.9 ft tall casts a shadow that is 21.4 ft long. Find the height of a man that casts a 27.6 ft shadow.
(Round to the nearest tenth)

A) 8.9 ft   B) 66.3 ft
C) 2.2 ft   D) 2.9 ft

74) A 7.4 ft tall tree standing next to an building casts a 23.7 ft shadow. If the building is 17.5 ft tall, then how long is its shadow?
(Round to the nearest tenth)

A) 5.5 ft   B) 2.3 ft
C) 56 ft   D) 75.8 ft

75) A light house that is 9.3 ft tall casts a shadow that is 26 ft long. Find the height of a tree that casts a 9 ft shadow.
(Round to the nearest tenth)

A) 72.8 ft B) 3.3 ft
C) 3.2 ft   D) 1.1 ft

76) If a 12.8 m tall ladder casts a 24.3 m long shadow, then how long is the shadow that a 15.5 m tall lighthouse casts?
(Round to the nearest tenth)

A) 6.7 m   B) 46.2 m
C) 8.2 m   D) 29.4 m

77) A tree that is 7 ft tall casts a shadow that is 39.9 ft long. Find the height of a tent that casts a 30.8 ft shadow. (Round to the nearest tenth)

A) 0.9 ft   B) 5.4 ft
C) 227.4 ft D) 1.2 ft

78) If a 14.7 m tall lighthouse casts a 82.3 m long shadow, then how tall is a roller coaster that casts a 101.9 m shadow?
(Round to the nearest tenth)

A) 18.2 m   B) 460.9 m
C) 3.3 m   D) 2.6 m

79) A lamp post that is 16 ft tall casts a shadow that is 44.8 ft long. Find the length of the shadow that a 3.9 ft boy casts. (Round to the nearest tenth)

A) 5.7 ft   B) 10.9 ft
C) 1.4 ft   D) 125.4 ft

80) If a 3.7 ft car casts a 20.7 ft long shadow, then how tall is a ladder that casts a 96.3 ft shadow?
(Round to the nearest tenth)

A) 3.1 ft   B) 115.9 ft
C) 17.2 ft D) 0.7 ft

©All rights reserved-Math-Knots LLC., VA-USA
www.math-knots.com | www.a4ace.com

Grade 7
Volume 1
Week 4

81) Find the distance between fairfax and Bethesda on a map with a scale of 1 cm : 11.6 km if they are actually 27.8 km apart. (Round to the nearest tenth)

A) 0.2 cm B) 2.2 cm
C) 322.5 cm D) 2.4 cm

82) A map has a scale of 1 in : 7.7 mi. If Ashburn and Richmond are 0.9 in apart on the map, then how far apart are the real cities? (Round to the nearest tenth)

A) 5.9 mi B) 0.1 mi
C) 53.1 mi D) 6.9 mi

83) Find the distance between Fair field and Fair city if they are 1.4 cm apart on a map with a scale of 1 cm : 14.7 km. (Round to the nearest tenth)

A) 23.1 km B) 302.8 km
C) 20.6 km D) 0.1 km

84) Pittsburg and Chicago are 65 km from be on a map that has a scale of 1 cm : 17.1 km ? (Round to the nearest tenth)

A) 3.8 cm B) 0.2 cm
C) 4.1 cm D) 1111.5 cm

85) Florida and South Carolina are 5.7 mi from each other. How far apart would the cities be on a map that has a scale of 1 in : 5.2 mi? (Round to the nearest tenth)

A) 1.1 in B) 1 in
C) 29.6 in D) 0.2 in

86) Find the distance between Dallas and Houston if they are 2.6 in apart on a map with a scale of 1 in : 18.2 mi. (Round to the nearest tenth)

A) 47.3 mi B) 860.9 mi
C) 0.1 mi D) 43 mi

87) Find the distance between San Fransisco and Sam Ramon if they are 2 in apart on a map with a scale of 1 in : 8.8 mi. (Round to the nearest tenth)

A) 154.9 mi B) 17.6 mi
C) 0.2 mi D) 19.9 mi

88) Alexandria and Falls Church are 3 cm apart on a map that has a scale of 1 cm : 17.7 km. How far apart are the real cities? (Round to the nearest tenth)

A) 939.9 km B) 0.2 km
C) 53.1 km D) 46.7 km

©All rights reserved-Math-Knots LLC., VA-USA   155   www.math-knots.com | www.a4ace.com

89) A map has a scale of 1 in : 19.7 mi. If Georgetown and Gainesville are 67 mi apart, then they are how far apart on the map ? (Round to the nearest tenth)

A) 3.1 in   B) 0.2 in
C) 1319.9 in   D) 3.4 in

90) Find the distance between Potomac and Rockville on a map with a scale of 1 cm : 10.4 km if they are actually 6.2 km apart. (Round to the nearest tenth)

A) 1.7 cm   B) 0.1 cm
C) 0.6 cm   D) 64.5 cm

91) A map has a scale of 1 in : 11.3 mi. If Baltimore and Bethesta are 14.7 mi apart, then they are how far apart on the map? (Round to the nearest tenth)

A) 1.3 in   B) 1.1 in
C) 166.1 in   D) 0.1 in

92) Aldie and Arlington are 2 in apart on a map that has a scale of 1 in : 9.9 mi. How far apart are the real cities? (Round to the nearest tenth)

A) 17.6 mi   B) 0.2 mi
C) 196 mi   D) 19.8 mi

93) Find the distance between Mountain view and San Ramon on a map with a scale of 1 in : 11.1 mi if they are actually 23.3 mi apart. (Round to the nearest tenth)

A) 0.2 in   B) 258.6 in
C) 2.4 in   D) 2.1 in

94) Find the distance between Arlington and South riding if they are 3.4 in apart on a map with a scale of 1 in : 17.1 mi. (Round to the nearest tenth)

A) 0.2 mi   B) 52.9 mi
C) 58.1 mi   D) 993.5 mi

95) Find the distance between Fort myers and Holly hill if they are 0.9 in apart on a map with a scale of 1 in : 12.8 mi. (Round to the nearest tenth)

A) 11.5 mi   B) 0.1 mi
C) 147.2 mi   D) 10 mi

96) Find the distance between Plant city and San ford on a map with a scale of 1 in : 20 mi if they are actually 74 mi apart. (Round to the nearest tenth)

A) 1480 in   B) 0.2 in
C) 3.3 in   D) 3.7 in

97) Find the distance between West Palm beach and Winter springs if they are 0.7 in apart on a map with a scale of 1 in : 7.4 mi. (Round to the nearest tenth)

A) 38.5 mi   B) 4.9 mi
C) 5.2 mi   D) 0.1 mi

98) Find the distance between Treasure island and Temple terrace if they are 1.6 in apart on a map with a scale of 1 in : 7.5 mi (Round to the nearest tenth).

A) 90 mi   B) 0.2 mi
C) 12.8 mi   D) 12 mi

99) Find the distance between Tampa and Seaside if they are 3.6 cm apart on a map with a scale of 1 cm : 13.8 km. (Round to the nearest tenth)

A) 42.2 km   B) 49.7 km
C) 0.3 km   D) 685.9 km

100) Panama city and Palm beach are 5.5 km from each other. How far apart would the cities be on a map that has a scale of 1 cm : 9.2 km ?
(Round to the nearest tenth)

A) 0.6 cm   B) 0.1 cm
C) 1.5 cm   D) 50.6 cm

Grade 7
Volume 1
Week 4

Grade 7
Volume 1
Week 5

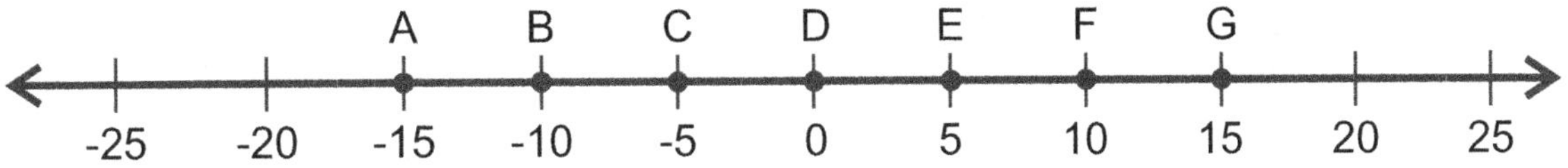
A B C D E F G
-25 -20 -15 -10 -5 0 5 10 15 20 25

1) Find the letter in the number line best represents the answer to below problem

5 + 15 =

2) Find the letter in the number line best represents the answer to below problem

(-10) + (-20) =

3) Find the letter in the number line best represents the answer to below problem

(-20) + 10 =

4) Find the letter in the number line best represents the answer to below problem

20 + (-10) =

5) Find the letter in the number line best represents the answer to below problem

10 + (-20) =

Grade 7
Volume 1
Week 5

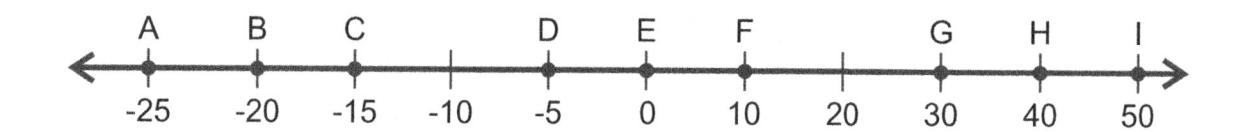

```
  A    B    C         D    E    F         G    H    I
<-•----•----•----|----•----•----•----|----•----•----•->
 -25  -20  -15  -10   -5   0   10   20   30   40   50
```

6) Find the letter in the number line best represents the answer to each problem

40 + (-10) =

7) Find the letter in the number line best represents the answer to below problem

(-40) + (-10) =

8) Find the letter in the number line best represents the answer to below problem

(-40) + 10 =

9) Find the letter in the number line best represents the answer to below problem

(-10) + 40 =

10) Find the letter in the number line best represents the answer to below problem

10 + (-40) =

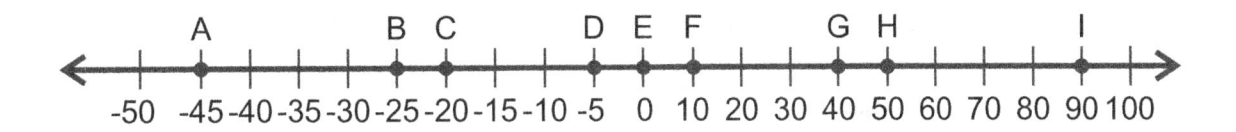

```
  A        B  C     D E F      G H        I
<-+--+--+--+--+--+--+--+--+--+--+--+--+--+--+->
 -50 -45-40-35-30-25-20-15-10 -5  0 10 20 30 40 50 60 70 80 90 100
```

11) Find the letter in the number line best represents the answer to below problem

50 + (-40) =

12) Find the letter in the number line best represents the answer to below problem

40 + (-50) =

13) Find the letter in the number line best represents the answer to below problem

(-50) + 40 =

14) Find the letter in the number line best represents the answer to below problem

50 + 40 =

15) Find the letter in the number line best represents the answer to below problem

(-40) + 50 =

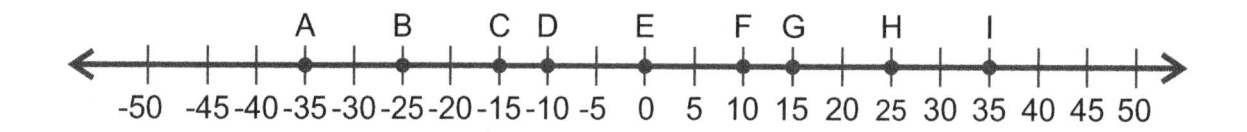

```
         A     B    C  D     E    F  G    H    I
←──┼──┼──●──┼──●──┼──●●──┼──●──┼──●●──┼──●──┼──●──┼──┼──→
  -50 -45-40-35-30-25-20-15-10 -5  0  5  10 15 20 25 30 35 40 45 50
```

16) Find the letter in the number line best represents the answer to below problem

    12 + (-15) =

17) Find the letter in the number line best represents the answer to below problem

    (-15) + (-12) =

18) Find the letter in the number line best represents the answer to below problem

    (-18) + 12 =

19) Find the letter in the number line best represents the answer to below problem

    15 + (-12) =

20) Find the letter in the number line best represents the answer to below problem

    (-16) + 20 =

```
        A   B  C         D E F         G H I
←---|---|--●--●--●--|--|--●-●-●--|--|--●-●-●--|--|--|--→
  -10 -9 -8 -7 -6 -5 -4 -3 -2 -1 0 1 2 3 4 5 6 7 8 9 10
```

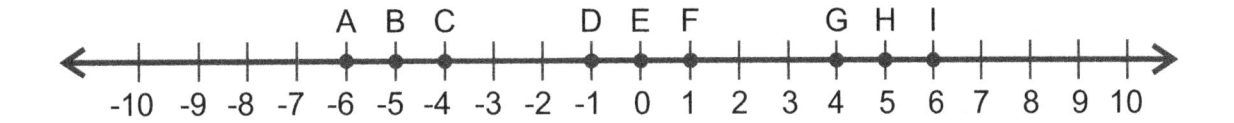

21) Find the letter in the number line best represents the answer to below problem

    11 + 5 =

22) Find the letter in the number line best represents the answer to below problem

    (-24) + 25 =

23) Find the letter in the number line best represents the answer to below problem

    25 + (-21) =

24) Find the letter in the number line best represents the answer to below problem

    (-21) + (-15) =

25) Find the letter in the number line best represents the answer to below problem

    (-15) + 11 =

26) Which expression(s) are equivalent to -88.8 + (-82.7)?

   A) -88.8 - (+82.7)

   B) 88.8 - (-82.7)

   C) 88.8 - (-82.7)

   D) 88.8 + (+82.7)

27) Which expression(s) are equivalent to 66 + (+16)?

   A) 66 + (16)

   B) -66 - (+16)

   C) -66 - (16)

   D) -66 + (+16)

28) Which expression(s) are equivalent to –79.3 – (–58.86)?

   A) 79.3 – (58.86)

   B) 79.3 + (+58.86)

   C) 79.3 – (–58.86)

   D) 79.3 – (+58.86)

29) Which expression(s) are equivalent to $-\frac{1}{5} - (-\frac{1}{7})$

   A) $-\frac{1}{5} - (+\frac{1}{7})$

   B) $\frac{1}{5} + (+\frac{1}{7})$

   C) $-(+\frac{1}{5}) + (+\frac{1}{7})$

   D) $-\frac{1}{5} + (+\frac{1}{7})$

30) Which expression(s) are equivalent to $\frac{4}{5} + (\frac{1}{5})$?

   A) $-(-\frac{4}{5}) - (-\frac{1}{5})$ B) $-\frac{4}{5} + (-\frac{1}{5})$

   C) $\frac{4}{5} + (+\frac{1}{5})$ D) $-\frac{4}{5} - (-\frac{1}{5})$

31) Which expression(s) are equivalent to 57.7 + (+64.4)?

   A) –57.7 + (–64.4)   B) –57.7 + (+64.4)

   C) 57.7 + (64.4)   D) 57.7 - (- 64.4)

32) Which expression(s) are equivalent to -70.06 + (+21.65)?

   A) 70.06 + (21.65)   B) – (70.06) + (21.65)

   C) 70.06 + (-21.65)   D) -70.06 - (-21.65)

33) Which expression(s) are equivalent to

$\frac{2}{9} - (+\frac{3}{6})$?

A) $-\frac{2}{9} - (-\frac{3}{6})$ B) $\frac{2}{9} - (\frac{3}{6})$

C) $\frac{2}{9} + (+\frac{3}{6})$ D) $\frac{2}{9} + (\frac{3}{6})$

34) Which expression(s) are equivalent to 4.54 + 6.17?

A) -4.54 + (-6.17)   B) -4.54 - (6.17)

C) 4.54 - (6.17)   D) 4.54 - (-6.17)

35) Which expression(s) are equivalent to 91 + (+62)?

A) 91 - (-62)   B) -91 + (-62)

C) 91 + (-62)   D) -91 + (+62)

36) Which expression(s) are equivalent to 88 − (−14) ?

A) −(-88) − (−14)   B) -88 − (+14)

C) 88 + (14)   D) 88 + (−14)

37) Which expression(s) are equivalent to

$\frac{3}{9} + (\frac{3}{11})$?

A) $-\frac{3}{9} - (-\frac{3}{11})$ B) $-\frac{3}{9} - (+\frac{3}{11})$

C) $\frac{3}{9} + (+\frac{3}{11})$ D) $-\frac{3}{9} - (\frac{3}{11})$

38) Which expression(s) are equivalent to 2.09 + (4.107)?

A) 2.09 - (-4.107)   B) −(-2.09) + (+4.107)

C) 2.09 + (-4.107)   D) 2.09 - (+4.107)

39) Determine if the values in the table are proportional (yes) or not (no).

X	Y
8	1
24	3
32	4
64	8

167

40) Determine if the values in the table are proportional (yes) or not (no).

X	Y
88	-8
44	-4
22	-2
11	-1

41) Determine if the values in the table are proportional (yes) or not (no).

X	Y
9	10
27	30
36	40
54	60

42) Determine if the values in the table are proportional (yes) or not (no).

X	Y
-50	-4
-60	-3
-75	-2
-85	-1

43) Determine if the values in the table are proportional (yes) or not (no).

X	Y
-18	-10
-7	-6
-6	-8
-15	-9

44) Determine if the values in the table are proportional (yes) or not (no).

X	Y
-10	-70
-9	-63
-6	-42
-3	-21

45) Determine if the values in the table are proportional (yes) or not (no).

X	Y
-100	20
-50	10
-40	8
-30	6

46) Determine if the values in the table are proportional (yes) or not (no).

X	Y
-8	-40
-6	-30
-4	-20
-2	-10

47) Determine if the values in the table are proportional (yes) or not (no).

X	Y
64	-48
24	-18
16	-12
4	-3

48) Determine if the values in the table are proportional (yes) or not (no).

X	Y
5	30
7	42
10	40
15	60

49) Determine if the values in the table are proportional (yes) or not (no).

X	Y
30	3
120	4
77	7
72	8

50) Determine if the values in the table are proportional (yes) or not (no).

X	Y
1	5
2	6
6	14
20	40

51) Determine if the values in the table are proportional (yes) or not (no).

X	Y
-10	-40
-8	-32
-6	-24
-2	-8

Grade 7
Volume 1
Week 5

52) Determine if the values in the table are proportional (yes) or not (no).

X	Y
10	16
70	112
80	128
90	144

53) Determine if the values in the table are proportional (yes) or not (no).

X	Y
-7	-5
-5	-6
-2	-8
-1	-8

54) Determine if the values in the table are proportional (yes) or not (no).

X	Y
-9	-13
-10	-12
-7	-11
-6	-10

55) Determine if the values in the table are proportional (yes) or not (no).

X	Y
50	-10
45	-9
25	-5
55	-11

56) Determine if the values in the table are proportional (yes) or not (no).

X	Y
4	-8
1	-3
0	-8
-15	-15

57) Determine if the values in the table are proportional (yes) or not (no).

X	Y
32	32
37	37
67	67
93	93

58) Determine the constant of proportionality for each table. Express your answer as y = kx

Number of Glasses (x)	120	195	220	325	390
Oranges Used (y)	24	39	44	65	78

For every glass of orange juice there were _____ oranges used.

59) Determine if the values in the table are proportional (yes) or not (no).

Bricks (x)	11	17	24	30	40
Weight in kilograms (y)	99	153	216	270	360

Every brick weighs _____ kilograms.

60) Determine if the values in the table are proportional (yes) or not (no).

Plants(x)	11	14	10	9	25
Flowers (y)	275	350	250	225	625

Every plant blooms _____ flowers.

61) Determine if the values in the table are proportional (yes) or not (no).

Hair pins (x)	17	50	96	121	654
Price ($y)	17	50	96	121	654

For each piece of hair pin it costs _____ dollars.

62) Determine if the values in the table are proportional (yes) or not (no).

Video games Sold (x)	16	24	25	34	40
Revenue Earned ($y)	256	384	400	544	640

Every video game earns _____ dollars.

63) Determine if the values in the table are proportional (yes) or not (no).

Pounds of Almonds (x lbs)	12	11	22	30	55
Price ($y)	168	154	308	420	770

For every pound of almonds it cost _____ dollars.

64) Determine if the values in the table are proportional (yes) or not (no).

Votes for Jack (x)	14	22	31	44	56
Votes for Jill (y)	252	396	558	792	1008

For Every vote for Jack there were _____ votes for Jill

65) Determine if the values in the table are proportional (yes) or not (no).

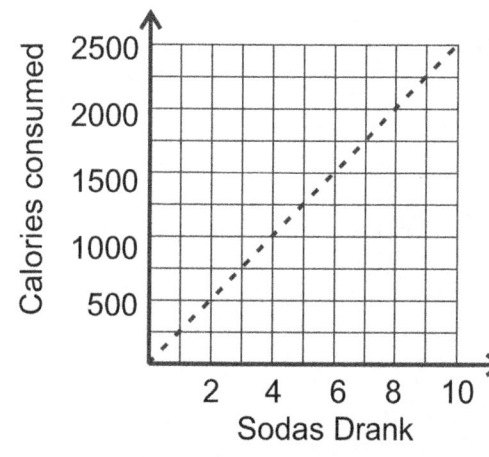

A) The point (2, 500) shows that drinking 2 sodas will give a total of 500 calories.

B) The point (1, 250) shows that drinking 1 soda will give a total of 250 calories.

C) The point (1000, 4) shows that drinking 4 sodas will give a total of 1000 calories.

D) The point (4, 1000) shows that 1000 calories you'd have to drink 4 sodas.

66) Determine if the values in the table are proportional (yes) or not (no).

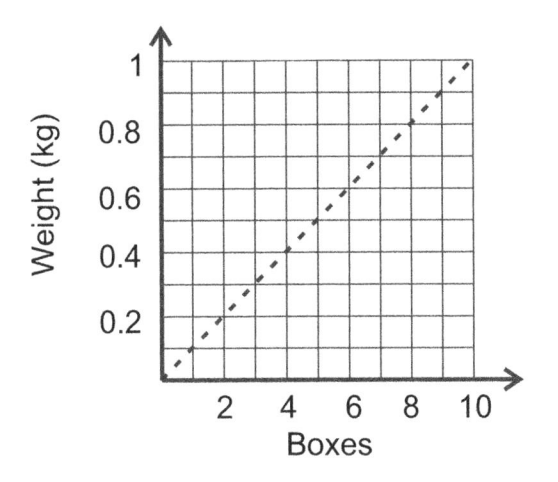

A) The point (8, 0.4) shows that 0.4 kilograms is the weight of 8 boxes.

B) The point (20, 1) shows that 20 boxes weigh 1 kilogram.

C) The point (0.5, 10) shows that 0.5 kilograms is the weight of 10 boxes.

D) The point (16, 0.8) shows that 16 boxes weigh 0.8 kilograms.

67) Determine which statements about the graph are true.

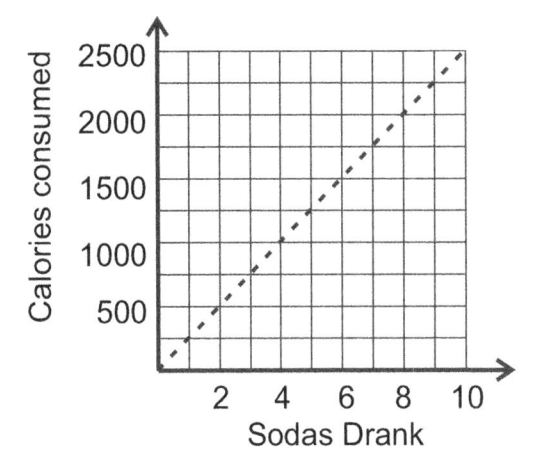

A) The point (6, 1500) shows that drinking 6 sodas will mean you consumed 1500 calories.

B) The point (10, 2500) shows that to consume 2500 calories you'd have to drink 10 sodas.

C) The point (250, 1) shows that drinking 1 soda will mean you consumed 250 calories.

D) The point (2000, 8) shows that drinking 8 sodas will mean you consumed 2000 calories.

68) Determine if the values in the table are proportional (yes) or not (no).

[Graph: Price vs Pieces of cupcakes, with dashed line from origin through points, y-axis labeled Price with values 3, 6, 9, 12, 15; x-axis labeled Pieces of cupcakes with values 2, 4, 6, 8, 10]

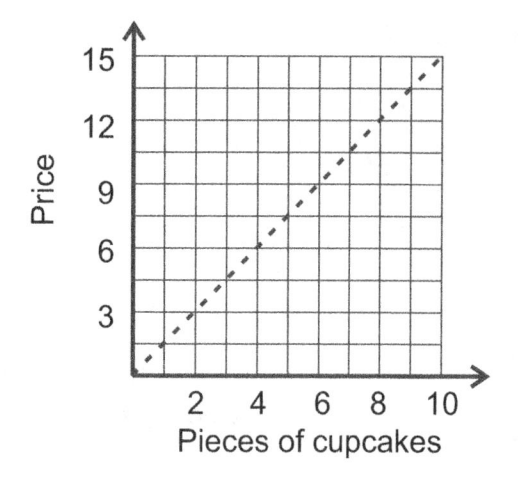

A) The point (3, 4.5) shows that with $4.5 you can buy 3 cupcakes.

B) The point (7, 10.5) shows that 7 cupcakes will cost $10.5.

C) The point (9, 12) shows that 9 cupcakes will cost $12.

D) The point (4, 6) shows that 4 cupcakes will cost $6.

69) Determine if the values in the table are proportional (yes) or not (no).

[Graph: Batches of cookies vs Cups of flour, with dashed line from origin through points, y-axis labeled Batches of cookies with values 3, 6, 9, 12, 15; x-axis labeled Cups of flour with values 2, 4, 6, 8, 10]

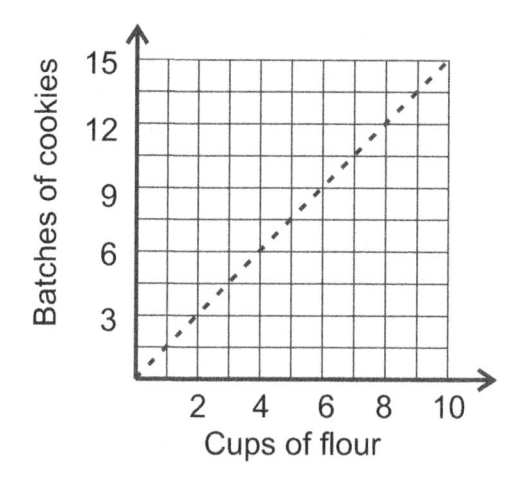

A) The point (5, 2) shows that 2 cups of flour will make 5 batches of cookies.

B) The point (7, 3) shows that 3 cups of flour will make 7 batches of cookies.

C) The point (4.5, 9) shows that 4.5 cups of flour will make 9 batches of cookies.

D) The point (8, 4.5) shows 8 batches of cookies require 3.5 cups of flour.

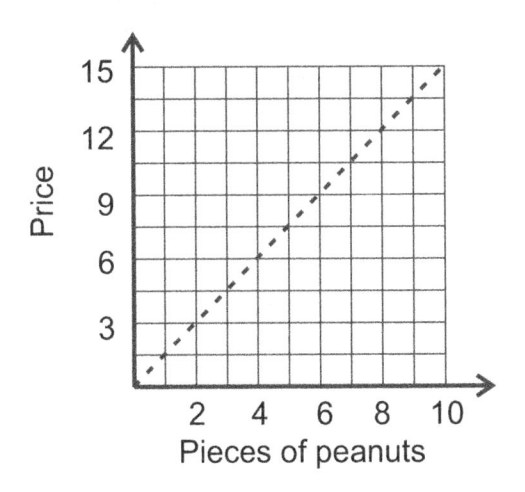

70) Determine if the values in the table are proportional (yes) or not (no).

Pieces of peanuts (x-axis) vs **Price** (y-axis)

A) The point (4, 6) shows that with $6 you can buy 4 packs of peanuts

B) The point (8, 12) shows that with $12 you can buy 8 packs of peanuts.

C) The point (9, 13.5) shows that 9 packs of peanuts will cost $13.5.

D) The point (10.5, 7) shows that with $10.5 you can buy 7 packs of peanuts

71) Determine which statements about the graph are true.

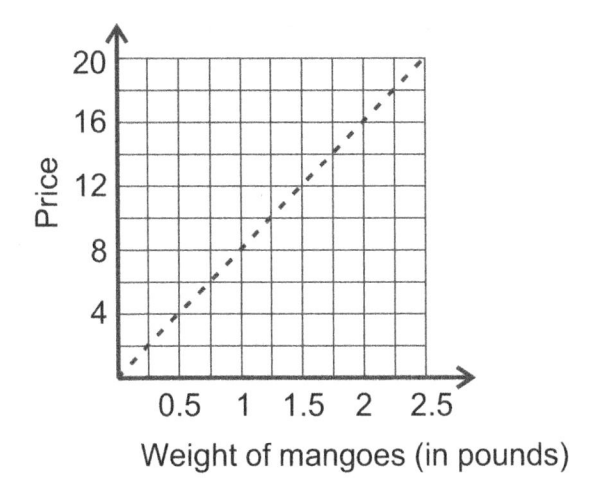

Weight of mangoes (in pounds) (x-axis) vs **Price** (y-axis)

A) The point (11, 1.25) shows that 1.25 pounds of mangoes will cost $10.

B) The point (1.5, 12) shows that 1.5 pound of mangoes will cost $12.

C) The point (2.5, 16) shows that it would cost $16 for 2.5 pounds of mangoes.

D) The point (14, 1.75) shows that mowing 1.75 pounds of mangoes will cost $14.

Grade 7
Volume 1
Week 5

72) A construction contractor used the equation Y = KX to determine it would cost him $11.91 to buy 3 boxes of nails. How much is each box?

73) The equation 77.07 = 7k shows that buying 7 bags of almonds would cost 11.01 dollars. How much is it for one bag?

74) An industrial printing machine printed 1570 pages in 3 minutes. How much would it have printed in 9 minutes?

75) An ice cream truck driver determined he had made $33.75 after selling 15 ice cream bars (using the equation y = kx). How much would he have earned if he sold 45 bars?

76) A movie theater used Y = 11.04X to calculate how much money they made selling buckets of popcorn where Y is the total and K is the price per bucket. How much would they make if they sold 9 buckets?

77) A grocery store paid $147.35 for 7 crates of orange juice. This can be expressed by the equation Y = KX. How much would they have paid for 15 crates?

78) The equation 704 = (176)4 is used to determine how many pages would be needed to make 4 books. How many pages are in one book?

79) Sally bought 4 boxes of hair pins for $15.92. This can be expressed by the equation 15.92 = (3.98)4. How much would it cost for 8 boxes?

80) Elsa walked $\frac{1}{2}$ of a mile in $\frac{1}{3}$ of an hour. At this rate, how far will she have travelled after an hour?

81) Sally needs $\frac{1}{2}$ of a bag of oranges to make $\frac{1}{3}$ of a gallon of juice. If she wanted to make a full gallon of juice how many bags of oranges would she need?

82) Ron spent $\frac{1}{2}$ of an hour playing on his phone. That used up $\frac{1}{3}$ of his battery. How long would he have to play on the phone to use the entire battery?

83) A bottle of gasoline contains $\frac{1}{2}$ of a liter to fill up $\frac{1}{3}$ of a motorcycle gas tank. How many bottles would you need to fill up the gas tank entirely?

84) Using a water hose Tom fills up $\frac{1}{3}$ of a pool in $\frac{1}{2}$ of an hour. At this rate, how many hours would it take to fill the pool?

85) Lola $\frac{1}{2}$ of a box of nails while working on a birdhouse. At this rate, how many boxes will he need to finish 5 bird houses?

86) A juicer was able to squeeze a pint of juice from $\frac{1}{2}$ bag of apples. This amount of juice filled up $\frac{1}{3}$ of a jug. At this rate, how many bags will it take to fill the entire jug?

87) A basket of lemons weighed $\frac{1}{2}$ of a pound and could make a jug of lemonade that was $\frac{1}{3}$ full. How many baskets of lemons would you need to fill up two jugs?

88) An ant going full speed was taking $\frac{1}{2}$ of a minute to move $\frac{1}{5}$ of a centimeter. At this rate, how long would it take the ant to travel a centimeter?

89) A bag of grass seeds weighed $\frac{1}{2}$ of a kilogram. That was enough to cover $\frac{1}{8}$ of a front lawn with seed. How many bags would it take to completely cover a lawn?

90) A bag of chocolate mix that weighed $\frac{1}{2}$ of a kilogram could make enough brownies to feed $\frac{1}{7}$ of the students in grade 7. How many bags of chocolate mix would be needed to feed all of the students in the class?

91) A car can travel 189 miles on 9 gallons of gasoline. How far can it travel on 14 gallons?

92) 49 lbs of peppers cost $98. How many lbs of peppers can you get with $30?

93) 8 kg of bananas cost $56. How many kilograms of bananas can you get with $189?

94) A boat can travel 72 miles on 18 gallons of gasoline. How far can it travel on 90 gallons?

95) A boat can travel 390 miles on 195 gallons of gasoline. How much gasoline will it need to go 174 miles?

96) A car travels 169 kilometers in 1 hours (with a constant speed). How much time will it take traveling 221 kilometers?

97) A boat can travel 23 kilometers on 23 liters of gasoline. How far can it travel on 16 liters?

98) A car travels 450 miles in 12 hours. How far can it travel in 6 hours?

99) A boat can travel 342 kilometers on 114 liters of gasoline. How much gasoline will it need to go 51 kilometers?

100) A car can travel 310 miles on 10 gallons of gasoline. How much gasoline will it need to go 465 miles?

101) 49 kg of bananas cost $294. How much would 11.

102) A car can travel 280 miles on 10 gallons of gasoline. How much gasoline will it need to 168 miles?

103) Helen covers a distance of 10.2 miles in 3 hours on her bicycle, find the distance covered by her in 10 hours.

**Grade 7
Volume 1
Week 5**

104) A train 360 m long is running at a speed of 90 km/ hr. What time will it take to cross 140 m long bridge?
(Given, 1 km = 1000 m)

105) Convert the speed of 36 miles/ hour to miles/ sec.

106) Convert 15 miles/sec to miles/hour.

107) A bus runs at 68.4 miles/hr. Calculate the distance traveled by the bus in 15 seconds.

108) Peter's car is running at 72 miles/hour. How much time will it take to cover 450 miles?

109) Ron walks at a speed of 8 miles/hr and covers a certain distance in 3 hours 15 minutes. If he covers the same distance on bike at the rate of 30 miles/hr, calculate the time taken by him?

110) A truck has traveled 90 miles in 1.5 hours. Which proportion can be used to determine how long, it would take for the truck to travel 450 miles, assuming it is moving at the same speed?

111) A car travels at a speed of 85 miles/hour. How much distance will it cover in 3.5 hours?

112) A shot is fired at a distance of 30 miles away from Mary's house and she hears the sound after 3 seconds. Find the speed at which the sound travelled?

113) John goes to his school walking at 5 miles/hr and reaches in 30 minutes. Calculate the time taken to reach his school if he runs at speed of 6 miles/hr.

114) A train leaves station A on Wednesday at 6:30 PM and reaches station B next day at 8:30 AM. If the distance between the two stations is 1484 miles, find the average speed of the train.

115) A car travels a distance of 258 miles in 6 hours. What is the speed of the car?

116) If a train covers a distance of 14 miles in 10 minutes, find the speed of the train.

117) A car is running at 40 miles/hour, takes 9 hours to reach destination. At what speed = it should travel so that the journey is completed in 6 hours?

118) A bus running at 36 miles/hour covers a certain distance in 40 minutes. How much time will it take to cover the same distance at a speed of 30 miles/hour?

119) A train is running at a speed of 60 km/hour. The length of the train is 150 m. How much time will it take to cross an electric pole? (Given, 1km = 1000 m)

120) A train 300 m long passes a tree in 12 seconds. What is the speed of the train in km/hour? (Given, 1 km = 1000 m)

121) A car covers 100 miles in the first two hours and 56 miles in the third hour. Find the average speed of the car.

122) An airplane travels the first 2500 miles at a speed of 500 miles/hour, the next 1500 miles at a speed of 300 miles/hour. Find the average speed of the airplane in the whole journey.

123) An aero plane travels at a speed of 900 miles/hour. Convert the speed into miles/sec.

124) Convert 30 m/sec into km/hour. (Given, 1km = 1000 m)

125) A car travels at a speed of 90 miles/ hour. Express this speed in miles/ second.

126) A train, 150 m long, is running at a speed of 54 km/hr. What time will it take to cross an electric pole? (Given, 1km = 1000 m)

127) Two trains of length 130 m and 95 m are traveling in opposite directions on parallel lines at 32 km/hr and 49 km/hr respectively. In what time will they pass each other from the moment they meet?

Grade 7
Volume 1
Week 5

128) Two trains, 125 m and 175 m in length, are running in the same direction on parallel lines, one at the rate of 40 km/hr and the other at 22 km/hr. In what time will they completely pass each other?

129) A car travels 60 miles in 50 minutes. Find the speed of the car in miles/hr.

130) Harris jogs 8 miles in 8 minutes. What is his speed in miles/hr?

131) If A, B and C can do a piece of work in 10 days, 12 days and 15 days respectively, how long will they take to finish it if they work together?

132) A cistern can be filled by two taps A and B in 12 minutes and 16 minutes respectively. The full cistern can be emptied by a third tap C in 8 minutes. If all taps are turned on at the same time, in how much time will the empty cistern be filled up completely?

133) Amy alone can paint a house in 20 days and Macy alone can paint the house in 30 days. How much time will they take to paint the house, if both Amy and Macy work together?

134) An IT company wants to buy 220 new cell phone cases at z dollars apiece for their employees. Because they were buying so many, they got 27% off the price. Which expression shows how much money they saved?

A) 0.27 × 220z

B) 220z + 1.27

C) 220z + 0.27

D) 220z − 0.27

135) A new box of cereal contains having 49% more almonds. The original cereal had c cups of almonds. Which expression shows total number of cups almonds in the new cereal?

A) y + 1.49

B) y × 0.49

C) y + (0.49 × y)

D) y 0.49

136) Wilson drew a square with each side being exactly 9 in long. If he wanted to make the square 15% larger which expression, can he use to find new side length?

A) 9 × 0.15

B) 9 + 1.15

C) 9 + 0.15

D) 9 × 1.15

137) A cell phone company dropped the prices on their phones by 11%. Which expression shows the new price of the phones (p)?

A) p × 0.11

B) p - 1.11

C) p - 0.11p

D) p - 0.11

138) A fruit seller raised the price on watermelons 8%. The original price for each was X dollars. Which expression shows the new price of the watermelons?

A) X + 0.8

B) X × 0.8

C) X + (0.8 × X)

D) X + 1.8

139) Joe was earning $18 an hour before his raise. After his 5% raise he was making $18.9 an hour. Which expression shows how his new hourly rate was calculated?

A) 18 + 1.05

B) 18 × 1.05

C) 18 × 0.05

D) 18 + 1.05

140) Over the winter gas prices dropped 3%. Which expression shows the new price of a gallon of gas? (the old price is represented by g)

A) g - 0.03

B) g - 1.03

C) g - 0.03g

D) g × 0.03

**Grade 7
Volume 1
Week 5**

141) Best electronics store has a black Friday sale of 11% off on all computer monitors. Which expression shows the amount of savings if you bought 25 monitors for z dollars a piece?

   A) 25z – 0.11

   B) 0.11 × 25z

   C) 25z + 0.11

   D) 25z + 1.11

142) Jade's house was on sale for $22,871. If you wanted to offer 16% less than the asking price (p) which expression shows how much you should offer?

   A) p – 1.16

   B) p – 0.16

   C) p × 0.16

   D) p – 0.16p

143) The regular price of a computer was $1899 dollars, but over the Christmas weekend it'll be on sale for 18% off. Which expression shows the difference in price from normal(n) to sale?

   A) 1899 – 10

   B) 1899 × 0.18

   C) 1899 – 0.1

   D) 1899 – 1.1

144) A farmer produced is 12% more crop last years. This year's crop is represented by c. Which expression can be used to calculate the last year's crop?

   A) c 1.12

   B) c × 0.12

   C) c – 0.12

   D) c – 1.12

145) A sandwich shop was charging $6.72 for a sandwich but raised the price 6% making them cost $7.06. Which expression shows how the new price was calculated?

   A) 6.72 + 1.06 = $7.60

   B) 1.72 × 0.06 = $0.34

   C) 6.72 + 0.06 = $7.06

   D) 6.72 × 1.06 = $7.06

146) A boutique is offering 25 % off any item(d) on the weekend sale. Which expression can be used to calculate the new cost of an item?

   A) d × 0.25

   B) d – 1.25

   C) d – 0.25d

   D) d – 0.25

147) A cell phone company dropped the prices on their phones by 8%. Which expression shows the new price of the phones(p)?

A) p – 1.08

B) p – 0.08

C) p – 0.08p

D) p × 0.08

148) Jolly was earning $16 an hour before his raise. After her 5% raise she was making $16.8 an hour. Which expression shows how her new hourly rate was calculated?

A) 16 + 0.05 = $16.05

B) 16 × 1.05 = $16.8

C) 16 × 0.05 = $16.8

D) 16 + 1.05 = $17.05

149) Over the winter gas prices dropped 6%. Which expression shows the new price of a gallon of gas? (the old price is represented by g)

A) g × 0.06

B) g – 0.06g

C) g – 1.06

D) g – 0.06

150) The regular price of a bike was $548, but over the weekend it'll be on sale for 20% discount. Which expression shows the difference in price?

A) b × 0.2

B) b – 1.2

C) b – 20

D) b – 0.2

151) John drew a picture with each side being exactly 7 in long. If he wanted to make the picture 9% larger which expression, can he use to find the new sides length?

A) 7 + 0.09

B) 7 + 1.09

C) 7 × 0.09

D) 7 × 1.09

152) A house was on sale for $435,788. If you wanted to offer 14% less than the asking price(p) which expression shows how much you should offer?

A) p – 0.14p

B) p – 1.14

C) p × 0.14

D) p – 0.14

Grade 7
Volume 1
Week 5

153) A Yummy bakery sells a cookie for $2.11, a brownie for $3.60 and a cake for $28.78. On Wednesday they sold 12 cookies, 8 brownies and 5 cakes. The cost in ingredients for each cookie is $0.85, for each brownie is $0.55 and for each cake is $4.70. How much profit did they make on Wednesday?

154) A recycling center in the city was offering cash for different types of recyclables. For each pound of paper recycled they offered $2.76, for each pound of aluminum they offered $2.21 and for each pound of plastic they offered $0.99. George took in 15 pounds of paper, 6 pounds of aluminum and 11 pounds of plastic. How much money did he earn?

155) Violet ordered 5 pounds of beef at $7.28 per pound, 7 pounds of potatoes at $11.78 a pound and 9 pounds of carrots at $4.44 per pound. She estimates that this will make 42 meals for a fundraiser. If she charges $32.55 per meal how much profit will he make?

156) A new pizza shop invested $12,056.92 to open it in the city mall. During first week they sold 162 pizzas ($7.58 profit each), 90 sodas ($2.65 profit each), 560 sodas ($2.07 profit each) and 275 burgers ($8.10 profit each). Their expenses for the first month were $2,956.82. How much money do they still need to make to break even?

157) At the local fair 9 employees were paid $21.09 an hour. They worked for 8 hours at regular price and then for 5 hours at overtime price (an extra $12.35 an hour). How much money did they make total?

Grade 7 Volume 1 Week 6

Grade 7
Volume 1
Week 6

1) Identify if the given pair of ratios form a proportion.

$$\frac{10}{6} \text{ and } \frac{20}{12}$$

A) Yes   B) No

2) Identify if the given pair of ratios form a proportion.

$$\frac{8}{10} \text{ and } \frac{56}{70}$$

A) Yes   B) No

3) Identify if the given pair of ratios form a proportion.

$$\frac{10}{7} \text{ and } \frac{50}{28}$$

A) No   B) Yes

4) Identify if the given pair of ratios form a proportion.

$$\frac{14.4}{10.2} \text{ and } \frac{3.6}{3.4}$$

A) Yes   B) No

5) Identify if the given pair of ratios form a proportion.

$$\frac{6.4}{9.6} \text{ and } \frac{3.2}{2.4}$$

A) No   B) Yes

6) Identify if the given pair of ratios form a proportion.

$$\frac{10}{6} \text{ and } \frac{70}{54}$$

A) Yes   B) No

7) Identify if the given pair of ratios form a proportion.

$$\frac{56}{42} \text{ and } \frac{8}{6}$$

A) No   B) Yes

8) Identify if the given pair of ratios form a proportion.

$$\frac{4.8}{4.4} \text{ and } \frac{2.4}{2.2}$$

A) Yes   B) No

9) Identify if the given pair of ratios form a proportion.

$$\frac{3.9}{3.6} \text{ and } \frac{15.6}{10.8}$$

A) Yes   B) No

10) Identify if the given pair of ratios form a proportion.

$$\frac{5.4}{7} \text{ and } \frac{2.7}{3.5}$$

A) Yes   B) No

11) Solve the below proportion.

$$\frac{10}{x} = \frac{15}{19}$$

A) 2 B) $\frac{38}{3}$

C) 25 D) $\frac{15}{13}$

12) Solve the below proportion.

$$\frac{32}{31} = \frac{25}{x}$$

A) $\frac{45}{23}$ B) 2

C) 23 D) $\frac{775}{32}$

13) Solve the below proportion.

$$\frac{n}{43} = \frac{29}{16}$$

A) 32 B) $19\frac{11}{12}$

C) $\frac{1247}{16}$ D) $6\frac{4}{37}$

14) Solve the below proportion.

$$\frac{v}{23} = \frac{33}{36}$$

A) $10\frac{11}{13}$ B) $\frac{11}{8}$

C) $\frac{19}{25}$ D) $\frac{253}{12}$

15) Solve the below proportion.

$$\frac{16}{b} = \frac{44}{8}$$

A) $24\frac{32}{39}$ B) 12

C) 2 D) $\frac{32}{11}$

16) Solve the below proportion.

$$\frac{x}{38} = \frac{36}{25}$$

A) $\frac{1368}{25}$ B) $8\frac{20}{33}$

C) $\frac{7}{10}$ D) 6

17) Solve the below proportion.

$$\frac{41}{12} = \frac{6}{p}$$

A) 43 B) $\frac{72}{41}$

C) $\frac{25}{38}$ D) 6

18) Solve the below proportion.

$$\frac{40}{10} = \frac{49}{a}$$

A) $17\frac{13}{43}$ B) $\frac{49}{4}$

C) 1 D) $23\frac{23}{35}$

**Grade 7
Volume 1
Week 6**

19) Solve the below proportion.

$$\frac{23}{k} = \frac{14}{48}$$

A) $\frac{53}{38}$ B) 47

C) $\frac{552}{7}$ D) 13

20) Solve the below proportion.

$$\frac{3.5}{r} = \frac{4.7}{4.4}$$

A) 5.8   B) 7.4

C) 3.28   D) 2.4

21) Solve the below proportion.

$$\frac{n}{7} = \frac{7.3}{7.89}$$

A) 7.8   B) 6.48

C) 2   D) 2.4

22) Solve the below proportion.

$$\frac{n}{7} = \frac{4.4}{4.7}$$

A) 6.55   B) 6.9

C) 1.9   D) 6.7

23) Solve the below proportion.

$$\frac{6}{p} = \frac{17}{4}$$

A) 1.41   B) 5.9

C) 5.5   D) 5.6

24) Solve the below proportion.

$$\frac{44}{45} = \frac{8}{x}$$

A) $\frac{90}{11}$ B) $\frac{16}{21}$

C) 26   D) $9\frac{9}{32}$

25) Solve the below proportion.

$$\frac{v}{3.3} = \frac{6.6}{7.3}$$

A) 2.98   B) 5

C) 3   D) 8

26) Solve the below proportion.

$$\frac{x}{4.9} = \frac{7}{7.3}$$

A) 1   B) 4.7

C) 3.4   D) 5.1

27) Solve the below proportion.

$$\frac{4}{k} = \frac{17}{5}$$

A) 6.7 B) 8
C) 15 D) 1.18

28) Solve the below proportion.

$$\frac{4}{x} = \frac{9}{7}$$

A) 3.11 B) 4
C) 17.9 D) 10.799

29) Solve the below proportion.

$$\frac{v}{11} = \frac{5}{3}$$

A) 7.5 B) 18.33
C) 8.4 D) 17.9

30) Solve the below proportion.

$$\frac{k}{17} = \frac{15}{12}$$

A) 3.4 B) 21.25
C) 5.4 D) 9.53

31) Solve the below proportion.

$$\frac{17}{6} = \frac{r}{15}$$

A) 42.5 B) 8
C) 10 D) 6.633

32) Solve the below proportion.

$$\frac{11}{19} = \frac{17}{k}$$

A) 17.7 B) 12.416
C) 12 D) 29.36

33) A photo is 130 cm wide and 60 cm tall. If it is reduced to a width of 13 cm, then how tall will it be?
(Round to the nearest whole number)

A) 130 cm B) 6 cm
C) 1 cm D) 600 cm

34) Solve the below proportion.

$$\frac{15}{x} = \frac{9}{11}$$

A) 13.2 B) 18.33
C) 1 D) 1.4

35) Solve the below proportion.

$$\frac{5}{4x} = \frac{16}{8}$$

A) 0.63 B) 6

C) 18.6 D) 12.5

36) Solve the below proportion.

$$\frac{x}{16} = \frac{11}{13}$$

A) 6 B) 2

C) 9.7 D) 13.54

37) The currency in Kuwait is the Kuwa. The exchange rate is approximately 7 Kuwas to $24. At this rate, how many Kuwas would you get if you exchanged $528?
(Round to the nearest whole number)

A) 193 Kuwas B) 154 Kuwas

C) 169 Kuwas D) 1810 Kuwas

38) A triangle is 4 in wide and 7 in tall. If it is enlarged to a height of 70 in, then how wide will it be?
(Round to the nearest whole number)

A) 42 in B) 40 in

C) 45 in D) 700 in

39) A frame is 196 in tall and 126 in wide. If it is reduced to a width of 9 in, then how tall will it be?
(Round to the nearest whole number)

A) 2744 in B) 1 in

C) 13 in D) 14 in

40) James took a trip to the Caribbean islands. Upon leaving he decided to convert all of his Caros back into dollars. How many Caros did he receive if he exchanged 132 Caros at a rate of $4 to 11 Caros ?
(Round to the nearest whole number)

A) $36 B) $48

C) $41 D) $363

41) A rectangle is 6 in tall and 5 in wide. If it is enlarged to a width of 60 in, then how tall will it be?
(Round to the nearest whole number)

A) 1 in B) 5 in

C) 72 in D) 720 in

42) The currency in Malaysia is the Riggs. The exchange rate is approximately 35 Riggs for $10. At this rate, how many dollars would you get if you exchanged 140 Riggs?
(Round to the nearest whole number)

A) $490 B) $34

C) $40 D) $38

43) Mary reduced the size of a photo to a height of 8 cm. What is the new width if it was originally 52 cm wide and 32 cm tall?
(Round to the nearest whole number)

A) 2 cm   B) 13 cm

C) 14 cm   D) 32 cm

44) Bella enlarged the size of a rectangle to a width of 28 in. What is the new height if it was originally 14 in tall and 7 in wide?
(Round to the nearest whole number)

A) 112 in   B) 4 in

C) 56 in   D) 7 in

45) Lola reduced the size of a photo to a width of 14 in. What is the new height if it was originally 140 in wide and 90 in tall?
(Round to the nearest whole number)

A) 8 in   B) 900 in

C) 1 in   D) 9 in

46) A rectangle is 13 in wide and 12 in tall. If it is enlarged to a height of 36 in, then how wide will it be?
(Round to the nearest whole number)

A) 37 in   B) 36 in

C) 39 in   D) 41 in

47) If you can buy one bunch of seedlees green grapes for $2, then how many can you buy with $34 ?
(Round to the nearest whole number)

A) 18   B) 17

C) 16   D) 68

48) Find the missing side from the given pair of similar figures

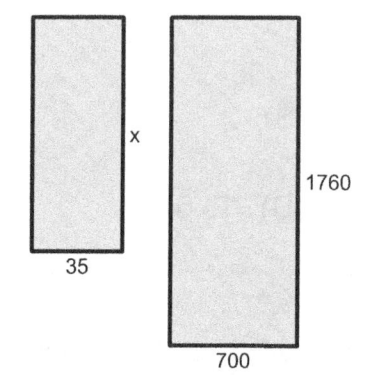

A) 35200   B) 88

C) 1760   D) 700

49) Find the missing side from the given pair of similar figures

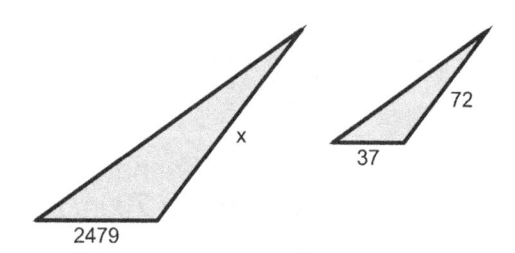

A) 166093 B) 323208

C) 4824   D) 1

Grade 7
Volume 1
Week 6

50) Find the missing side from the given pair of similar figures

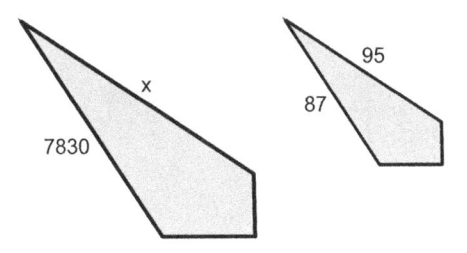

A) 90 B) 8550
C) 170100 D) 75

51) Find the missing side from the given pair of similar figures

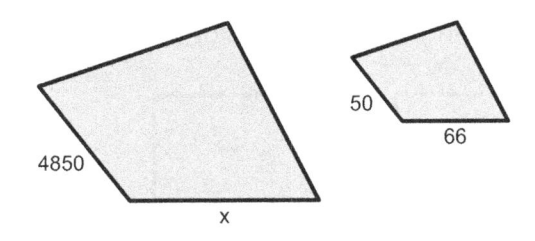

A) 470450 B) 6402
C) 0.6 D) 6693

52) Find the missing side from the given pair of similar figures

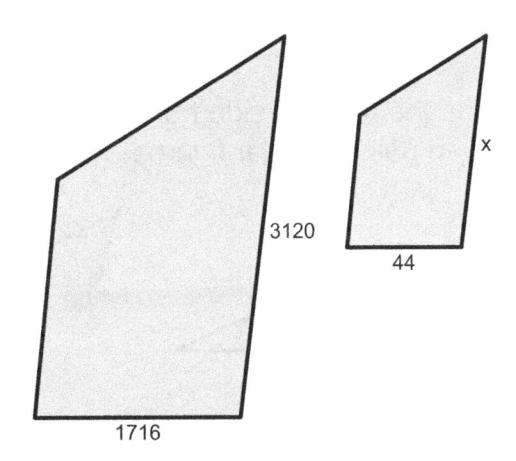

A) 66924 B) 1950
C) 3120 D) 80

53) Find the missing side from the given pair of similar figures

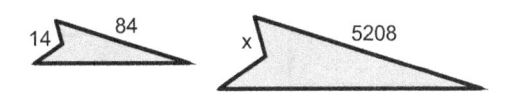

A) 84 B) 0.2
C) 868 D) 1240

54) Find the missing side from the given pair of similar figures

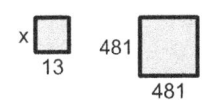

A) 943 B) 17797
C) 481 D) 13

55) Find the missing side from the given pair of similar figures

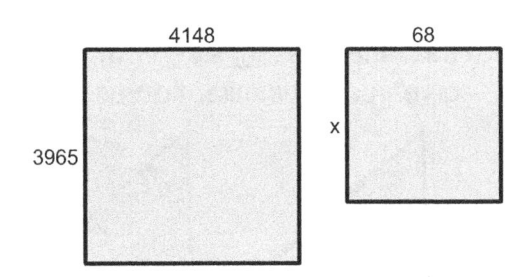

A) 3965 B) 4148
C) 65 D) 61

**Grade 7
Volume 1
Week 6**

56) Find the missing side from the given pair of similar figures

8.3
2.6
x
13

A) 41.5   B) 207.5
C) 65   D) 13

57) Find the missing side from the given pair of similar figures

90
64
8190
x

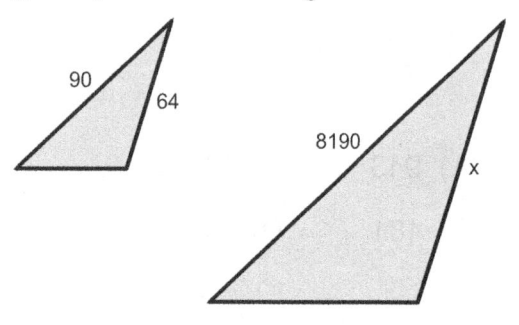

A) 0.7   B) 606
C) 64   D) 5824

58) Find the missing side from the given pair of similar figures

74
2146
x
1421

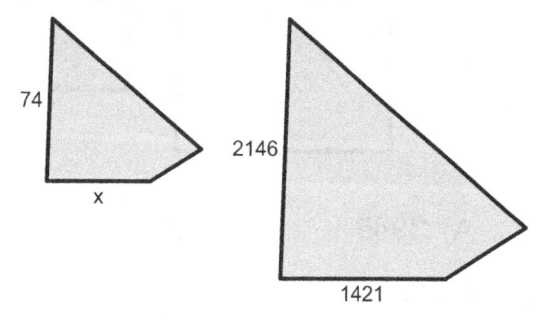

A) 74   B) 812
C) 49   D) 1421

59) Find the missing side from the given pair of similar figures

103.2
8.6
24
x

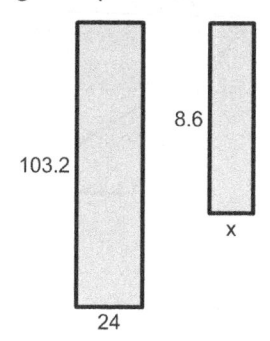

A) 288   B) 1238.3
C) 24   D) 2

60) Find the missing side from the given pair of similar figures

6.2
5.1
51
x

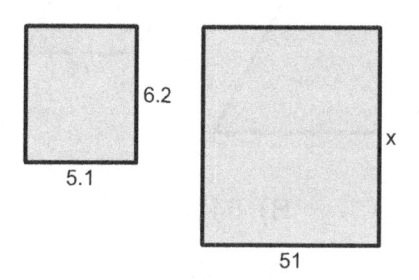

A) 510   B) 6.2
C) 620   D) 62

61) Find the missing side from the given pair of similar figures

30.4
55.2
3.8
x

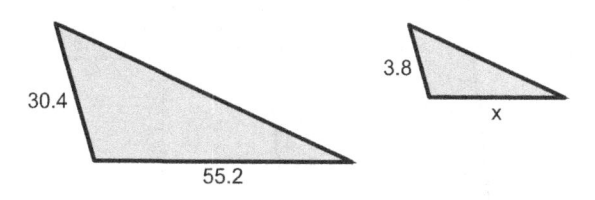

A) 6.9   B) 563.2
C) 55.2   D) 70.4

©All rights reserved-Math-Knots LLC., VA-USA   198   www.math-knots.com | www.a4ace.com

Grade 7
Volume 1
Week 6

62) Find the missing side from the given pair of similar figures

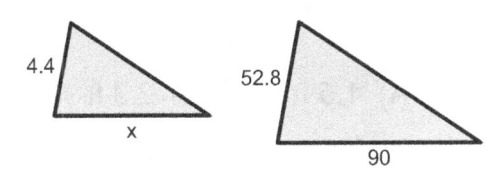

4.4 — x — 52.8 — 90

A) 7.5   B) 90
C) 96   D) 0.6

63) Find the missing side from the given pair of similar figures

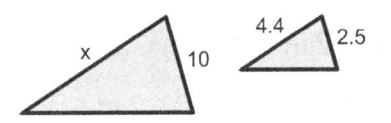

x, 10, 4.4, 2.5

A) 70.4   B) 17.6
C) 4   D) 40

64) Find the missing side from the given pair of similar figures

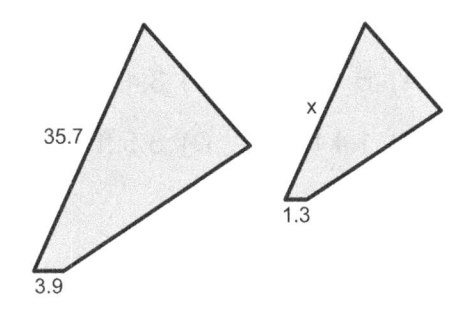

35.7, 3.9, x, 1.3

A) 107.1   B) 10
C) 11.9   D) 3.9

65) Find the missing side from the given pair of similar figures

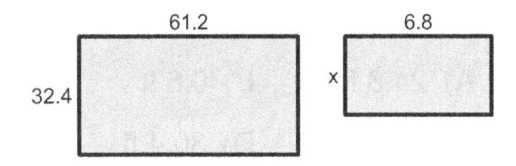

61.2, 32.4, 6.8, x

A) 550.8   B) 61.2
C) 9   D) 3.6

66) A model plane is 7 cm long. If it was built with a scale of 1 cm : 2.5 m, then how long is the real plane?

A) 2.1 m   B) 17.5 m
C) 9.5 m   D) 2.8 m

67) A christmas tree that is 1.5 ft tall casts a shadow that is 3.2 ft long. Find the length of the shadow that a 7.2 ft vase casts.

A) 6.7 ft   B) 15.4 ft
C) 0.7 ft   D) 3.4 ft

68) A model plane has a scale of 1 cm : 3.5 m. If the real plane is 17.5 m tall, then how tall is the model plane?

A) 14 cm   B) 61.3 cm
C) 5 cm   D) 0.9 cm

©All rights reserved-Math-Knots LLC., VA-USA   199   www.math-knots.com | www.a4ace.com

**Grade 7
Volume 1
Week 6**

69) A globe that is 2.1 ft tall casts a shadow that is 8 ft long. Find the length of the shadow that a 6.5 ft woman casts.

   A) 24.8 ft   B) 0.6 ft
   C) 1.7 ft   D) 30.4 ft

70) A train is 12 ft tall. A model of it was built with a scale of 1 in : 2 ft. How tall is the model?

   A) 24 in   B) 1.1 in
   C) 10 in   D) 6 in

71) Find the distance between Chantilly and Aldie if they are 1.7 in apart on a map with a scale of 1 in : 5.6 mi.

   A) 8.4 mi   B) 9.5 mi
   C) 0.3 mi   D) 53.2 mi

72) A SUV is 11.2 ft long. A model of it was built with a scale of 1 in : 2.8 ft. How long is the model?

   A) 31.4 in   B) 8.4 in
   C) 2.2 in   D) 4 in

73) A tree that is 7.1 ft tall casts a shadow that is 22 ft long. Find the height of a tent bath that casts a 14 ft shadow.

   A) 1.5 ft   B) 2.3 ft
   C) 4.5 ft   D) 68.2 ft

74) A pond is 12 ft long. A model of it was built with a scale of 1 in : 6 ft. How long is the model?

   A) 2.2 in   B) 72 in
   C) 6 in   D) 2 in

75) A 2.6 ft tall car standing next to a tree casts a 4.7 ft shadow. If the tree casts a shadow that is 11 ft long, then how tall is it?

   A) 6.1 ft   B) 3.4 ft
   C) 1.4 ft   D) 8.5 ft

76) A map has a scale of 1 cm : 15.9 km. If Austin and Irving are 2.2 cm apart on the map, then how far apart are the real cities?

   A) 37.1 km   B) 0.1 km
   C) 556.5 km D) 35 km

**Grade 7
Volume 1
Week 6**

77) A 8.4 ft tall lamp post booth standing next to a ladder casts a 44.5 ft shadow. If the ladder is 10.3 ft tall, then how long is its shadow?

A) 235.9 ft B) 1.9 ft

C) 1.6 ft D) 54.6 ft

81) Evaluate the below

$$191 - (-273) - (-10)$$

A) 947 B) 509

C) 474 D) 899

78) A model rocket has a scale of 1 cm : 6.2 m. If the model rocket is 3 cm tall, then how tall is the real rocket?

A) 9.2 m B) 0.9 m

C) 0.5 m D) 18.6 m

82) If a 7.3 m tall statue casts a 28.5 m long shadow, then how tall is a lighthouse that casts a 41 m shadow?

A) 1.9 m B) 2.7 m

C) 10.5 m D) 111.2 m

79) Evaluate the below

$$208 + (-319) + 453$$

A) 154 B) 293

C) 342 D) 160

83) A model plane has a scale of 1 in : 2 ft. If the model plane is 5 in tall, then how tall is the real plane?

A) 1.7 ft B) 10 ft

C) 2.5 ft D) 7 ft

80) Evaluate the below

$$176 - 170 + 32$$

A) 146 B) −163

C) 38 D) 241

84) A statue that is 15.9 ft tall casts a shadow that is 68.4 ft long. Find the height of an tree that casts a 75.3 ft shadow.

A) 3.7 ft B) 294.1 ft

C) 4.1 ft D) 17.5 ft

85) Evaluate the below

$$301 + (-499) - (-65)$$

A) 318 B) −286

C) −168 D) −133

86) Evaluate the below

$$(-244) + 178 - (-223)$$

A) −152 B) 268

C) 471 D) 157

87) Evaluate the below

$$(-117) + (-117) + 478$$

A) 244 B) −144

C) 516 D) −65

88) Find the product of the below.

$$-14 \times 15 \times 3$$

A) −630 B) −643

C) 6664 D) −6664

89) Find the product of the below.

$$10 \times -27 \times -28$$

A) 756 B) 7563

C) 7560 D) −756

90) Find the product of the below.

$$-10 \times 15 \times -30$$

A) 4500 B) 4520

C) −25 D) 4477

91) Find the quotient of the below.

$$-105 \div -5$$

A) 21 B) −525

C) 7 D) −37

92) Find the quotient of the below.

$$192 \div -6$$

A) −1 B) −32

C) 1152 D) 186

Grade 7
Volume 1
Week 6

93) Find the quotient of the below.

$1140 \div -38$

A) 14   B) −43320

C) 1102   D) −30

94) Evaluate the below expression.

$\frac{19}{18} - 12\frac{4}{9} - 12\frac{5}{7}$

A) $-24\frac{797}{1008}$ B) $-26\frac{13}{126}$

C) $-25\frac{47}{252}$ D) $-24\frac{13}{126}$

95) Find the product of the below.

$-30 \times 15 \times -35$

A) 15750   B) −15750

C) −756   D) −50

96) Find the product of the below.

$-29 \times -27 \times -14$

A) −10985   B) −10942

C) −10990   D) −10962

97) Find the product of the below.

$27 \times -9 \times 13$

A) −3159   B) 31

C) −3163   D) −3192

98) Find the quotient of the below.

$423 \div -47$

A) −9   B) 19881

C) 42   D) −47

99) Find the quotient of the below.

$1131 \div -29$

A) −34   B) −19

C) −39   D) −32799

100) Find the quotient of the below.

$-253 \div -23$

A) −45   B) −230

C) 11   D) −32

©All rights reserved-Math-Knots LLC., VA-USA

101) Evaluate the below expression.

$$\left(-\frac{1}{15}\right) + \frac{8}{9} - \left(-\frac{36}{25}\right)$$

A) $1\frac{461}{900}$ B) $2\frac{59}{225}$

C) $-6\frac{2099}{3150}$ D) $\frac{1951}{3150}$

102) Evaluate the below expression.

$$\left(-\frac{13}{11}\right) - \left(-\frac{29}{21}\right) - \left(-2\frac{1}{13}\right)$$

A) $2\frac{610}{1001}$ B) $2\frac{829}{3003}$

C) $2\frac{21647}{24024}$ D) $-\frac{2203}{6006}$

103) Evaluate the below expression.

$$\frac{3}{10} - \frac{17}{18} - \left(-\frac{22}{15}\right)$$

A) $-\frac{8}{45}$ B) $-6\frac{67}{630}$

C) $\frac{37}{45}$ D) $\frac{22}{45}$

104) Find the product of the below.

$$-2 \times \frac{3}{10} \times -\frac{16}{11}$$

A) $\frac{48}{55}$ B) $-1\frac{7}{55}$

C) $\frac{137}{220}$ D) $-2\frac{47}{110}$

105) Find the product of the below.

$$4\frac{1}{2} \times \frac{2}{3} \times -\frac{1}{2}$$

A) $-1\frac{1}{2}$ B) $1\frac{1}{2}$

C) $-4\frac{2}{3}$ D) $-7\frac{1}{10}$

106) Evaluate the below expression.

$$\left(-\frac{1}{4}\right) + 15 - 3\frac{1}{8}$$

A) $7\frac{37}{40}$ B) $13\frac{7}{88}$

C) $17\frac{39}{40}$ D) $11\frac{5}{8}$

107) Evaluate the below expression.

$$2\frac{3}{25} + \frac{8}{23} + \left(-3\frac{16}{25}\right)$$

A) $-1\frac{99}{575}$ B) $-\frac{731}{10925}$

C) $\frac{1313}{1725}$ D) $-2\frac{2041}{5175}$

108) Find the product of the below.

$$-1\frac{3}{5} \times -\frac{7}{6} \times \frac{3}{7}$$

A) $2\frac{2}{5}$ B) $-\frac{13}{15}$

C) $\frac{4}{5}$ D) $-2\frac{71}{210}$

Grade 7
Volume 1
Week 6

109) Find the product of the below.

$$2\frac{7}{12} \times -\frac{5}{3} \times -\frac{1}{2}$$

A) $1\frac{5}{18}$ B) $3\frac{3}{8}$

C) $2\frac{11}{72}$ D) $\frac{5}{12}$

110) Find the product of the below.

$$-2\frac{3}{5} \times -3\frac{9}{10} \times -\frac{8}{5}$$

A) $5\frac{59}{80}$ B) $-8\frac{1}{10}$

C) $-16\frac{28}{125}$ D) 0

111) Find the product of the below.

$$0 \times -\frac{1}{9} \times -\frac{10}{11}$$

A) 0   B) $-\frac{3}{7}$

C) $-1\frac{2}{99}$ D) $2\frac{1}{2}$

112) Find the quotient of the below.

$$\frac{-7}{13} \div -2\frac{8}{11}$$

A) $\frac{77}{390}$ B) $3\frac{7}{10}$

C) $2\frac{27}{143}$ D) $-\frac{77}{390}$

113) Find the quotient of the below.

$$1\frac{17}{18} \div \frac{-14}{13}$$

A) $2\frac{11}{117}$ B) $3\frac{5}{234}$

C) $-1\frac{29}{36}$ D) $-\frac{36}{65}$

114) Find the product of the below.

$$-1\frac{5}{7} \times 0 \times \frac{7}{6}$$

A) $\frac{1}{6}$ B) 0

C) $-\frac{4}{11}$ D) 2

115) Find the product of the below.

$$0 \times 2\frac{1}{12} \times -\frac{10}{7}$$

A) $17\frac{7}{25}$ B) $-17\frac{7}{25}$

C) 0   D) $\frac{55}{84}$

116) Find the quotient of the below.

$$-30 \div \frac{8}{29}$$

A) $8\frac{8}{29}$ B) $-\frac{1}{2}$

C) $\frac{4}{435}$ D) $-108\frac{3}{4}$

117) Find the quotient of the below.

$$-3\frac{4}{7} \div \frac{23}{15}$$

A) $-5\frac{10}{21}$ B) $1\frac{1}{2}$

C) $-2\frac{53}{161}$ D) $5\frac{10}{21}$

118) Find the quotient of the below.

$$\frac{-19}{17} \div 2\frac{11}{15}$$

A) $-\frac{285}{697}$ B) $-3\frac{217}{255}$

C) $2\frac{127}{285}$ D) $-\frac{3}{4}$

119) Find the quotient of the below.

$$\frac{-5}{8} \div 9\frac{9}{25}$$

A) $\frac{125}{1872}$ B) $-\frac{125}{1872}$

C) $-9\frac{197}{200}$ D) $8\frac{147}{200}$

120) Seven jars of peanuts cost $14. How many jars can you buy for $42? (Round to the nearest whole number)

A) 19 B) 84

C) 21 D) 20

121) Helen took a trip to Panama islands. Upon leaving she decided to convert all of her Panas back into dollars. How many dollars did she receive if she exchanged 78 Panas at a rate of $7 to 39 Panas? (Round to the nearest whole number)

A) $13 B) $434.57

C) $15 D) $14

122) Find the quotient of the below.

$$6\frac{3}{20} \div \frac{-2}{7}$$

A) -14 B) $-21\frac{21}{40}$

C) $-2\frac{15}{23}$ D) $5\frac{121}{140}$

123) Find the quotient of the below.

$$\frac{-1}{2} \div \frac{-7}{6}$$

A) $15\frac{1}{6}$ B) $\frac{3}{7}$

C) $9\frac{6}{11}$ D) $-2\frac{1}{3}$

124) A painting is 18 in tall and 24 in wide. If it is reduced to a width of 8 in, then how tall will it be?
(Round to the nearest whole number)

A) 7 in B) 5 in

C) 3 in D) 6 in

**Grade 7
Volume 1
Week 6**

125) Julia can finish a piece of work in 6 days while Kelly takes 12 days to do it alone. If both Julia and Kelly work together, how long will they take to finish off the work?

126) Fiona and Zara together can finish a piece of work in   days. Fiona alone can finish it in 12 days. In how many days can Zara alone finish the work?

127) Convert 45 m/sec into km/hr

128) Convert 36 km/hr into m/sec.

129) Which expression(s) are equivalent to $\frac{7}{9} - (\frac{2}{9})$

A) $\frac{7}{9} + (+\frac{2}{9})$

B) $-\frac{7}{9} - (-\frac{2}{9})$

C) $-\frac{7}{9} - (-\frac{2}{9})$

D) $\frac{7}{9} + (-\frac{2}{9})$

130) A car travels a distance of 540 miles in 6 hours. What is the speed of the car?

131) Convert 35 m/sec into km/hr.

132) Convert 40 m/sec into km/hr.

©All rights reserved-Math-Knots LLC., VA-USA   www.math-knots.com | www.a4ace.com

Grade 7
Volume 1
Week 6

133) Convert 45 km/hr into m/sec

134) A local florist used the equation Y = KX to determine how many flowers she'd need for 15 bouquets. She determined she'd need 315 flowers. How many flowers were in each bouquet?

135) Sally used the equation Y = KX to calculate the profit made $127.89 after selling 7 boxes of his cookies for $p each. How much would he have made had he sold 15 boxes?

136) Convert 48.6 km/hr into m/sec.

137) Convert 48.6 km/hr into m/sec.

138) Which expression(s) are equivalent to 13 + (-44)?

A) 13 - (44)

B) 13 + (-14)

C) -13 + (+44)

D) -13 - (+44)

139) Convert 72 km/hr into m/sec.

Grade 7
Volume 1
Week 6

140) Peter started the year with $6072.85 in his bank account. In January, he got 2 paychecks. One for $1883.94 and the other was for $1651.75. He ended up spending $701.28 on food, $546.24 on gas, $152.40 and on car insurance and $902.40 for other things. Then he decided to buy a new iPad which cost him $605.63. Find the balance amount in Peter's account?

141) Convert 4 miles/sec into miles/hr.

142) Convert 0.75 miles/sec into miles/hr.

143) A bus covers 8 miles in 15 minutes. How far does it go in 2 hours 30 minutes?

144) Beth and Dora can do a piece of work in 12 hours. Dora and Jolly can do the work in 15 hours. Beth and Jolly can finish the work in 20 hours. In how many hours will they finish the work if they work together?

145) Dora and Tara when working together complete the work in 6 hours. If Dora works alone, she can do it in 9 hours. How much time will Tara alone take to finish it?

146) Convert 108 km/hr into m/sec.

Grade 7
Volume 1
Week 6

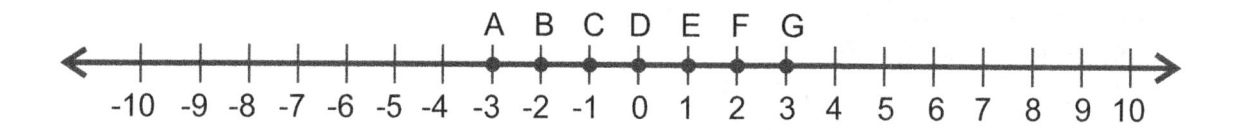

```
           A  B  C  D  E  F  G
<--+--+--+--+--+--+--+--+--+--+--+--+--+--+--+--+--+--+--+--+-->
  -10 -9 -8 -7 -6 -5 -4 -3 -2 -1  0  1  2  3  4  5  6  7  8  9 10
```

147) Find the letter in the number line best represents the answer to below problem

(-1) + 2 =

148) Find the letter in the number line best represents the answer to below problem

1 + (-2) =

149) Find the letter in the number line best represents the answer to below problem

2 + 1 =

150) Find the letter in the number line best represents the answer to below problem

2 + (-1) =

151) Find the letter in the number line best represents the answer to below problem

(-2) + 1 =

152) Determine if the values in the table are proportional (yes) or not (no).

X	Y
11	1
12	2
11	3
18	4

153) Determine if the values in the table are proportional (yes) or not (no).

X	Y
-4	-10
-80	-100
-40	-50
-88	-110

154) Convert 54 km/hr into m/sec

155) A car travels 120 km, 180 km and 150 km at the rate of 48 km/hr, 45 km/hr and 60 km/hr respectively. Find the average speed of the whole journey.

156) The equation 28.25 = 5a shows that buying 5 bags of apples would cost 18.25 dollars. How much is it for one bag?

157) A car travels at a speed of 90 miles/hour. How much distance will it cover in 8 hours?

158) A bus travels 240 km, 225 km and 175 km at the rate of 60 km/hr, 45 km/hr and 50 km/hr respectively. Find the average speed of the whole journey.

159) A construction contractor used the equation Y = KX to determine it would cost him $1868.04 to buy 9 boxes of wood. Find the cost of the box of wood?

162) An ice cream truck driver used the equation Y = KX to calculate the profit made by selling 7 ice cream bars. He determined he'd make $28.56. How much did he make per bar sold?

160) A train travels 98 km, 124 km and 80 km at the rate of 28 km/hr, 31 km/hr and 32 km/hr respectively. Find the average speed of the whole journey.

163) Lisa and her 3 friends were going to the movie. Each ticket cost $15.00 and then they bought 4 popcorns at $5.50 a piece and 4 soda's at $1.00 each. How much money did they spend total for their trip to the movies?

161) Which expression(s) are equivalent to 54.78 − (−79.95)?

A) -54.78 + (-79.95)

B) 54.78 + (+79.95)

C) 54.78 − (79.95)

D) -54.78 − (+79.95)

164) Convert 1.2 miles/sec into miles/hr.

165) A train covers 14 miles in 12 minutes. How far does it go in 3 hours?

166) A baker used the equation Y = KX to calculate that he had made $589.90 after selling 34 boxes of his cookies for $17.35 each. How much would he have made had he sold 4 boxes?

167) Determine which statements about the graph are true

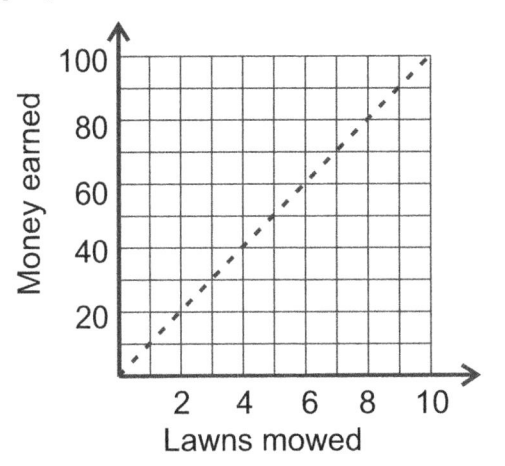

[Graph: Money earned (y-axis, 0 to 100) vs Lawns mowed (x-axis, 2 to 10), showing a linear relationship through the origin]

A) The point (3, 30) shows that you would earn $30 for working 3 hours.

B) The point (90, 9) shows that you would earn $90 for working 9 hours.

C) The point (1, 10) shows that working 1 lawn will earn you $10.

D) The point (10, 1) shows that working 1 lawn will earn you $10.

168) Dora was playing a trivia game she gained points for correct answers and lost points for incorrect answers. At the start of round 2 she was at -800 points. During the round she answered 750 questions correct and she answered 450 points questions incorrect. What was her score at the end of the round?

169) The equation Y = KX shows you would make $7.18 for recycling 2 pounds of cans. How much would you make if you recycled 7 pounds?

170) Vanessa used the equation Y = KX to determine she would need 136 beads to create 4 necklaces. How many beads did she use per necklace?

171) Determine the constant of proportionality for each table. Express your answer as y = kx

Concert Tickets Sold (x)	10	15	21	33	45
Revenue ($y)	90	135	189	297	405

Every ticket sold _____ dollars are earned.

172) Determine the constant of proportionality for each table. Express your answer as y = kx

Boxes of Candy (x)	13	20	36	40	56
Pieces of Candy (y)	260	400	720	800	1120

For every box of candy, you get _____ pieces.

173) To determine how many pages would be needed to make 9 books you can use the equation, 459 = (561)9. How many pages would be in 8 books?

174) The equation 99.63 = (11.07)9 shows how much it cost for a company to buy 9 new uniforms. How much does it cost per uniform?

175) An industrial printing machine printed 824 pages in 8 minutes. How many pages did it print in one minute?

176) A florist used the equation 128 = (16)8 to determine how many flowers she'd need for 8 bouquets. How many flowers would she need for 9 bouquets?

Grade 7 Volume 1 Week 6

177) At a restaurant a meal used to cost $9, but a new owner decided to raise the price by 13 percent. How much does the meal cost now?

178) An older TV screen was 26 centimeters wide. The newer model increased the screen size by 6 percent. How wide is the new TV screen?

179) A restaurant bill was $28 after tax. If you wanted to leave a 17 percent tip, how much should you leave total?

180) Carol ordered a shirt online that cost her $26 total. The package arrived late so the seller took 14 percent off the price. How much did she end up paying?

181) A plant was 41 centimeters tall. After a month it had grown 20 percent taller. How tall was the plant after a month?

182) A loan company charged 25 percent interest on every dollar borrowed. If a person borrowed 186 dollars, how much would they end up paying total?

183) A video game was $31. This weekend the game will be marked 20 percent off. How much is the game this weekend?

184) At the hardware store you can buy 4 boxes of bolts for $16.52. This can be expressed by the equation 16.50 = (4.13)4. How much would it cost for 8 boxes?

**Grade 7
Volume 1
Week 6**

185) A small bag of flour weighed 26 ounces. A large bag was 15 percent heavier. How much does the large bag weigh?

186) Gwen's hair was 40 centimeters long. Over summer though she cut off 18 percent. How long was her hair after the haircut?

187) The regular price of a computer was $1899 dollars, but over the Christmas weekend it'll be on sale for 18% off. Which expression shows the difference in price from normal(n) to sale?

A) n – 10

B) n × 0.18

C) n – 0.1

D) n – 1.1

188) Lana was using a container to fill up a fishbowl. The container held $\frac{1}{2}$ of a gallon of water and filled $\frac{1}{3}$ of the fishbowl. At this rate, how many containers will it take to fill the fishbowl?

189) A discount bottle of perfume was $\frac{1}{2}$ of a liter. That was enough to fill $\frac{1}{3}$ of a jug. How many bottles of perfume would you need to fill the entire jug?

190) At noon the temperature was -2 C. Over the next 5 hours the temperature dropped another 2 degrees. Then every hour until 4:00 the temperature rose 3 degrees. What was the temperature at 4:00?

191) Determine the constant of proportionality for each table. Express your answer as y = kx

Phone Sold (x)	2	5	3	6	4
Money Earned (y)	94	235	141	282	188

Every phone sold earns _____ dollars.

192) Determine the constant of proportionality for each table. Express your answer as y = kx

Pounds of Beef Jerky (x)	2	4	5	8	9
Price in dollars (y)	20	40	50	80	90

For every pound of beef jerky it cost _____ dollars.

193) Determine the constant of proportionality for each table. Express your answer as y = kx

Tickets Sold (x)	4	9	8	5	7
Money Earned (y)	48	108	96	60	84

Every ticket sold _____ dollars are earned.

194) Determine the constant of proportionality for each table. Express your answer as y = kx

Cans of Paint (x)	2	5	6	9	7
Bird Houses Painted (y)	8	20	24	36	28

For every can of paint you could paint _____ bird houses.

195) Determine the constant of proportionality for each table. Express your answer as y = kx

Time in minute (x)	4	3	10	7	9
Distance traveled in meters (y)	76	57	190	133	171

Every minute _____ meters are travelled.

196) Determine the constant of proportionality for each table. Express your answer as y = kx

Time in minute (x)	8	3	6	4	10
Gallons of Water Used (y)	240	90	180	120	300

Every minute _____ gallons of water are used.

197) Determine the constant of proportionality for each table. Express your answer as y = kx

Boxes of Candy (x)	5	9	3	2	6
Pieces of Candy (y)	90	162	54	36	108

For every box of candy you get _____ pieces.

198) Determine the constant of proportionality for each table. Express your answer as y = kx

Pieces of Chicken (x)	3	10	7	9	4
Price in dollars (y)	6	20	14	18	8

For each piece of chicken it costs _____ dollars.

199) Determine the constant of proportionality for each table. Express your answer as y = kx

Lawns Mowed (x)	7	6	2	9	3
Dollars Earned (y)	294	252	84	378	126

For every lawn mowed _____ dollars were earned.

Grade 7
Volume 1
Week 7

1) Convert the below fraction to percent

$$8\frac{1}{2}$$

2) Convert the below fraction to percent

$$\frac{39}{40}$$

3) Convert the below fraction to percent

$$\frac{2}{333}$$

4) Convert the below fraction to percent

$$\frac{23}{33}$$

5) Convert the below fraction to percent

$$1\frac{61}{100}$$

6) Convert the below fraction to percent

$$\frac{10}{33}$$

7) Convert the below fraction to percent

$$\frac{1}{33}$$

8) Convert the below fraction to percent

$$\frac{71}{100}$$

Grade 7
Volume 1
Week 7

9) Convert the below fraction to percent

$$9\frac{1}{3}$$

10) Convert the below fraction to percent

$$\frac{9}{10}$$

11) Convert the below fraction to percent

$$\frac{1}{10}$$

12) Convert the below fraction to percent

$$\frac{2}{3}$$

13) Convert the below fraction to percent

$$\frac{4}{5}$$

14) Convert the below fraction to percent

$$\frac{1}{2}$$

15) Convert the below fraction to percent

$$\frac{7}{1000}$$

16) Convert the below fraction to percent

$$\frac{1}{4}$$

©All rights reserved-Math-Knots LLC., VA-USA   www.math-knots.com | www.a4ace.com

17) Convert the below fraction to percent

$$\frac{8}{11}$$

18) Convert the below fraction to percent

$$\frac{7}{8}$$

19) Convert the below decimal to percent

0.0014

A) 0.7 %   B) 0.14 %

C) 0.07 %   D) 0.0014 %

20) Convert the below decimal to percent

0.6

A) 0.6 %   B) 150 %

C) 60 %   D) 65 %

21) Convert the below fraction to percent

$$\frac{29}{100}$$

22) Convert the below fraction to percent

$$\frac{11}{16}$$

23) Convert the below decimal to percent

0.49

A) 4.9 %   B) 49.1 %

C) 49 %   D) 0.49 %

24) Convert the below decimal to percent

0.0072

A) 0.0072 % B) 0.72 %

C) 72 %   D) $1388.\overline{8}$ %

25) Convert the below decimal to percent

0.5

A) 1.2 %   B) 5000 %
C) 0.5 %   D) 50 %

26) Convert the below decimal to percent

0.3

A) 31 %   B) 300 %
C) 0.3 %   D) 30 %

27) Convert the below decimal to percent

0.06

A) 3.5 %   B) 6 %
C) 1666.$\overline{6}$ %   D) 0.06 %

28) Convert the below decimal to percent

0.888

A) 88.8%   B) 1.$\overline{126}$ %
C) 0.888 %   D) 108.8 %

29) Convert the below decimal to percent

0.46

A) 2350 %   B) 0.46 %
C) 23.5 %   D) 46 %

30) Convert the below decimal to percent

0.837

A) 837 %   B) 83.7 %
C) 0.837 %   D) 93.7 %

31) Convert the below decimal to percent

0.926

A) 92.6 %   B) 463.5 %
C) 0.926 %   D) 94.6 %

32) Convert the below decimal to percent

0.04

A) 1.25 %   B) 0.4 %
C) 4 %   D) 0.04 %

33) Convert the below decimal to percent

0.67

A) 0.67 % B) 67 %
C) 6.7% D) 671 %

34) Convert the below decimal to percent

0.0089

A) 20.89 % B) 0.89 %
C) 0.0089 % D) 4.45 %

35) Convert the below decimal to percent

0.0013

A) 14 % B) 0.0013 %
C) 0.13 % D) $76923.\overline{076923}$ %

36) Convert the below decimal to percent

0.0032

A) 0.32 % B) 60.32 %
C) 0.0032 % D) 90.32 %

37) Convert the below decimal to percent

0.009

A) 0.45 % B) $11111.\overline{1}$ %
C) 0.9 % D) 0.009 %

38) Convert the below decimal to percent

0.367

A) 0.367 % B) 36.7 %
C) 36.71 % D) 3.67 %

39) Convert the below decimal to percent

0.08

A) 4 % B) 9 %
C) 0.08 % D) 8 %

40) Convert the below decimal to percent

0.0043

A) 44 % B) 0.43 %
C) 4.3 % D) 0.0043 %

41) What percent of 191 ft is 33.4 ft?

A) 105.1%   B) 17.5%
C) 571.9%   D) 677.8%

42) What percent of 287 hours is 183 hours?

A) 1.57%   B) 63.8%
C) 156.8%   D) 0.64%

43) 98 hours is what percent of 394 hours?

A) 3.2%   B) 127.3%
C) 315.8%   D) 24.9%

44) What percent of $307 is $58?

A) 2612.5%   B) 18.9%
C) 26.1%   D) 529.3%

45) What percent of 187 inches is 25 inches?

A) 748%   B) 0.13%
C) 7.5%   D) 13.4%

46) 73 minutes is what percent of 216 minutes?

A) 832%   B) 33.8%
C) 8.3%   D) 0.34%

47) 324 ft is what percent of 338 ft?

A) 2.5%   B) 1.04%
C) 95.9%   D) 0.96%

48) 248 minutes is what percent of 432 minutes?

A) 0.57%   B) 1.74%
C) 57.4%   D) 174.2%

49) What percent of 484 miles is 268.5 miles?

A) 0.55%   B) 477.8%
C) 55.5%   D) 180.3%

50) What percent of 440 tons is 233 tons?

A) 0.53%   B) 1.89%
C) 53%   D) 188.8%

51) 414 minutes is what percent of 374 minutes?

A) 110.7% B) 1.11%
C) 90.3%   D) 71.7%

52) What is 73% of 57 km?

A) 41.6 km B) 78.1 km
C) 4161 km D) 10962 km

53) What percent of 318 minutes is 463 minutes?

A) 253.4% B) 1.78%
C) 68.7%   D) 145.6%

54) What percent of 311 minutes is 137 minutes?

A) 44.1%   B) 159%
C) 2.3%   D) 4383.3%

55) 33% of 1 minute is what?

A) 0.33 minutes
B) 3 minutes
C) 26.4 minutes
D) 33 minutes

Grade 7
Volume 1
Week 7

56) What is 195% of 151 hours?

A) 294.5 hours   B) 152.8 hours

C) 77.4 hours   D) 50 hours

57) 230% of 193 miles is what?

A) 443.9 miles

B) 44390 miles

C) 83.9 miles

D) 92.9 miles

58) What is 160% of 186 grams?

A) 116.3 grams

B) 57.8 grams

C) 297.6 grams

D) 29760 grams

59) What is 120% of 30 hours?

A) 25 hours   B) 3600 hours

C) 180.2 hours   D) 36 hours

60) 82% of 111 cm is what?

A) 107.9 cm   B) 8544 cm

C) 91 cm   D) 135.4 cm

61) 35% of 92 grams is what?

A) 263.6 grams

B) 3220 grams

C) 262.9 grams

D) 32.2 grams

62) 97% of 28 minutes is what?

A) 15810 minutes

B) 27.2 minutes

C) 28.9 minutes

D) 2716 minutes

63) 60% of 186 grams is what?

A) 310 grams

B) 27.6 grams

C) 11160 grams

D) 111.6 grams

64) What is 2% of 22 tons?

A) 0.44 tons   B) 320.6 tons
C) 1100 tons   D) 44 tons

65) 71% of $88 is what?

A) $123.94   B) $1043.68
C) $62.48   D) $3767.70

66) 30 ft is 106% of what?

A) 3180 ft   B) 100.7 ft
C) 31.8 ft   D) 28.3 ft

67) 192.5 minutes is 89% of what?

A) 216.3 minutes
B) 8514 minutes
C) 85.1 minutes
D) 396.3 minutes

68) 173% of 94 km is what?

A) 54.3 km   B) 212.1 km
C) 16262 km   D) 162.6 km

69) 180.6 minutes is 31% of what?

A) 582.6 minutes
B) 5598.6 minutes
C) 21.4 minutes
D) 56 minutes

70) 28.6 minutes is 60% of what?

A) 47.7 minutes
B) 1716 minutes
C) 17.2 minutes
D) 170.2 minutes

Grade 7
Volume 1
Week 7

71) 3 grams is 68% of what?

A) 6.1 grams   B) 2 grams

C) 609 grams   D) 4.4 grams

75) 300% of what is 78 grams?

A) 84.6 grams   B) 73.3 grams

C) 26 grams   D) 8463 grams

72) $82 is 32% of what?

A) $256.25   B) $26.24

C) $2476.60   D) $24.77

76) 280% of what is 113 hours?

A) 1738 hours   B) 28.1 hours

C) 2813.9 hours   D) 40.4 hours

73) 47% of what is 176 cm?

A) 82.7 cm   B) 374.5 cm

C) 8272 cm   D) 3.2 cm

77) 156.6 grams is 97% of what?

A) 161.4 grams   B) 44.7 grams

C) 151.9 grams   D) 13.6 grams

74) 65% of what is 56 inches?

A) 6075 inches   B) 60.8 inches

C) 49.7 inches   D) 86.2 inches

78) $58 is 7% of what?

A) $23.41   B) $406

C) $4.06   D) $828.57

©All rights reserved-Math-Knots LLC., VA-USA   www.math-knots.com | www.a4ace.com

Grade 7
Volume 1
Week 7

79) 78.4 hours is 65% of what?

  A) 120.6 hours   B) 5096 hours
  C) 81 hours   D) 51 hours

83) 47% of what is 12 grams?

  A) 57 grams   B) 5.6 grams
  C) 564 grams   D) 25.5 grams

80) Find the percentage increase or decrease from 242 cm to 388 cm

  A) 137.6% increase
  B) 160.3% increase
  C) 37.6% increase
  D) 60.3% increase

84) Find the percentage increase or decrease from 276 minutes to 341 minutes

  A) 23.6% increase
  B) 123.6% increase
  C) 65% increase
  D) 19.1% decrease

81) Find the percentage increase or decrease from 126 inches to 214 inches

  A) 41.1% decrease
  B) 69.8% increase
  C) 88% decrease
  D) 88% increase

85) Find the percentage increase or decrease from 123 ft to 297 ft

  A) 174% decrease
  B) 141.5% increase
  C) 158.6% increase
  D) 241.5% increase

82) Find the percentage increase or decrease from 147 tons to 40 tons

  A) 367.5% decrease
  B) 107% decrease
  C) 72.8% decrease
  D) 72.8% increase

86) Find the percentage increase or decrease from 146.4 hours to 285 hours

  A) 48.6% decrease
  B) 138.6% increase
  C) 148.6% increase
  D) 94.7% increase

87) Find the percentage increase or decrease from 112 miles to 117 miles

A) 4.5% increase

B) 94.3% increase

C) 4.3% decrease

D) 5% increase

88) Find the percentage increase or decrease from 350 cm to 31 cm

A) 319% decrease

B) 1129% decrease

C) 8.9% decrease

D) 91.1% decrease

89) Find the percentage increase or decrease from 328 cm to 82 cm

A) 246% decrease

B) 400% decrease

C) 300% decrease

D) 75% decrease

90) Find the percentage increase or decrease from 296 km to 397 km

A) 101% increase

B) 34.1% increase

C) 25.4% decrease

D) 25.4% increase

91) Find the percentage increase or decrease from 91 inches to 234 inches

A) 157.1% increase

B) 161.1% increase

C) 157.1% decrease

D) 61.1% increase

92) Find the percentage increase or decrease from 352 m to 294 m

A) 16.5% increase

B) 16.5% decrease

C) 119.7% decrease

D) 19.7% decrease

93) Find the percentage increase or decrease from 275 tons to 288 tons

A) 13% increase

B) 4.7% increase

C) 4.5% decrease

D) 104.7% increase

94) Find the percentage increase or decrease from 224 ft to 193 ft

A) 13.8% increase

B) 16.1% increase

C) 13.8% decrease

D) 86.2% decrease

Grade 7
Volume 1
Week 7

95) Find the percentage increase or decrease from 241 m to 264 m

A) 8.7% decrease

B) 9.5% increase

C) 8.7% increase

D) 109.5% increase

96) Find the percentage increase or decrease from 285 km to 48 km

A) 83.2% decrease

B) 493.8% decrease

C) 593.8% decrease

D) 83.2% increase

97) Find the percentage increase or decrease from 239 km to 87 km

A) 274.7% decrease

B) 36.4% decrease

C) 174.7% increase

D) 63.6% decrease

98) Find the percentage increase or decrease from 132 tons to 18 tons

A) 114% decrease

B) 633.3% increase

C) 86.4% decrease

D) 633.3% decrease

99) Find the percentage increase or decrease from 87 hours to 293 hours

A) 206% increase

B) 170.3% increase

C) 236.8% decrease

D) 236.8% increase

100) Find the percentage increase or decrease from 104 hours to 39 hours

A) 166.7% decrease

B) 266.7% decrease

C) 62.5% decrease

D) 37.5% decrease

101) Find the percentage increase or decrease from 392 grams to 136 grams

A) 65.3% increase

B) 188.2% increase

C) 34.7% decrease

D) 65.3% decrease

102) Find the percentage increase or decrease from 313 tons to 350 tons

A) 11.8% decrease

B) 111.8% increase

C) 10.6% decrease

D) 11.8% increase

103) Find the percentage increase or decrease from $78 to $149.60

A) 47.9% increase

B) 71.6% decrease

C) 47.9% decrease

D) 91.8% increase

104) Find the percentage increase or decrease from 205.8 km to 351 km

A) 145.2% decrease

B) 41.4% decrease

C) 141.4% increase

D) 70.6% increase

105) Find the percentage increase or decrease from 371 minutes to 329 minutes

A) 42% decrease

B) 88.7% decrease

C) 12.8% increase

D) 11.3% decrease

106) Find the percentage increase or decrease from 163 km to 367 km

A) 125.2% increase

B) 225.2% increase

C) 55.6% increase

D) 55.6% decrease

107) Find the percentage increase or decrease from 240 grams to 158 grams

A) 34.2% decrease

B) 51.9% decrease

C) 34.2% increase

D) 82% decrease

108) Find the percentage increase or decrease from 145 inches to 200 inches

A) 27.5% increase

B) 37.9% increase

C) 137.9% increase

D) 55% decrease

109) Find the percentage increase or decrease from 344 m to 157 m

A) 54.4% decrease

B) 187% decrease

C) 54.4% increase

D) 45.6% decrease

110) Find the percentage increase or decrease from 144 km to 374 km

A) 230% increase

B) 161.5% increase

C) 75.6% increase

D) 159.7% increase

111) A pair of socks costs $25.95. The store is selling them at 20% discount. Find the selling price.

   A) $20.76   B) $5.19
   C) $24.65   D) $31.14

112) A toy car costs $10.50. The store is selling them at 42% discount. Find the selling price.

   A) $14.91   B) $12.07
   C) $12.60   D) $6.09

113) A hat costs $33.50. The store is selling them at 50% discount. Find the selling price.

   A) $35.18   B) $50.25
   C) $28.47   D) $16.75

114) A computer game costs $22.50. The store is selling them at 50% discount. Find the selling price.

   A) $33.75   B) $19.12
   C) $11.25   D) $24.75

115) A CD costs $12.99. The store is selling them at 40% discount. Find the selling price.

   A) $13.64   B) $7.79
   C) $18.19   D) $12.34

116) A bicycle costs $600.00. The store is selling them at 30% discount. Find the selling price.

   A) $180.00   B) $780.00
   C) $690.00   D) $420.00

117) A pair of socks costs $140.00. The store is selling them at 60% discount. Find the selling price.

   A) $56.00   B) $224.00
   C) $119.00   D) $84.00

118) A concert tickets costs $99.95. The store is selling them at 28% discount. Find the selling price.

   A) $127.94   B) $89.95
   C) $27.99   D) $71.96

Grade 7
Volume 1
Week 7

119) A bicycle costs $1,000.00. The store is selling them at 50% discount. Find the selling price.

A) $800.00   B) $500.00
C) $1,500.00   D) $1,050.00

120) A cell phone costs $249.95. The store is selling them at 25% discount. Find the selling price.

A) $187.46   B) $312.44
C) $62.49   D) $274.94

121) A cell phone costs $250.00. The store is selling them at 55% discount. Find the selling price.

A) $112.50   B) $387.50
C) $237.50   D) $137.50

122) A telescope costs $604.50. The store is selling them at 50% discount. Find the selling price.

A) $544.05   B) $302.25
C) $695.17   D) $906.75

123) A sled costs $99.99. The store is selling them at 50% discount. Find the selling price.

A) $50.00   B) $49.99
C) $149.98   D) $84.99

124) A sweater costs $21.00. The store is selling them at 50% discount. Find the selling price.

A) $6.30   B) $27.30
C) $18.90   D) $14.70

125) A kitten costs $134.95. The store is selling them at 36% discount. Find the selling price.

A) $107.96   B) $183.53
C) $86.37   D) $48.58

126) A car costs $16,500.00. The store is selling them at 30% discount. Find the selling price.

A) $11,550.00   B) $4,950.00
C) $17,325.00   D) $19,800.00

127) A pair of socks costs $22.99. The store is selling them at 35% discount. Find the selling price.

A) $31.04   B) $8.05
C) $14.94   D) $25.29

128) A concert tickets costs $179.50. The store is selling them at 10% discount. Find the selling price.

A) $197.45   B) $17.95
C) $170.53   D) $161.55

129) A cell phone costs $90.00. The store is marked up the price by 60%. Find the selling price.

A) $144.00   B) $108.00
C) $36.00   D) $54.00

130) A microphone costs $9.50. The store is marked up the price by 85%. Find the selling price.

A) $11.40   B) $17.57
C) $1.43   D) $17.58

131) A telescope costs $579.95. The store is selling them at 10% discount. Find the selling price.

A) $637.95   B) $492.96
C) $58.00   D) $521.96

132) A sled costs $59.95. The store is selling them at 10% discount. Find the selling price.

A) $68.94   B) $6.00
C) $65.95   D) $53.96

133) A pair of socks costs $10.00. The store is marked up the price by 10%. Find the selling price.

A) $9.00   B) $1.00
C) $11.00   D) $12.00

134) A hat costs $13.50. The store is marked up the price by 50%. Find the selling price.

A) $20.25   B) $11.47
C) $6.75   D) $6.41

Grade 7
Volume 1
Week 7

135) A jacket costs $430.00.
The store is marked up the price by 5%. Find the selling price.

A) $21.50   B) $451.50
C) $408.50   D) $365.50

136) A parrot costs $599.99.
The store is marked up the price by 35%. Find the selling price.

A) $809.99   B) $389.99
C) $689.99   D) $210.00

137) A sled costs $99.50.
The store is marked up the price by 90%. Find the selling price.

A) $9.95   B) $119.40
C) $189.05   D) $89.55

138) A pants costs $64.50.
The store is marked up the price by 70%. Find the selling price.

A) $109.65   B) $70.95
C) $19.35   D) $77.40

139) A purse costs $199.50.
The store is marked up the price by 25%. Find the selling price.

A) $249.38   B) $149.62
C) $179.55   D) $49.88

140) A lizard costs $39.99.
The store is marked up the price by 25%. Find the selling price.

A) $49.99   B) $33.99
C) $43.99   D) $29.99

141) A microphone costs $150.00.
The store is marked up the price by 30%. Find the selling price.

A) $105.00   B) $195.00
C) $45.00   D) $172.50

142) A book costs $13.50.
The store is marked up the price by 70%. Find the selling price.

A) $4.05   B) $15.52
C) $22.95   D) $16.20

143) A pen costs $1.45.
The store is marked up the price by 90%. Find the selling price.

A) $1.30   B) $1.59
C) $2.75   D) $2.76

144) A purse costs $69.95.
The store is marked up the price by 30%. Find the selling price.

A) $20.98   B) $62.96
C) $90.94   D) $66.45

145) A radio costs $74.99.
The store is marked up the price by 70%. Find the selling price.

A) $127.48   B) $63.74
C) $22.50   D) $52.49

146) A SUV costs $3,000.00.
The store is marked up the price by 87%. Find the selling price.

A) $390.00   B) $2,610.00
C) $3,150.00   D) $5,610.00

147) An oil costs $34.95.
The store is marked up the price by 90%. Find the selling price.

A) $66.41   B) $31.46
C) $3.49   D) $27.96

148) A goldfish costs $3.95.
The store is marked up the price by 70%. Find the selling price.

A) $3.56   B) $2.77
C) $6.72   D) $1.19

149) A shoes costs $115.00.
The store is marked up the price by 16%. Find the selling price.

A) $133.40   B) $120.75
C) $126.50   D) $96.60

150) A telescope costs $399.50.
The store is marked up the price by 65%. Find the selling price.

A) $259.68   B) $659.17
C) $419.48   D) $659.18

151) A pants costs $5.99.
The store charges 6% tax on every item. Find the selling price.

A) $6.35 B) $6.59
C) $5.63 D) $0.36

152) A pen costs $109.99.
The store charges 6% tax on every item. Find the selling price.

A) $116.59 B) $6.60
C) $103.39 D) $126.49

153) A pen costs $2.95.
The store charges 5% tax on every item. Find the selling price.

A) $3.10 B) $2.36
C) $0.15 D) $3.25

154) A pen costs $49.99.
The store charges 2% tax on every item. Find the selling price.

A) $1.00 B) $50.99
C) $48.99 D) $44.99

155) A pen costs $49.50.
The store charges 2% tax on every item. Find the selling price.

A) $39.60 B) $51.98
C) $0.99 D) $50.49

156) A pen costs $149.95.
The store charges 1% tax on every item. Find the selling price.

A) $164.94 B) $1.50
C) $148.45 D) $151.45

157) A pen costs $65.50.
The store charges 6% tax on every item. Find the selling price.

A) $69.43 B) $3.93
C) $68.78 D) $61.57

158) A pen costs $4.95.
The store charges 1% tax on every item. Find the selling price.

A) $4.90 B) $0.05
C) $4.70 D) $5.00

Grade 7
Volume 1
Week 7

159) A microphone costs $49.99. The store charges 6% tax on every item. Find the selling price.

A) $46.99   B) $52.99
C) $47.49   D) $3.00

160) A comic book costs $1.35. The store charges 2% tax on every item. Find the selling price.

A) $1.08   B) $0.03
C) $1.38   D) $1.28

161) A SUV costs $33,000.00. The store charges 3% tax on every item. Find the selling price.

A) $990.00   B) $32,010.00
C) $29,700.00   D) $33,990.00

162) An oil costs $23.50. The store charges 6% tax on every item. Find the selling price.

A) $1.41   B) $22.09
C) $25.85 D) $24.91

163) A computer game costs $37.95. The store charges 4% tax on every item. Find the selling price.

A) $39.47 B) $1.52
C) $36.43 D) $34.16

164) A radio costs $69.50. The store charges 5% tax on every item. Find the selling price.

A) $72.98 B) $3.48
C) $66.02 D) $83.40

165) A comb costs $0.95. The store charges 6% tax on every item. Find the selling price.

A) $0.89   B) $1.01
C) $0.06   D) $0.81

166) A microscope costs $149.50. The store charges 3% tax on every item. Find the selling price.

A) $145.01 B) $4.48
C) $153.99 D) $156.97

167) The original price of an SUV is $57,800. There is a state tax of 3% on every vehicle purchased. Find the selling price of the SUV.

A) $56,066.00   B) $69,360.00

C) $1,734.00   D) $59,534.00

168) The original price of an oil is $30.99. There is a state tax of 3% on every item. Find the selling price of the oil.

A) $26.34   B) $29.44

C) $31.92   D) $30.06

169) The original price of a comic book is $2.95. The store offers a discount of 50% with a tax of 2% on every item. Find the selling price of the comic book.

A) $0.03   B) $0.06

C) $1.50   D) $3.01

170) The original price of a comb is $2.10. The store offers a discount of 50% with a tax of 2% on every item. Find the selling price of the comb.

A) $3.15   B) $0.02

C) $3.09   D) $1.07

171) The original price of a pair of shoes is $119.50. There is a state tax of 4% on every item purchased. Find the selling price of the shoes.

A) $4.78   B) $114.72

C) $124.28   D) $101.58

Original price of a kitten: $160.00
Tax: 1%

172) The original price of a kitten is $160.00. There is a state tax of 1% on every item purchased. Find the selling price of the kitten.

A) $161.60   B) $184.00

C) $1.60   D) $158.40

173) The original price of a microphone is $49.99. The store offers a discount of 55% with a tax of 4% on every item. Find the selling price of the microphone.

A) $0.90   B) $26.39

C) $3.10   D) $23.40

174) The original price of a computer is $3,800.00. The store offers a discount of 20% with a tax of 2% on every item. Find the selling price of the computer.

A) $4,651.20   B) $3,724.00

C) $3,100.80   D) $15.20

175) The original price of a tie is $39.50. The store offers a discount of 25% with a tax of 3% on every item. Find the selling price of the tie.

A) $10.17   B) $50.86

C) $40.69   D) $30.51

176) The original price of a truck is $46,000. The store offers a discount of 20% with a tax of 3% on every item. Find the selling price of the truck.

A) $1,656.00   B) $37,904.00

C) $43,700.00   D) $36,800.00

177) The original price of a microphone is $259.99. The store offers a discount of 45% with a tax of 6% on every item. Find the selling price of the microphone.

A) $376.99   B) $151.57

C) $7.02   D) $142.99

178) The original price of a sweater is $46.50. The price of the sweater is marked up by 85% and then a discount of 40% was given. Find the selling price of the sweater.

A) $6.98   B) $51.62

C) $39.52   D) $65.10

179) The original price of a cell phone is $150. The store offers a discount of 54% with a tax of 6% on every item. Find the selling price of the cell phone.

A) $85.86   B) $4.86

C) $64.86   D) $73.14

180) The original price of a camera is $799.50. The store offers a discount of 10% with a tax of 5% on every item. Find the selling price of the camera.

A) $755.53   B) $83.95

C) $79.95   D) $839.48

181) The original price of a comb is $3.50. The store offers a discount of 18% with a tax of 3% on every item. Find the selling price of the comb.

A) $0.65   B) $4.25

C) $2.96   D) $4.01

182) The original price of a computer is $799.99. The price of the computer is marked up by 60% and then a discount of 20% was given. Find the selling price of the computer.

A) $256.00   B) $575.99

C) $959.99   D) $1,023.99

183) The original price of a radio is $63.50. The price of the radio is marked up by 90% and then a discount of 27% was given. Find the selling price of the radio.

   A) $88.07   B) $76.20
   C) $4.64   D) $6.35

184) The original price of a kitten is $210. The price of the kitten is marked up by 50% and then a discount of 5% was given. Find the selling price of the kitten.

   A) $330.75   B) $199.50
   C) $110.25   D) $299.25

185) The original price of a pants is $51.50. The price of the pants is marked up by 5% and then a discount of 40% was given. Find the selling price of the pants.

   A) $1.55   B) $1.03
   C) $32.45   D) $3.60

186) The original price of a sled is $49.50. The price of the sled is marked up by 36% and then a discount of 50% was given. Find the selling price of the sled.

   A) $15.84   B) $33.66
   C) $31.68   D) $67.32

187) The original price of a tie is $9.50. The price of the tie is marked up by 70% and then a discount of 50% was given. Find the selling price of the tie.

   A) $8.07   B) $9.97
   C) $1.43   D) $8.08

188) The original price of a bicycle is $1,100. The price of the bicycle is marked up by 30% and then a discount of 20% was given. Find the selling price of the bicycle.

   A) $1,144.00   B) $1,430.00
   C) $66.00   D) $616.00

189) The original price of a CD is $15.99. The price of the CD is marked up by 37% and then a discount of 20% was given. Find the selling price of the CD.

   A) $21.91   B) $8.06
   C) $17.53   D) $5.92

190) The original price of a microscope is $249.99. The price of the microscope is marked up by 39% and then a discount of 15% was given. Find the selling price of the microscope.

   A) $129.62   B) $295.36
   C) $22.87   D) $347.49

Grade 7
Volume 1
Week 7

191) The original price of a car is $36,000. The price of the car is marked up by 60% and then a discount of 6% was given. The state collects tax of 3% on every purchase. Find the selling price of the car.

A) $13,942.08 B) $1,334.88
C) $55,768.32 D) $14,832.00

192) The original price of a dress is $73. The price of the dress is marked up by 60% and then a discount of 45% was given. The state collects tax of 6% on every purchase. Find the selling price of the dress.

A) $12.35 B) $39.80
C) $68.09 D) $25.54

193) The original price of a video game is $19.99. The price of the video game is marked up by 70% and then a discount of 10% was given. The state collects tax of 1% on every purchase. Find the selling price of the video game.

A) $0.14 B) $12.59
C) $2.02 D) $30.89

194) The original price of a van is $55,000. The price of the van is marked up by 30% and then a discount of 50% was given. The state collects tax of 3% on every purchase. Find the selling price of the van.

A) $36,822.50 B) $104,032.50
C) $26,675.00 D) $3,217.50

195) The original price of a book is $11.50. The price of the book is marked up by 35% and then a discount of 30% was given. The state collects tax of 1% on every purchase. Find the selling price of the book.

A) $8.05 B) $0.05
C) $10.98 D) $14.80

196) The original price of a TV is $210.00. The price of the TV is marked up by 56% and then a discount of 30% was given. The state collects tax of 5% on every purchase. Find the selling price of the TV.

A) $240.79 B) $103.19
C) $82.32 D) $343.98

197) The original price of a scarf is $3.95. The price of the scarf is marked up by 80% and then a discount of 40% was given. The state collects tax of 4% on every purchase. Find the selling price of the scarf.

A) $1.97 B) $4.42
C) $4.44 D) $4.11

198) The original price of a bike is $24,500. The price of the bike is marked up by 60% and then a discount of 50% was given. The state collects tax of 5% on every purchase. Find the selling price of the bike.

A) $39,200.00 B) $20,580.00
C) $9,310.00 D) $19,600.00

Grade 7 Volume 1 Week 8

Grade 7
Volume 1
Week 8

1) The original price of a diamond ring is $10,500. The price of the diamond ring is marked up by 20% and then a discount of 30% was given. The state collects tax of 5% on every purchase. Find the selling price of the diamond ring.

   A) $9,261.00   B) $10,920.00
   C) $420.00   D) $9,975.00

2) The original price of a shirt is $43.99. The price of the shirt is marked up by 20% and then a discount of 60% was given. The state collects tax of 6% on every purchase. Find the selling price of the shirt.

   A) $27.98   B) $14.92
   C) $89.53   D) $22.38

3) An alloy contains 28% of aluminum, 40% copper and the rest of it is zinc. Find the quantity of zinc in a sample of 1000 oz of the alloy.

4) Celes saves 15% of her income. If her monthly saving is $600, what is her monthly income?

5) 40% of the total number of students of a school did not go for a picnic on a certain day. If the number of students who went for a picnic is 240, find the total number of students in the school.

6) In an examination, Ava secured 378 marks. If she secured 63% marks, find the maximum marks.

7) The price of a commodity rose by 50%. By what percent must a person reduce his consumption so that his expenditure does not increase?

8) A fruit seller brought some fruits to market. He sells 75% of them and still has 100 fruits. Find the number of fruits the fruit seller originally had?

Grade 7
Volume 1
Week 8

9) A number is increased by 20% and then decreased by 20%. Find the net increase or decrease percent.

10) Alice scored 432 points out of 600 in her school examination. Find her percentage of points.

11) Fill in the blank with suitable symbol

   25% of 500 _____ 20% of 600

12) In an election the winner won with 56% of votes. If the total number of votes polled were 45,800, find the number of votes his opponent secured.

13) Mr. Austen spends 35% of his monthly earnings on daily expenses, 20% on the rent of his house and 25% on studies of his children. He saves the rest of his salary. If his savings per month are $245, find his monthly salary.

14) In a city 40% of the population are men and 30% of the population are women. The rest of them are children. If the total population of the city is 12,890, find the population of children.

15) There were 850 trees in an orchard. In these 35% of trees are apple, 25% of trees are oranges and the rest are mango. How many mango trees are there in the orchard?

16) Robert and Edward started a business together investing 36% and 74% of the capital respectively. They share the profit also in the same proportion. If the profit in a particular year was $2,980, how much money will Edward get as his share?

**Grade 7
Volume 1
Week 8**

17) A school worked 260 days in a particular year. Lydia attended school for 247 days. What was the percentage of her attendance?

18) For a weekend project, Jane was given 40 questions. If she has completed 28 questions, what percent of the given work has been done?

19) In a laboratory, a mixture of 2 oz salt, 5 oz chalk and 3 oz sand was prepared. What percent of mixture was salt?

20) In a monthly test Daisy secured $\frac{45}{50}$ in mathematics, $\frac{64}{75}$ in science and $\frac{82}{100}$ in English. In which subject did she get the highest percentage?

21) At a quiz competition, Elizabeth answered 65% of the questions correctly. If the quiz master asked 40 questions, how many questions did she get correct?

22) Convert $60 of $240 into percent.

23) Find 23% of 356 gallons.

Ans: 81.88 gallons.

24) If 24% of a number is 498, find the number.

25) In an annual examination out of 60 students, $83\frac{1}{3}$ % of students were promoted to a higher grade. What number of students is promoted to higher grade?

26) In an election the number of votes the three contestants secured are 23,475; 43,456 and 34,985. If the total votes polled were 120,000, what is the percent of invalid votes?

27) Mrs. Williams bought 6 lb flour to make cakes. She used 4 lb flour. What percentage of flour did she use?

28) There were 36 apples, 34 oranges and 50 mangoes in a basket. What percent of fruits in the basket are oranges?

29) Eighty percent of the cost of a washing machine is $680.00. What is the total cost of the washing machine?

30) Express 8 oz as a percent of 5 pounds.

31) Paul saves 8% of his income. If his annual savings is $4000, what is his annual income?

Ans: $50,000.

32) How much is 15% of $3.60?

Ans: 54 cents.

Grade 7
Volume 1
Week 8

33) Mr. Joseph's monthly income is $5,680. If he saves 12.5% of his income, how much money does he save?

34) In an examination, the maximum points were 450. Kelly got 32% points and failed by 16 points. What are the passing points?

35) A number when increased by 5% equals 168. What is the original number?

36) The price of a television set increased from $450 to $495. What is the increase in its price?

37) There are 56 boys and 84 girls in a class. What is the percentage of boys in the class?

38) There are 56 boys and 84 girls in a class. What is the percentage of girls in the class?

39) There are 42 apples and 63 pears in a basket. What is the percentage of apples in the basket?

40) There are 42 apples and 63 pears in a basket. What is the percentage of pears in the basket?

41) There are 17 roses and 33 daisies in a bouquet. What is the percentage of roses in the bouquet?

42) There are 17 roses and 33 daisies in a bouquet. What is the percentage of daisies in the bouquet?

43) There are 15 red marbles, 18 green marbles and 12 blue marbles in a box. What is the percentage of red marbles in the box?

44) There are 15 red marbles, 18 green marbles and 12 blue marbles in a box. What is the percentage of blue marbles in the box?

45) There are 15 red marbles, 18 green marbles and 12 blue marbles in a box. What is the percentage of green marbles in the box?

46) Harry and his team won 10 basket ball matches and lost 4 matches. 2 matches were declared draw. What percent of matches were won by Harry's team?

47) Harry and his team won 10 basket ball matches and lost 4 matches. 2 matches were declared draw. What percent of matches were lost by Harry's team?

48) Harry and his team won 10 basket ball matches and lost 4 matches. 2 matches were declared draw. What percent of matches were declared draw?

49) In a pile of 80 clothes, 20% were skirts. How many skirts are there in all?

50) In a class of 45 students, 40% are boys. How many boys are there in all?

51) In a class of 55 students, 40% are girls. How many girls are there in all?

52) In a class of 60 students, 40% were absent on a rainy day. How many students were present in the class on that day?

53) The monthly income of Mr. David is $4,800. He saves 13% of his income and spends the rest. How much money does he spend every month?

54) The monthly income of Mr. David is $4,800. He saves 13% of his income and spends the rest. How much money does he save every month?

55) The monthly income of Mrs. Ruby is $5,250. Every month she deposits 12% of her income in a savings account. She spends 14% of her income on the education of her children, pays 20% for the rent of her apartment and the rest she spends on miscellaneous items. How much money does she save every month?

56) The monthly income of Mrs. Ruby is $5,250. Every month she deposits 12% of her income in a savings account. She spends 14% of her income on the education of her children, pays 20% for the rent of her apartment and the rest she spends on miscellaneous items. How much money does she pay as rent?

57) The monthly income of Mrs. Ruby is $5,250. Every month she deposits 12% of her income in a savings account. She spends 14% of her income on the education of her children, pays 20% for the rent of her apartment and the rest she spends on miscellaneous items. How much money does she spend on the education of her children?

58) The monthly income of Mrs. Ruby is $5,250. Every month she deposits 12% of her income in a savings account. She spends 14% of her income on the education of her children, pays 20% for the rent of her apartment and the rest she spends on miscellaneous items. How much money does she spend on miscellaneous items?

59) Mr. Robert earns $2,260 a month. He spends 54% of his income for his regular expenses every month and donates 18% of his income for charity. He saves the rest of his income. How much money does he save every month?

60) Mr. Bob earns $2,260 a month. He spends 54% of his income for his regular expenses every month and donates 18% of his income for charity. He saves the rest of his income. How much money does he donate every month?

61) Mr. Smith earns $2,260 a month. He spends 54% of his salary for his regular expenses every month, donates 18% to charity and saves the rest. How much does he spend on his monthly expenses?

Grade 7
Volume 1
Week 8

62) Ms. Lucy's runs a garment business. The turn over of her business is $32,500 per month. She spends 24% of her turn over on salaries for her staff, 32% on raw material and 8% on the rent for her office. The rest is her profit. How much is her profit per month

63) Two candidates, Kevin and Rachel, contested an election. The total votes polled were 4650. If Rachel wins with 62% of votes, find the number of votes received by each.

64) The maximum points in an examination are 500. Aidan gets 28% points and fails by 15 points. How many points should he get to pass the examination?

65) The maximum points in an examination are 500. Aidan gets 28% points and fails by 15 points. What is the minimum percentage to pass the examination?

66) A whole sale dealer charges 2% handling charges on first $5,000 and 1.8% on the remainder. How much handling charges does he charge for goods worth $8,000?

67) Bob saves 28% of his pocket money every month. If he saves $44.80 every month, how much is his pocket money?

68) 12% of some amount of money is $84. What is 43% of the same amount?

69) 18% of a number is 56. What is 45% of that number?

Grade 7
Volume 1
Week 8

70) 12% of certain weight is 45 kg. What is 108% of this weight?

71) $13\frac{1}{3}$% of a certain length is 30 miles.

Find 45% of the same length.

72) A piece of cotton cloth 120 inches long shrinks by $1\frac{2}{3}$ % after the first wash.

Find the length of the cloth after washing.

73) An ore contains 24% iron. How many pounds of ore is required to get 57 pounds of iron?

74) 3% of the value of a car is $240. What is the value of the car?

75) 1% of the cost of a book is $1.20. What is the cost of the book?

76) Luke's savings account has $80. At a later time he withdrew 5% of the amount from his account. How much money is left in his account?

77) The cost of 100 apples was $45. How much should you pay if the cost increases by 20%?

78) There are 60 oranges in a basket. They are increased by 150%. How many oranges are there in the basket now?

79) The number 500 is increased by 3.4%. What is the new number?

80) The number 70 is decreased by 40%. What is the new number?

81) There were 25 mangoes in a basket. 8% of them were rotten. How many good mangoes are there in the basket?

82) There are 24 pencils with Ella. 25% of them were broken. How many good pencils are there with Ella?

83) John has some kites. If he can make 10% more, he will have 66 kites. How many kites does John have now?

84) Allison has some money with her. If she had 15% more money, she would have had $207. How much money does she have now?

85) By what number must a given number 'x' be multiplied so that its value decreases by 15%?

86) The price of eggs increase from $3 for two to $3.80 for two. Find the percentage increase in the egg's price.

87) The number of students in a school increased from 480 to 600 during the academic year 2005 - 2006. What is the percentage of increase in number of students?

88) A person receives 12% increase in his salary because of which there is an increment of $1200 in his payment. What was his original salary?

89) Jacob bought a second hand car for $3000. He spent $500 for repairs and then sold it for $4200. What is his gain percent?

90) Ethan sold a fan for $52 and gained 4%. For how much should he sell the fan as to gain 6%?

91) Anthony sold his car for $5,320 at a loss of 5%. For how much should he sell the car as to gain 5%?

92) A time piece is sold for $49 at a gain of 12%. What is the cost price of the time piece?

93) Find the cost price when selling price is $851 and loss is 8%.

94) Mr. Ronald sold a room heater at a loss of 1%. If he had sold it for $75 more, he would have gained 3%. For how much did Mr. Ronald purchase the room heater?

95) Three ipods are purchased for $450 each. One of the ipod is sold at a loss of 12%. At what price should the other two ipods be sold so as to gain 10% on the whole transaction?

96) Ella bought a machine for $368 and spent $32 on its repairs. Then she sold it for $380. What was the gain/loss percent?

97) Isaac bought 100 books for $950. 5 books were torn and so he could not sell them. If he sells the remaining books at $8 per book, find his gain or loss percent?

98) John buys some toffees at the rate of 4 for a dollar and sells them at the rate of 3 for a dollar. Find his gain percent.

99) An electrician sells a room heater for $342, gaining $\left(\frac{1}{5}\right)$ th of its cost price. Find his gain percent.

100) A bike dealer sells a bike at a gain of 8%. If he had sold it for $75 less, he would have lost 2%. Find the cost price of the bike.

101) Find the loss or gain percent if an article is bought for $360 and sold for $450.

Grade 7
Volume 1
Week 8

102) By selling a refrigerator for $1260, Mr. Brown lost 10% of his out lay. For how much did he buy the refrigerator?

103) Susie bought 200 oranges for $400. Eight oranges were rotten and could not be sold. She sold the others at the rate of $24 per dozen. What is her gain/loss percent?

104) Ms. Julia bought 10 dozen apples at the rate of $8 per dozen. Due to a forecast of heavy snow fall, she had to sell these apples at 10% loss. At what price did she sell each apple?

105) Kevin bought a digital camera for $655 and sold it to Jose at a loss of 3% as the model was outdated. How much was Kevin's loss in this transaction?

106) Evan sells his television for $160 and loses $\left(\frac{1}{5}\right)$ of his outlay. For how much did he buy the television? buy the refrigerator?

107) A carpet was sold for $360 at a loss of 4%. At what price should it be sold to lose only 2%?

108) By selling a gold chain for $368 Peter lost 8%. For how much did he buy the gold chain?

109) By selling a mathematics book for $120, a publisher incurs a loss of 20%. How much did the book cost to the publisher?

©All rights reserved-Math-Knots LLC., VA-USA   www.math-knots.com | www.a4ace.com

110) A mp3 player is bought for $240 and sold at a loss of 5%. What is the selling price of the mp3 player?

111) An old house is sold for $\left(\frac{4}{5}\right)$ of its cost price. Find the loss percent.

112) A sofa set is sold for $120 at a gain of 25%. At what price should it be sold to gain 30%?

113) James bought a dining table for $180 and paid $20 for its transportation. If he sells it for $236, what is the gain percent?

114) An article was sold for $550 and the gain is 10%. At what price should it be sold to gain 12%?

115) Matt sold his motor bike for $552, losing 8%. For how much did he purchase the motor bike?

116) Eric sold two play-stations for $264 each, gaining 10% on one and losing 12% on the other. What is his gain or loss percent on the whole transaction?

117) Find the loss percent when the cost price of an article is $940 and the selling price is $893.

118) Find the selling price of an article when its cost price is $96 and the loss is 6.25%.

119) Find the selling price of an article if its cost price is $480 and the loss is 17.5%.

120) Find the cost price of an article when selling price is $1,470 and loss is 2%.

121) Grace bought apples at $28.80 per dozen and sold them at $42.00 per score. Find her loss or gain percent.

122) Ruby bought roses at $5 per pair and sold them at $8 per four. Find her loss or gain percent.

123) Olivia bought oranges at the rate of 6 for $8 and sold them at the rate of 8 for $10. Find her loss or gain percent.

124) Lemons are bought at the rate of 4 for $1.50. They were sold at a loss of 20%. What was the selling price of the lemons?

125) Tim sold a book for $213.20, gaining 4%. For how much did he buy the book?

Grade 7
Volume 1
Week 8

126) On selling his car for $6,083, Mr. William gains 10%. For how much should he sell it to gain 15%?

127) While selling his house for $65,780, Mr. Daniel gains 10%. What will be his gain percent if he sells the same for $76,544?

128) Milk was bought at 3 gallons for $14, and was sold at a loss of 10%. What was the selling price of milk?

129) Pears are bought at the rate of 50 for $20. They were sold at a loss of 5%. What was the selling price of the pears?

130) School bags are bought at the rate of 10 for $55. They were sold at a loss of 8%. What was the selling price of school bags?

131) While selling his bike at $861, David gains 5%. At what price must he sell to gain 10%?

132) A grocer bought 1 ton wheat at $4 per pound. He sold 1500 pounds at $5 perpound. At what rate per pound should he sell the remaining wheat so as to gain 20% on the whole? (Given, 1 ton = 2000 pounds)

133) A boy bought a bicycle for $310. He spent $54 on repairs and then sold it for $370. Find the loss or gain percent.

Grade 7
Volume 1
Week 9

1) An investment of $420 is made at 2% for 2 years at a bank. Find the ending balance.

   A) $436.80   B) $16.80
   C) $428.40   D) $436.97

2) An investment of $46,000 is made at 11% for 3 years at a bank. Find the ending balance.

   A) $16,911.03   B) $61,180.00
   C) $62,911.03   D) $51,060.00

3) An investment of $500 is made at 2% for 7 years at a bank. Find the ending balance.

   A) $510.00   B) $574.34
   C) $70.00   D) $570.00

4) An investment of $7,000 is made at 12% for 5 years at a bank. Find the ending balance.

   A) $11,200.00   B) $12,336.39
   C) $7,840.00   D) $4,200.00

5) An investment of $490 is made at 14% for 3 years at a bank. Find the ending balance.

   A) $235.96   B) $558.60
   C) $695.80   D) $205.80

6) An investment of $48,300 is made at 5% for 2 years at a bank. Find the ending balance.

   A) $50,715.00   B) $53,250.75
   C) $4,950.75   D) $53,130.00

7) An investment of $1,340 is made at 10% for 4 years at a bank. Find the ending balance.

   A) $536.00   B) $1,876.00
   C) $1,474.00   D) $1,961.89

8) An investment of $110 is made at 13% for 2 years at a bank. Find the ending balance.

   A) $28.60   B) $140.46
   C) $138.60   D) $124.30

9) An investment of $40,000 is made at 7% for 3 years at a bank. Find the ending balance.

A) $42,800.00   B) $48,400.00
C) $8,400.00   D) $49,001.72

10) An investment of $44,900 is made at 13% for 5 years at a bank. Find the ending balance.

A) $74,085.00   B) $50,737.00
C) $82,725.34   D) $29,185.00

11) An investment of $56,800 is made at 5% for 2 years at a bank. Find the ending balance.

A) $2,840.00   B) $62,480.00
C) $59,640.00   D) $5,680.00

12) An investment of $38,000 is made at 15% for 2 years at a bank. Find the ending balance.

A) $43,700.00   B) $12,255.00
C) $49,400.00   D) $50,255.00

13) An investment of $230 is made at 13% for 4 years at a bank. Find the ending balance.

A) $375.01   B) $145.01
C) $349.60   D) $259.90

14) An investment of $1,160 is made at 1% for 4 years at a bank. Find the ending balance.

A) $1,171.60   B) $1,206.40
C) $1,199.21   D) $1,207.10

15) An investment of $2,000 is made at 1% for 5 years at a bank. Find the ending balance.

A) $2,020.00   B) $2,100.00
C) $102.02   D) $2,102.02

16) An investment of $48,600 is made at 2% for 6 years at a bank. Find the ending balance.

A) $54,432.00   B) $54,731.49
C) $49,572.00   D) $5,832.00

17) An investment of $50,600 is made at 15% for 5 years at a bank. Find the ending balance.

A) $58,190.00   B) $88,569.27

C) $88,569.96   D) $88,550.00

18) An investment of $32,400 is made at 5% for 6 years at a bank. Find the ending balance.

A) $42,120.00   B) $43,419.10

C) $1,620.00   D) $34,020.00

19) An investment of $30 is made at 8% for 4 years at a bank. Find the ending balance.

A) $40.81   B) $32.40

C) $39.60   D) $10.81

20) An investment of $41,700 is made at 13% for 6 years at a bank. Find the ending balance.

A) $74,226.00   B) $86,817.39

C) $32,526.00   D) $47,121.00

21) An investment of $380 is made at 6% for 2 years at a bank. Find the ending balance.

A) $46.97   B) $22.80

C) $425.60   D) $426.97

22) An investment of $35,500 is made at 14% for 4 years at a bank. Find the ending balance.

A) $59,113.61   B) $54,600.00

C) $19,600.00   D) $39,900.00

23) An investment of $56,200 is made at 16% for 8 years at a bank. Find the ending balance.

A) $71,936.00

B) $128,136.00

C) $184,246.92

D) $128,046.92

24) An investment of $28,300 is made at 1% for 7 years at a bank. Find the ending balance.

A) $28,583.00   B) $1,981.00

C) $30,281.00   D) $30,341.43

25) An investment of $1,630 is made at 7% for 7 years at a bank. Find the ending balance.

A) $2,428.70   B) $1,744.10

C) $798.70   D) $2,617.42

26) Write the below number in standard notation.

3×10^{-2}

A) 0.003 B) 300
C) 0.0003 D) 0.03

27) Write the below number in standard notation.

9.74×10^{1}

A) 0.0974 B) 0.974
C) 97.4 D) 9.74

28) Write the below number in standard notation.

5.2×10^{0}

A) 5.2 B) 520
C) 5200 D) 0.52

29) Write the below number in standard notation.

4.2×10^{-5}

A) 0.000042 B) 4200000
C) 42000000 D) 420000

30) Write the below number in standard notation.

1.9×10^{-6}

A) 1900000 B) 19000
C) 0.0000019 D) 190000

31) Write the below number in standard notation.

2.9×10^{-4}

A) 290000 B) 0.000029
C) 29000 D) 0.00029

32) Write the below number in standard notation.

5×10^{-4}

A) 500 B) 0.005
C) 0.0005 D) 5000

33) Write the below number in standard notation.

8.02×10^{3}

A) 802 B) 80.2
C) 8.02 D) 8020

34) Write the below number in standard notation.

9.4×10^0

A) 940 B) 94
C) 0.94 D) 9.4

35) Write the below number in standard notation.

7.57×10^2

A) 0.757 B) 757
C) 7.57 D) 75.7

36) Write the below number in standard notation.

8.92×10^5

A) 0.00000892 B) 89200000
C) 8920000 D) 892000

37) Write the below number in standard notation.

7.32×10^6

A) 0.00732 B) 7320
C) 7320000 D)

38) Write the below number in standard notation.

5.8×10^0

A) 580 B) 5800
C) 5.8 D) 0.58

39) Write the below number in standard notation.

1.9×10^{-1}

A) 1.9 B) 19
C) 0.19 D) 190

40) Write the below number in standard notation.

2×10^0

A) 2 B) 0.2
C) 20 D) 0.02

41) Write the below number in scientific notation.

4850000

A) 0.0485×10^{-4}

B) 4.85×10^{-5}

C) 4.85×10^{6}

D) 4.85×10^{-4}

42) Write the below number in scientific notation.

40000

A) 4×10^{3} B) 4×10^{4}

C) 4×10^{1} D) 4×10^{0}

43) Write the below number in scientific notation.

80

A) 8×10^{0} B) 8×10^{1}

C) 8×10^{-1} D) 8×10^{2}

44) Write the below number in scientific notation.

0.000008

A) 8×10^{0} B) 8×10^{6}

C) 8×10^{1} D) 8×10^{-6}

45) Write the below number in scientific notation.

9400

A) 9.4×10^{3} B) 9.4×10^{0}

C) 94×10^{0} D) 9.4×10^{-1}

46) Write the below number in scientific notation.

410

A) 4.1×10^{2} B) 4.1×10^{3}

C) 4.1×10^{-3} D) 0.41×10^{-3}

47) Write the below number in scientific notation.

4970000

A) 4.97×10^{-6} B) 4.97×10^{-7}

C) 4.97×10^{6} D) 4.97×10^{7}

48) Write the below number in scientific notation.

7700000

A) 7.7×10^{-6} B) 7.7×10^{6}

C) 7.7×10^{-5} D) 0.77×10^{6}

49) Write the below number in scientific notation.
0.000732

A) 7.32×10^4 B) 7.32×10^{-4}
C) 0.732×10^4 D) 0.732×10^6

50) Write the below number in scientific notation.
7000000

A) 7×10^1 B) 7×10^6
C) 7×10^{-6} D) 0.7×10^1

51) Write the below number in scientific notation.
0.00000187

A) 1.87×10^2 B) 1.87×10^{-7}
C) 1.87×10^{-2} D) 1.87×10^{-6}

52) Write the below number in scientific notation.
0.098

A) 9.8×10^0 B) 9.8×10^{-1}
C) 9.8×10^{-2} D) 98×10^0

53) Write the below number in scientific notation.
0.000157

A) 15.7×10^{-4} B) 1.57×10^{-5}
C) 157×10^{-4} D) 1.57×10^{-4}

54) Write the below number in scientific notation.
8300

A) 8.3×10^3 B) 0.83×10^3
C) 0.83×10^1 D) 0.83×10^{-1}

55) Write the below number in scientific notation.
14700000

A) 1.47×10^5 B) 1.47×10^6
C) 0.147×10^5 D) 1.47×10^7

56) The distance between Earth and Sun is 93,000,000 miles. What is this distance when expressed in scientific notation?

Grade 7
Volume 1
Week 9

57) Earth's circumference is 4×10^4 km. What is its circumference in meters?

58) The population of 4 cities is given in the table below :
   city a 3.5×10^5
   city b 4.1×10^6
   city c 1.2×10^6
   city d 6.5×10^5
Order the cities according to the population from greatest to least.

59) Express 11.2×10^{-6} in usual form.

60) Express 3.2256×10^{-4} in usual form.

61) Express the following in scientific notation: 0.8

62) The planet Jupiter has orbit of length 1.12×10^6 km. Express this distance in usual form.

63) Express the following in usual form :

5.5874×10^{-3}

64) Express the following in usual form :

0.8×10^{-1}

65) Express the following in usual form :

$(-4.4) \times 10^{-3}$

66) Express the following in scientific notation:
0.00000072

67) Express the following in scientific notation:
5600000000000

68) The speed of light is 300000000 m/sec. Express this speed in scientific notation.

69) Earth is about 149000000 km from the Sun. What is this distance expressed in scientific notation ?

70) Express in scientific notation
(-0.0000000052)

71) Express the following numbers in order from greatest to least:

$3.1 \times 10^{5}, 2.4 \times 10^{-3}, 2.5 \times 10^{-1}$

72) What is the usual form of 1.364×10^{-3} ?

73) The length of a river is 238,500,000,000 miles. How do you represent this length in scientific notation?

91) Order the following numbers in order from least to greatest:

5.12×10^{-3}, 3.72×10^{-8},

7.45×10^{-5}, 6.38×10^{-4}

75) How do you represent 23,000,000,000 in scientific notation?

76) Is 2.349×10^{-9} greater than 1.98×10^{-4}?

77) The atomic radius of a gold atom is 1.3×10^{-10} m. Express this in usual form.

78) The radius of Sun is 432,000 miles. Express this length in scientific notation.

79) One light year is 5,899,000,000,000 miles. Express this distance in scientific notation.

80) The temperature at the centre of the Sun is 15,000,000 K. Express the temperature in scientific notation.

81) The size of a plant cell is approximately 12.76×10^{-6} m. Express the size in usual form.

82) Express 2.46×10^6 in usual form.

83) Express 3.4×10^8 in usual form.

84) What is the usual form of 1.45×10^{-4}?

85) Express 40,720,000 in scientific notation.

86) Express 670,000 in scientific notation.

87) Express 0.0000345 in scientific notation.

88) Is 6.23×10^8 greater than 6.23×10^6?

89) Is 2.678×10^{-7} greater than 2.678×10^{-5}?

90) Order the following numbers in order from greatest to least:

2.34×10^5, 4.18×10^6,

1.78×10^8, 6.29×10^4.

91) At what rate per cent per annum will a sum triple itself in 20 years?

92) Divide $18,000 into two parts such that the simple interest on the first part for 2 years at 6% per annum is equal to the simple interest on the second part for 3 years at 8% per annum.

93) Feng borrowed $2620 from his friend Smith. He returned $3537 to Smith after 1 year. Calculate the rate of simple interest.

94) Clark deposited $3500 at the rate of 8% per annum for 4 years. Find the simple interest.

95) In how much time will $5600 amount to $6720 at 8% simple interest per annum?

96) Find the principal that will yield an interest of $306 in 3 years at 10% per annum.

97) At what rate per cent per annum will the simple interest on $3750 be $1200 in 4 years?

98) Find the time in which the interest on $4260 is $284, reckoned at 5% per annum.

99) How much money will amount to $4230 at $7\frac{1}{2}$ % per annum simple interest in 2 years 4 months?

100) At what percent per annum will the simple interest on $2345 be $281.40 in 2 years?

101) In how many years will the simple interest on $3860 be $579 at $7\frac{1}{2}$ % per annum?

102) Find the simple interest on $6,372 at $6\frac{2}{3}$ % per annum for 9 months.

103) At what rate per cent per annum will $500 amount to $605 in 3 years.

104) Find the simple interest on $500 at 12% per annum for 3 years.

105) At what rate per cent per annum will $500 amount to $650 in 3 years.

106) In what time will $720 amount to $792 if simple interest is calculated at 4% per annum ?

107) Find the simple interest on $2,850 at 3.5% per annum for 8 months.

108) Jenna invested $1,825 in a bank for 6 months at 14% per annum simple interest. How much interest would she receive at the end of 6 months?

109) Mia invested $300 at 11% per annum simple interest which amounted to $498 in a certain time. For how long did Mia make the investment?

110) If $240 amounts to $360 in 4 years, how much will $450 amount to in 6 years at the same rate of interest per annum?

111) A sum of money lent at simple interest amounts to $2,592 in 2 years and $3,888 in 5 years. Find the sum and the rate of interest?

112) The simple interest on a certain sum for 4 years at 6% per annum is $36 more than the simple interest on the same sum for 2 years at 8% per annum. Find the sum.

113) Divide $6,000 in two parts such that the simple interest on the first part for 3 years at 5% per annum is equal to the simple interest on the second part for 5 years at 6% per annum.

114) A sum of money invested at 5% per annum amounts to $644 in 3 years. What will it amount to in 1 year 3 months at the same rate?

115) A sum of money invested at 4% per annum amounts to $1044 in 5 years. What will it amount to in 6 years 9 months at 10% per annum?

116) sum of money invested at 6% per annum amounts to $840 in 2 years. What will it amount to in 2 years 9 months at 8% per annum?

117) A sum when reckoned at 7.5% per annum amounts to $3921.50 in 2 years. Find the sum.

Grade 7 Volume 1 Week 10

Grade 7
Volume 1
Week 10

1) Fill in the blanks with correct symbol:

$$\left(\frac{-3}{7}\right) \_\_\_\_ \left(\frac{-6}{13}\right)$$

2) Arrange the following rational numbers in ascending order:

$$\left(\frac{-5}{11}\right), \left(\frac{1}{-3}\right), \left(\frac{-6}{7}\right), \left(\frac{-3}{4}\right)$$

3) State True or False

$$\left(\frac{-12}{5}\right)$$

lies to the right of 0 on the number line.

4) Compare the following rational numbers using the correct symbol

$$\left(\frac{5}{9}\right) \_\_\_\_ \left(\frac{3}{8}\right)$$

5) Compare the following rational numbers using the correct symbol

$$\left(\frac{3}{11}\right) \_\_\_\_ \left(\frac{6}{11}\right)$$

6) Compare the following rational numbers using the correct symbol

$$\left(\frac{-2}{3}\right) \_\_\_\_ \left(\frac{4}{-5}\right)$$

7) Compare the following rational numbers using the correct symbol

$$\left(\frac{6}{7}\right) \_\_\_\_ 0$$

8) Compare the following rational numbers using the correct symbol

$$\left(\frac{2}{5}\right) \_\_\_\_ \left(\frac{4}{7}\right)$$

9) Arrange the following rational numbers in ascending order:

$\left(\dfrac{5}{-6}\right), \left(\dfrac{1}{4}\right), \left(\dfrac{-1}{-8}\right), \left(\dfrac{-11}{4}\right)$

10) Arrange the following rational numbers in ascending order:

$\left(\dfrac{-4}{5}\right), \left(\dfrac{-3}{4}\right), \left(\dfrac{17}{-20}\right), (-1)$

11) Arrange the following rational numbers in descending order:

$\left(\dfrac{1}{-2}\right), \left(\dfrac{-4}{5}\right), 0, \left(\dfrac{4}{7}\right)$

12) Arrange the following rational numbers in descending order:

$\left(\dfrac{7}{-9}\right), \left(\dfrac{-5}{8}\right), \left(\dfrac{-8}{7}\right), \left(\dfrac{4}{-3}\right)$

13) State True or False:

The rational numbers $\left(\dfrac{3}{7}\right), \left(\dfrac{-3}{8}\right)$ are on the opposite sides of 0 on the number line.

14) State True or False:

$\left(\dfrac{-12}{7}\right)$ lies to the right of 0 on the number line.

15) Express $\left(\dfrac{143}{25}\right)$ in decimal form.

16) Write the following number in its standard form:

$\left(\dfrac{-7}{-63}\right)$

Grade 7
Volume 1
Week 10

17) Arrange the following rational numbers in ascending order:

$\left(\dfrac{-3}{5}\right), \left(\dfrac{-8}{12}\right), \left(\dfrac{-6}{20}\right), \left(\dfrac{-13}{6}\right)$

18) Compare the following rational numbers:

$\left(\dfrac{-4}{9}\right)$ _____ $\left(\dfrac{-5}{12}\right)$

19) Which of the following rational numbers is less than $\left(\dfrac{-7}{8}\right)$?

A) $\left(\dfrac{7}{8}\right)$

B) $\left(\dfrac{-9}{8}\right)$

C) $\left(\dfrac{-5}{8}\right)$

D) $\left(\dfrac{-3}{8}\right)$

20) State True or False:

$\left(\dfrac{13}{5}\right)$ lies to the left of $\left(\dfrac{-13}{5}\right)$ on the number line.

21) Pick the greater rational number.

A) $\left(\dfrac{-2}{5}\right)$ B) $\left(\dfrac{-4}{5}\right)$

22) Simplify and express the result as a rational number

28.796 - 2.755 - 13.22

23) Use appropriate symbol to fill in the blank:

$\left(\dfrac{-3}{8}\right)$ _____ (0.375)

Grade 7
Volume 1
Week 10

24) State True or False:

$\left(\dfrac{-7}{4}\right)$ is greater than $\left(\dfrac{-4}{7}\right)$

25) Fill in the blank with suitable symbol

$\left(\dfrac{2}{7}\right)$ _____ $\left(\dfrac{7}{2}\right)$

26) Fill in the blank with suitable symbol

$\left(\dfrac{-1}{2}\right)$ _____ $\left(\dfrac{1}{-2}\right)$

27) State True or False:

$\left(\dfrac{-3}{5}\right) > \left(\dfrac{-5}{3}\right)$

28) State True or False:

$\left(\dfrac{2}{5}\right)$ lies to the right of $\left(\dfrac{1}{5}\right)$ on the number line.

29) State True or False:

$\left(\dfrac{13}{3}\right)$ lies to the left of $\left(\dfrac{-13}{3}\right)$ on the number line.

30) Use appropriate symbol to fill in the blank:

0.4576 _____ $\left(\dfrac{14}{35}\right)$

31) Use appropriate symbol to fill in the blank:

$\left(\dfrac{64}{125}\right)$ _____ 0.6

32) Compare the following rational numbers using the correct symbol

$\left(\dfrac{-5}{7}\right)$ _____ $\left(\dfrac{5}{-7}\right)$

33) Compare the following rational numbers using the correct symbol

$\left(\dfrac{8}{5}\right)$ _____ $\left(\dfrac{-8}{5}\right)$

34) Compare the following rational numbers using the correct symbol

$\left(\dfrac{35}{11}\right)$ _____ $\left(\dfrac{-35}{-11}\right)$

35) Arrange the following rational numbers in descending order:

$\left(\dfrac{2}{5}\right), \left(\dfrac{5}{7}\right), \left(\dfrac{-7}{8}\right), \left(\dfrac{3}{7}\right)$

36) Arrange the following rational numbers in descending order:

$\left(\dfrac{12}{13}\right)$ _____ $\left(\dfrac{-21}{19}\right)$

37) Arrange the following rational numbers in ascending order:

$\left(\dfrac{3}{-4}\right), \left(\dfrac{-5}{-7}\right), \left(\dfrac{4}{-9}\right), \left(\dfrac{-2}{-3}\right)$

38) Express $\dfrac{7}{33}$ in its decimal form.

39) Represent 0.2777.......... as a rational number.

$\dfrac{5}{18}$

**Grade 7
Volume 1
Week 10**

40) Which of the following rational numbers lies between $\left(\dfrac{5}{8}\right)$ and $\left(\dfrac{8}{9}\right)$?

A) $\left(\dfrac{3}{7}\right)$

B) $\left(\dfrac{6}{7}\right)$

C) $\left(\dfrac{2}{7}\right)$

D) $\left(\dfrac{3}{8}\right)$

41) Fill in the blank with appropriate symbol

$\left(\dfrac{-7}{8}\right)$ ____ $\left(\dfrac{-3}{8}\right)$

42) Fill in the blank with appropriate symbol

$\left(\dfrac{-8}{5}\right)$ ____ $\left(\dfrac{-1}{5}\right)$

43) Fill in the blank with appropriate symbol

$\left(\dfrac{0}{7}\right)$ ____ $\left(\dfrac{1}{7}\right)$

44) Fill in the blank with appropriate symbol

$\left(\dfrac{5}{6}\right)$ ____ $\left(\dfrac{-5}{6}\right)$

45) State True or False

The sum of two rational numbers is not always a rational number.

46) State True or False

The sum of any rational number and zero is the rational number itself.

Grade 7
Volume 1
Week 10

47) Evaluate

$$\left(\frac{11}{-16}\right)+\left(\frac{3}{-8}\right)+\left(\frac{1}{4}\right)$$

48) Simplify

$$(2)+\left(\frac{-1}{3}\right)+\left(\frac{-3}{4}\right)$$

49) State True or False

Addition is commutative on rational numbers.

50) When we add numbers with equal denominators, we

A) add the numerators

B) add the numerators and add denominators

C) subtract the numerators

D) subtract the numerators and add the denominators

51) State True or False:

$$\left(\frac{-24}{25}\right)$$ is in its standard form.

52) State True or False:

$$\left(\frac{-28}{35}\right)$$ in its standard form.

53) State True or False:

The standard form of $\left(\frac{40}{-52}\right)$ is $\left(\frac{-4}{13}\right)$

54) State True or False:

The standard form of $\left(\frac{-48}{-52}\right)$ is $\left(\frac{-12}{-13}\right)$

55) Find the sum of: $\dfrac{3}{10}, \dfrac{7}{10}, \dfrac{7}{10}$

56) You're making a peach and strawberry smoothie. You add $\dfrac{7}{12}$ cup of water, $\dfrac{7}{3}$ cup of orange juice, and $\dfrac{2}{9}$ cup of milk. How much liquid have you added?

57) Evaluate

$$\left(\dfrac{-19}{-6}\right)+\left(\dfrac{-22}{5}\right)+\left(\dfrac{13}{-15}\right)$$

58) Evaluate

$$\left(\dfrac{-3}{11}\right)+\left(\dfrac{7}{11}\right)$$

59) Evaluate

$$\left(\dfrac{-18}{37}\right)+\left(\dfrac{-35}{37}\right)$$

60) Evaluate

$$\left(\dfrac{-9}{13}\right)+(-2)+\left(\dfrac{11}{12}\right)$$

61) Evaluate

$$\dfrac{2}{3}+\left(\dfrac{-19}{6}\right)+\dfrac{34}{9}+\left(\dfrac{-5}{3}\right)$$

62) Fill in the blank:

$$\dfrac{14}{15}+\left(\dfrac{-18}{20}\right)+\left(\dfrac{-4}{5}\right) = \underline{\qquad}$$

Grade 7
Volume 1
Week 10

63) Four boxes are $\frac{5}{9}, \frac{7}{12}, \frac{11}{18}$ and $\frac{5}{18}$ ft high respectively. If these boxes were stacked one on top of another, what is the total height of the boxes?

64) Simplify

$$\left(\frac{3}{8}\right) + \left(\frac{-5}{6}\right) + \left(\frac{3}{-4}\right) + \left(\frac{1}{2}\right)$$

65) Fill in the blank:

$$\left(\frac{-5}{7}\right) + \left(\frac{3}{5}\right) + \left(\frac{-3}{7}\right) = \underline{\qquad}$$

66) Four electricity poles are in a row. The distance between the consecutive poles is $\frac{3}{4}$ miles, $\frac{5}{6}$ miles and $\frac{3}{5}$ miles, respectively. What is the total distance between the first electric pole and the fourth electric pole?

67) Find the sum of $\frac{1}{5}$, $\left(\frac{-1}{3}\right)$ and $\left(\frac{-1}{6}\right)$

68) The sum of two rational numbers is (-1). If one of the numbers is $\left(\frac{-6}{19}\right)$, find the other.

69) Fill in the blank:

$$2 + \left(\frac{-12}{15}\right) + \left(\frac{-7}{10}\right) = \underline{\qquad}$$

70) Simplify

$$\left(\frac{-11}{12}\right) + \left(\frac{-5}{18}\right) + \frac{1}{4}$$

$$\frac{-17}{18}$$

Grade 7
Volume 1
Week 10

71) The weight of an empty container is $\frac{5}{6}$ pounds. If $\frac{8}{5}$ pounds of grapes are kept in the container, what is its total weight of the container?

72) Add $\left(\frac{5}{-21}\right)$ and $\left(\frac{67}{21}\right)$

73) Add $\frac{-2}{5}$ and $\frac{3}{4}$

74) Find the sum of $\frac{-5}{11}, \frac{11}{5}, \frac{-7}{11}$

75) Evaluate

$$\frac{1}{8} + 1 + \left(\frac{-2}{5}\right)$$

76) Evaluate

$$\left(\frac{9}{5}\right) + \left(\frac{-2}{3}\right) + \left(\frac{-6}{5}\right)$$

77) Fill in the blank:

$$\left(\frac{-3}{8}\right) + \left(\frac{-5}{12}\right) = \underline{\qquad}$$

78) State True or False

$$\left(\frac{5}{-3}\right) + \left(\frac{-3}{5}\right) = \left(\frac{-34}{15}\right)$$

Grade 7
Volume 1
Week 10

79) Evaluate

$$\left(\frac{3}{7}\right)+\left(\frac{-2}{7}\right)+\left(\frac{12}{7}\right)$$

80) Evaluate

$$\left(\frac{3}{-8}\right)+(4)+\left(\frac{-5}{4}\right)$$

81) Evaluate

$$\left(\frac{-3}{10}\right)+\left(\frac{7}{15}\right)+\left(\frac{-8}{5}\right)$$

82) Add

$$\left(\frac{-4}{5}\right)+\left(\frac{-3}{20}\right)$$

83) Add $\left(\frac{-4}{5}\right)$ and $\left(\frac{-2}{7}\right)$

84) Add $\frac{3}{4}$ and $\left(\frac{-3}{6}\right)$

85) Add

$$\left(\frac{3}{4}\right)+\left(\frac{-2}{3}\right)+\left(\frac{2}{-5}\right)$$

86) Find the sum:

$$\left(\frac{-5}{9}\right)+\left(\frac{2}{3}\right)+\left(\frac{1}{-18}\right)$$

87) Find the sum:

$$\left(\frac{12}{7}\right)+\left(\frac{-2}{5}\right)+\left(\frac{5}{-7}\right)$$

88) Evaluate

$$\left(\frac{1}{4}\right)+\left(\frac{-3}{5}\right)+\left(\frac{3}{20}\right)$$

89) Add $\left(\frac{-12}{35}\right)$ and $\left(\frac{4}{7}\right)$

$$\frac{8}{35}$$

90) Simplify

$$(-1)+\left(\frac{-5}{4}\right)+\left(\frac{3}{4}\right)$$

91) Subtract $\left(\frac{-23}{27}\right)$ from 1.

92) Subtract $\left(\frac{-3}{11}\right)$ from $\left(\frac{-17}{8}\right)$.

93) Subtract $\left(\frac{1}{3}\right)$ from $\left(\frac{3}{4}\right)$.

94) Subtract (-14) from $\left(\frac{-3}{14}\right)$

95) Add the following:

$$\frac{9}{25} + \frac{27}{250}$$

96) Simplify:

$$\frac{8}{3} + \frac{5}{9}$$

97) State True or False:

Subtraction is associative on rational numbers

98) The sum of two rational numbers is (-3). If one of the numbers is $\left(\frac{-12}{5}\right)$, find the other.

99) What number should be added to $\left(\frac{-6}{5}\right)$ so as to get $\left(\frac{-2}{3}\right)$?

100) A bush in Olivia's garden was $\frac{27}{4}$ inches long, She trimmed $\frac{15}{8}$ inches off of the top of the bush. What is the height of the bush now?

101) Subtract $\frac{5}{8}$ from $\frac{3}{8}$

102) Subtract $\left(\frac{-15}{11}\right)$ from 0.

103) The sum of two rational numbers is $\left(\dfrac{5}{14}\right)$ If one of the numbers is $\left(\dfrac{-5}{7}\right)$, find the other.

104) Subtract $\left(\dfrac{-2}{5}\right)$ from 7.

105) The sum of two rational numbers is $\left(\dfrac{74}{75}\right)$ If one of the numbers is $\left(\dfrac{18}{25}\right)$, find the other.

106) The sum of two rational numbers is $\left(\dfrac{18}{25}\right)$ If one of the numbers is $\left(\dfrac{-9}{15}\right)$, find the other.

107) Evaluate

$\left(\dfrac{3}{7}+\dfrac{4}{9}\right)-\left(\dfrac{5}{21}\right)=$ _____

108) Subtract $\left(\dfrac{7}{8}\right)$ from $\left(\dfrac{21}{4}\right)$

109) Subtract $\left(\dfrac{-5}{2}\right)$ from $\left(\dfrac{-7}{5}\right)$

110) The sum of two rational numbers is 9. If one of the numbers is $\left(\dfrac{41}{6}\right)$, find the other rational number.

111) What rational number should be added to $\left(\dfrac{1}{8}\right)$ so as to get $\left(\dfrac{4}{9}\right)$?

112) Fill in the blank:

$\left(\dfrac{2}{7}\right) - \left(\dfrac{3}{5}\right) =$ _____

113) Subtract $\left(\dfrac{-17}{9}\right)$ from 0.

114) State True or False:

$\left(\dfrac{7}{11}\right) - \left(\dfrac{5}{9}\right) = \left(\dfrac{8}{99}\right)$

115) What rational number should be added to $\left(\dfrac{2}{3}\right)$ so as to get $\left(\dfrac{5}{4}\right)$?

116) The sum of $\left(\dfrac{-1}{2}\right)$ two rational numbers is $\left(\dfrac{5}{7}\right)$. If one of the rational numbers is , find the other rational number.

117) Find the additive inverse of $\left(\dfrac{-11}{13}\right)$

118) Find the additive inverse of $\left(\dfrac{-12}{17}\right)$

Grade 7
Volume 1
Week 10

119) Find the additive inverse of $\left(\dfrac{4}{9}\right)$

120) Find the additive inverse of $\left(\dfrac{-13}{-15}\right)$

121) The sum of two rational numbers is $\left(\dfrac{-4}{9}\right)$. If one of the numbers is $\left(\dfrac{3}{5}\right)$, find the other.

122) Fill in the blank

$\left(\dfrac{-5}{7}\right) - \left(\dfrac{-4}{5}\right) =$ _____

123) Subtract $\left(\dfrac{-4}{5}\right)$ from $\left(\dfrac{6}{8}\right)$

124) Subtract $\left(\dfrac{-4}{5}\right)$ from $\left(\dfrac{-6}{15}\right)$

125) Evaluate

$(-5) - \left(\dfrac{-18}{11}\right)$

126) Evaluate

$\left(\dfrac{-6}{8}\right) - \left(\dfrac{3}{4}\right)$

127) Evaluate

$$\left(\frac{14}{15}\right) - \left(\frac{5}{6}\right)$$

128) Evaluate

$$\left(\frac{-5}{-7}\right) - \left(\frac{-4}{7}\right)$$

129) Evaluate

$$\left(\frac{7}{-12}\right) - \left(\frac{19}{4}\right)$$

130) What should be subtracted from $\left(\frac{-5}{6}\right)$ to get (-1) ?

131) What should be added from $\frac{-4}{5}$ to get (-1) ?

132) Multiply:

$$\left(\frac{-13}{5}\right) \text{ by } 15$$

133) Simplify :

$$\left[\left(\frac{-12}{7}\right) \times \left(\frac{-14}{27}\right)\right] + \left[\left(\frac{-8}{45}\right) \times \left(\frac{9}{16}\right)\right]$$

134) A car is moving at an average speed of $40\frac{2}{5}$ miles/ hr. How much distance will it cover in $17\frac{1}{2}$ hours?

135) State True or False:

The product of a rational number and 1 is 1.

136) State True or False:

Every rational number when multiplied by 0 gives 0.

137) State True or False:

The product of two rational numbers is always a rational number.

138) State True or False:

Zero has no reciprocal.

139) Multiply:

$\dfrac{5}{7}$ by $\dfrac{5}{6}$

140) Multiply:

$\left(\dfrac{25}{-10}\right)$ by $\left(\dfrac{-80}{-50}\right)$

141) Multiply:

$\left(\dfrac{-36}{4}\right)$ by $\left(\dfrac{20}{-3}\right)$

142) Multiply:

$\left(\dfrac{-40}{21}\right)$ by 7

143) Multiply:

$$\left(\frac{-14}{91}\right) \text{ by } \left(\frac{13}{-21}\right)$$

144) Multiply:

$$\left(\frac{64}{-27}\right) \text{ by } \left(\frac{-3}{-8}\right)$$

145) Multiply:

$$\left(\frac{5}{18}\right) \text{ by } \left(\frac{-36}{5}\right)$$

146) Multiply:

$$\left(\frac{16}{-21}\right) \text{ by } \left(\frac{21}{-16}\right)$$

147) Simplify:

$$\left[\left(\frac{7}{24}\right) \times (-48)\right] + \left[\left(\frac{-4}{9}\right) \times \left(\frac{3}{-2}\right)\right]$$

148) Simplify:

$$\left[\left(\frac{-9}{8}\right) \times \left(\frac{-64}{27}\right)\right] - \left[\left(\frac{-13}{5}\right) \times (-15)\right]$$

149) Simplify:

$$\left[\left(\frac{-13}{-15}\right) \times \left(\frac{25}{-26}\right)\right] - \left[\left(\frac{-7}{30}\right) \times \left(\frac{-5}{14}\right)\right]$$

150) What number should be added to $\left(\frac{-7}{13}\right)$ so as to get 3 ?

151) The height of an electric pole is $\frac{32}{3}$ ft.

A rose bush is about three-eights the height of the electric pole. What is the height of the rose bush?

152) Find the product of

$\left(\frac{-18}{22}\right)$, $\left(\frac{42}{63}\right)$ and $\left(\frac{-9}{15}\right)$

153) Fill in the blank

$\left[\left(\frac{-34}{18}\right)+\left(\frac{5}{9}\right)\right] \times \left[\left(\frac{3}{8}\right)-\left(\frac{2}{7}\right)\right]$ = _____

154) Evaluate

$\left[\left(\frac{-13}{20}\right)+\left(\frac{7}{10}\right)\right] \times \left(\frac{5}{6}\right)$

155) Simplify

$\left(\frac{13}{2}\right) \times \left(\frac{5}{-26}\right) \times (-3)$

156) Find the multiplicative inverse of $\frac{12}{17}$

157) Divide $\left(\frac{-3}{8}\right)$ by $\left(\frac{-7}{9}\right)$

158) Find the denominator of the rational number -11

159) Multiply $\left(\dfrac{-8}{5}\right)$ by $\left(\dfrac{2}{-7}\right)$

160) Simplify

$$\left(\dfrac{7}{6}\right) \times \left(\dfrac{5}{-3}\right) \times (-8)$$

161) State True or False:

$$\dfrac{7}{9} \times \dfrac{2}{11} = \dfrac{2}{11} \times \dfrac{7}{9}$$

162) Evaluate

$$\left[\left(\dfrac{5}{6}\right) + \left(\dfrac{7}{-5}\right)\right] \times \left(\dfrac{3}{2}\right)$$

163) Evaluate

$$\left[\left(\dfrac{-5}{8}\right) + \left(\dfrac{3}{-7}\right)\right] \times \left(\dfrac{5}{9}\right)$$

164) Multiply

$$\left(\dfrac{-7}{-11}\right) \times \left(\dfrac{-9}{2}\right)$$

165) Simplify

$$\left[\left(\dfrac{-2}{3}\right) + \left(\dfrac{-5}{-7}\right)\right] \times \left[\left(\dfrac{-8}{3}\right) + \left(\dfrac{7}{2}\right)\right]$$

166) Fill in the blank :

$$\left(\dfrac{16}{-3}\right) \times \left(\dfrac{5}{7}\right) = \left(\dfrac{5}{7}\right) \times \underline{}$$

167) Find the product

$$\left(\frac{-4}{7}\right) \times \left(\frac{-5}{8}\right)$$

168) Find the product

$$\left(\frac{-5}{6}\right) \times \left(\frac{-3}{5}\right)$$

169) Find the product

$$\left(\frac{7}{18}\right) \times (-12)$$

170) Find the product

$$(-6) \times \left(\frac{-5}{9}\right)$$

171) Find the product

$$\left(\frac{6}{13}\right) \times \left(\frac{-26}{36}\right)$$

172) Multiply :

$$\left(\frac{5}{24}\right) \text{ by } \left(\frac{-18}{20}\right)$$

173) Multiply :

$$\left(\frac{-19}{28}\right) \text{ by } \left(\frac{21}{-19}\right)$$

174) Multiply :

$$\left(\frac{-12}{7}\right) \text{ by } \left(\frac{9}{16}\right)$$

175) Multiply :

$$\left(\frac{-13}{7}\right) \text{ by } \left(\frac{6}{7}\right)$$

176) Multiply :

$$\left(\frac{16}{15}\right) \text{ by } \left(\frac{-25}{36}\right)$$

177) Simplify

$$\left[\left(\frac{2}{16}\right)+\left(\frac{-3}{4}\right)\right]\times\left(\frac{8}{15}\right)$$

178) Simplify

$$\left[\left(\frac{6}{-77}\right)\times\left(\frac{88}{3}\right)\right]-\left[\left(\frac{13}{55}\right)\times\left(\frac{-33}{91}\right)\right]$$

179) Simplify

$$\left[\left(\frac{-12}{7}\right)\times\left(\frac{-14}{36}\right)\right]-\left[\left(\frac{-6}{7}\right)\times\left(\frac{14}{15}\right)\right]$$

180) Simplify

$$\left[\left(\frac{7}{24}\right)\times(-48)\right]+\left[\left(\frac{-3}{4}\right)\times\left(\frac{4}{3}\right)\right]$$

181) State True or False:

$$\left[(-16)\div\left(\frac{6}{5}\right)\right]\div\left(\frac{-9}{10}\right)=(-16)\div\left[\left(\frac{6}{5}\right)\div\left(\frac{-9}{10}\right)\right]$$

182) By what rational number should we multiply $\left(\frac{-15}{56}\right)$ to get $\left(\frac{-5}{8}\right)$?

Grade 7
Volume 1
Week 10

183) Divide the sum of $\frac{72}{12}$ and $\frac{8}{3}$ by their difference.

184) State True or False:

Reciprocal of 1 is -1

185) Fill in the blank:

$\frac{32}{45} \div \frac{18}{25} =$ _____

186) The product of two rational numbers is $\left(\frac{-65}{12}\right)$. If one of the numbers is $\left(\frac{5}{12}\right)$, find the other number.

187) Fill in the blank:

$\left(\frac{21}{11}\right) \times \left(\frac{3}{7}\right) \div \left(\frac{-15}{11}\right) =$ _____

188) Divide the product of $\left(\frac{-3}{5}\right)$ and $\left(\frac{-3}{2}\right)$ by the sum of $\left(\frac{-3}{4}\right)$ and $\left(\frac{-3}{2}\right)$.

189) **By what rational number should** $\left(\frac{-55}{56}\right)$ be divided to get $\left(\frac{5}{8}\right)$?

$\left(\frac{-11}{7}\right)$

190) Fill in the blank:

$\left(\frac{32}{25}\right) \div \left(\frac{18}{45}\right) =$ _____

©All rights reserved-Math-Knots LLC., VA-USA   www.math-knots.com | www.a4ace.com

191) The product of two rational numbers is $\left(\dfrac{5}{21}\right)$. If one number is $\left(\dfrac{5}{9}\right)$, find the other number.

192) Simplify

$$\left(\dfrac{9}{4}\right) \div \left(\dfrac{7}{2}\right)$$

193) Simplify

$$(-9) \div \left(\dfrac{5}{-7}\right)$$

194) Divide $\left(\dfrac{-8}{11}\right)$ by $\left(\dfrac{7}{3}\right)$

195) Which of the following numbers are negative rational numbers?

A) $\left(\dfrac{-5}{-9}\right)$ B) $\left(\dfrac{2}{3}\right)$

C) $\left(\dfrac{-11}{15}\right)$ D) $\left(\dfrac{0}{8}\right)$

196) Which of the following numbers are positive rational numbers?

A) $\left(\dfrac{-3}{7}\right)$ B) $\left(\dfrac{-2}{-2}\right)$

C) $\left(\dfrac{4}{-9}\right)$ D) (-8)

197) Find the numerator of the below rational number:

$$\left(\dfrac{-2}{19}\right)$$

Grade 7
Volume 1
Week 10

198) Divide the sum of $\dfrac{25}{2}$ and $\dfrac{7}{3}$ by their difference.

199) Divide the sum of $\dfrac{7}{8}$ and $\dfrac{1}{3}$ by their difference.

200) Find the reciprocal of $\left(\dfrac{-1}{9}\right)$

201) Find the reciprocal of $\left(\dfrac{-6}{7}\right)$

202) Find the multiplicative inverse of $\left(\dfrac{9}{-11}\right)$

203) Find the multiplicative inverse of -7

204) Find the multiplicative inverse of $\left(\dfrac{-7}{-12}\right)$

205) By what rational number should $\left(\dfrac{-8}{9}\right)$ be divided to get $\left(\dfrac{-4}{15}\right)$?

206) The product of two rational numbers is -15. If one of the numbers is -6, what is the other number?

207) State True or False:

The quotient of two integers is always an integer.

208) State True or False:

All integers are rational numbers.

209) Express $\left(\dfrac{-12}{13}\right)$ as a rational number with numerator 48.

210) State True or False:

Every integer is a rational number.

211) State True or False:

Zero is not a rational number.

212) State True or False:

Every natural number is a rational number.

213) The quotient of two integers is always a rational number.

214) State True or False:

Every terminating decimal is a rational number.

215) State True or False:

Every repeating decimal is a rational number.

216) State True or False:

A negative rational number always lies to the right of 0 on number line.

217) State True or False:

A rational number can be represented on the number line.

218) Is -1 a rational number?

219) Which of the following numbers are negative rational numbers?

A) $\left(\dfrac{-5}{8}\right)$ B) $\left(\dfrac{-6}{-7}\right)$

C) $\left(\dfrac{1}{2}\right)$ D) $\left(\dfrac{3}{7}\right)$

220) Which of the following numbers are positive rational numbers?

A) $\left(\dfrac{-2}{3}\right)$ B) (- 5)

C) $\left(\dfrac{-3}{-7}\right)$ D) $\left(\dfrac{1}{-7}\right)$

221) Pick the positive rational number.

A) $\left(\dfrac{-4}{5}\right)$ B) $\left(\dfrac{-12}{56}\right)$

C) $\left(\dfrac{13}{-47}\right)$ D) $\left(\dfrac{-1}{-14}\right)$

222) Express $\left(\dfrac{-3}{-8}\right)$ with a positive denominator.

223) Express $\left(\dfrac{3}{-4}\right)$ with a positive denominator.

224) Express $\left(\dfrac{6}{10}\right)$ with a denominator equal to 50.

225) Express $\left(\dfrac{4}{7}\right)$ with a denominator equal to -49.

226) Express $\left(\dfrac{-5}{8}\right)$ with a denominator equal to -48.

227) Express $\left(\dfrac{-4}{15}\right)$ with a denominator equal to -135.

228) Express the following rational number in its decimal form:

$$\dfrac{122}{9}$$

229) Express 0.161616...... as a fraction.

230) Express $\frac{35}{80}$ as a decimal number.

231) Represent 0.583333........as a rational number.

232) Find the numerator of the following rational number 3.

233) Convert 0.433433433

234) Convert $\frac{5}{18}$ into its equivalent decimal form.

235) Express $\frac{2}{33}$ in its decimal form.

236) Does $\frac{7}{25}$ equals 0.27?

237) Does $\frac{7}{18}$ equals 0.383838..........?

238) State True or False:

$\frac{3}{18}$ is equal to 0.16666......

239) Express 2.564 as a rational number.

240) Express 4.05 as a rational number.

241) Convert 0.3525252........... into its equivalent rational number.

242) Convert 0.007007007...........into its equivalent rational number.

243) State True or False:

$\left(\frac{-13}{7}\right)$ and $\left(\frac{39}{-21}\right)$ are equivalent fractions.

244) State True or False:

$\left(\frac{5}{-7}\right)$ and $\left(\frac{-20}{28}\right)$ are equivalent fractions.

Grade 7
Volume 1
Week 10

245) Express $\dfrac{27}{30}$ as a decimal number.

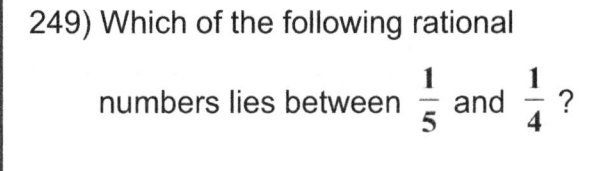

246) Convert 1.1666........as a rational number.

247) Convert 0.242424........into a rational number.

248) Express $\dfrac{13}{3}$ as a decimal number.

249) Which of the following rational numbers lies between $\dfrac{1}{5}$ and $\dfrac{1}{4}$?

A) $\dfrac{7}{25}$ B) $\dfrac{6}{25}$

C) $\dfrac{1}{45}$ D) $\dfrac{1}{35}$

250) Express 0.02222.....as a rational number.

251) Is $\dfrac{5}{7}$ greater than 0.8 ?

252) Convert $\left(\dfrac{57}{12}\right)$ into decimal form.

**Grade 7
Volume 1
Week 10**

253) Is $\frac{6}{18}$ equal to 0.3333......... ?

254) Express $\frac{71}{12}$ as a decimal number.

255) Express $\frac{3}{11}$ as a decimal number.

256) Is $\left(\frac{5}{8}\right)$ greater than 0.58?

257) Is $\left(\frac{16}{99}\right)$ greater than 0.2626......?

258) Express $\frac{25}{11}$ as a decimal number.

259) Does $\left(\frac{8}{11}\right)$ equal to 0.525?

260) Does $\left(\frac{23}{25}\right)$ equal to 0.92?

261) Which of the following is a pair of equivalent fractions?

A) $\left(\dfrac{7}{15}\right), \left(\dfrac{28}{-60}\right)$

B) $\left(\dfrac{3}{12}\right), \left(\dfrac{1}{-4}\right)$

C) $\left(\dfrac{-2}{3}\right), \left(\dfrac{3}{2}\right)$

D) $\left(\dfrac{-9}{-4}\right), \left(\dfrac{36}{16}\right)$

262) Find x such that : $\dfrac{16}{x} = -8$

263) Find the value of 'a' such that

$\left(\dfrac{1}{5}\right) = \left(\dfrac{-8}{a}\right)$

264) Pick an equivalent rational number for $\left(\dfrac{25}{35}\right)$

A) $\dfrac{50}{70}$ B) $\dfrac{35}{49}$

C) $\dfrac{5}{7}$ D) All options

265) Which of the following rational numbers is in its standard form?

A) $\left(\dfrac{-12}{39}\right)$ B) $\left(\dfrac{-13}{39}\right)$

C) $\left(\dfrac{-3}{39}\right)$ D) $\left(\dfrac{-2}{39}\right)$

266) Which of the following rational numbers is in its standard form?

A) $\left(\dfrac{-3}{-4}\right)$ B) $\left(\dfrac{-12}{24}\right)$

C) $\left(\dfrac{-17}{19}\right)$ D) $\left(\dfrac{-2}{-10}\right)$

267) Find 'a' such that :

$$\frac{16}{a} = -2$$

268) Fill in the blanks:

$$\left(\frac{36}{-24}\right) = \left(\frac{6}{}\right)$$

269) What is the standard form of the rational number $\frac{54}{75}$?

270) Is $\frac{11}{15}$ equal to $\frac{55}{95}$?

271) Which of the following rational numbers is equivalent to $\left(\frac{-6}{13}\right)$?

A) $\left(\frac{12}{23}\right)$ B) $\left(\frac{-18}{39}\right)$

C) $\left(\frac{3}{4}\right)$ D) $\left(\frac{-30}{75}\right)$

272) Reduce the rational number $\frac{36}{90}$ to its standard form.

273) Fill in the blank

$$\left(\frac{-12}{23}\right) = \left(\frac{}{138}\right)$$

274) Is $\dfrac{76}{100}$ equivalent to $\dfrac{19}{25}$?

275) Find 'x' such that

$$\left(\dfrac{-17}{24}\right) = \left(\dfrac{x}{-72}\right)$$

276) Reduce $\left(\dfrac{56}{98}\right)$ in its standard form.

277) Reduce $\left(\dfrac{-90}{126}\right)$ in its standard form.

278) Which of the following rational numbers is equivalent to $\dfrac{6}{13}$?

A) $\left(\dfrac{36}{52}\right)$ B) $\left(\dfrac{18}{52}\right)$

C) $\left(\dfrac{24}{78}\right)$ D) $\left(\dfrac{24}{52}\right)$

279) Reduce $\left(\dfrac{-20}{35}\right)$ in its standard form.

280) Fill in the blank

$$\left(\dfrac{5}{-9}\right) = \left(\dfrac{}{72}\right)$$

Grade 7
Volume 1
Week 10

281) Reduce $\dfrac{24}{42}$ in its standard form.

282) Is $\dfrac{5}{11}$ equal to $\dfrac{15}{21}$?

283) Find the equivalent form of rational number $\dfrac{3}{4}$

A) $\left(\dfrac{9}{10}\right)$ B) $\left(\dfrac{6}{12}\right)$

C) $\left(\dfrac{12}{15}\right)$ D) $\left(\dfrac{9}{12}\right)$

284) Express $\left(\dfrac{65}{35}\right)$ in standard form.

285) Express $\left(\dfrac{1}{-7}\right)$ in standard form.

286) Find the equivalent form of rational number $\dfrac{2}{5}$

A) $\left(\dfrac{5}{2}\right)$ B) $\left(\dfrac{10}{4}\right)$

C) $\left(\dfrac{4}{10}\right)$ D) $\left(\dfrac{-2}{5}\right)$

287) Reduce $\frac{72}{63}$ in its standard form.

288) Is $\frac{7}{3}$ equal to $\frac{42}{18}$?

289) Which of the following rational numbers is equivalent to $\frac{7}{11}$?

A) $\left(\frac{7}{22}\right)$ B) $\left(\frac{11}{7}\right)$

C) $\left(\frac{21}{33}\right)$ D) $\left(\frac{33}{21}\right)$

290) Fill in the missing numeral

$$\left(\frac{7}{8}\right) = \left(\frac{}{40}\right)$$

291) State True or False:

$\left(\frac{2}{7}\right)$ and $\left(\frac{6}{21}\right)$ are equivalent rational numbers.

292) Pick a pair of equivalent rational numbers.

A) $\frac{3}{8}, \frac{3}{4}$ B) $\frac{2}{7}, \frac{7}{2}$

C) $\frac{5}{2}, \frac{10}{4}$ D) $\frac{5}{2}, \frac{4}{10}$

293) Fill in the missing numeral

$$\left(\frac{5}{-9}\right) = \left(\frac{}{54}\right)$$

294) Pick an equivalent rational number for $\left(\frac{-11}{24}\right)$

295) Pick an equivalent rational number for $\left(\frac{7}{10}\right)$

A) $\left(\frac{27}{20}\right)$ B) $\left(\frac{28}{40}\right)$

C) $\left(\frac{21}{100}\right)$ D) $\left(\frac{-21}{30}\right)$

296) Pick an equivalent rational number for $\left(\frac{-9}{15}\right)$

A) $\left(\frac{-18}{85}\right)$ B) $\left(\frac{-3}{7}\right)$

C) $\left(\frac{-3}{5}\right)$ D) $\left(\frac{-27}{42}\right)$

297) Pick an equivalent rational number for $\left(\frac{3}{14}\right)$

A) $\left(\frac{-3}{14}\right)$ B) $\left(\frac{3}{-14}\right)$

C) $\frac{3}{4}$ D) $\left(\frac{-3}{-14}\right)$

298) Which of the following rational numbers is in its standard form?

A) $\left(\dfrac{25}{45}\right)$ B) $\left(\dfrac{15}{50}\right)$

C) $\left(\dfrac{-12}{15}\right)$ D) $\left(\dfrac{12}{25}\right)$

299) State True or False:

The standard form of $\left(\dfrac{-36}{-48}\right)$ is $\left(\dfrac{3}{4}\right)$

Grade 7
Volume 1
Week 11

Grade 7
Volume 1
Week 11

1) Convert the below fraction to a percentage.

$$8\frac{7}{10}$$

A) 876 %   B) 8.7 %

C) 870 %   D) 880 %

2) Convert the below fraction to a percentage.

$$\frac{28}{33}$$

A) 848.$\overline{48}$ %   B) 0.$\overline{84}$ %

C) 84.$\overline{84}$ %   D) 28.33 %

3) Convert the below decimal to a percentage.

0.0006

A) 3.5 %   B) 0.0006 %

C) 0.06 %   D) 3333.$\overline{3}$ %

4) Convert the below decimal to a percentage.

0.65

A) 0.65 %   B) 72 %

C) 65 %   D) 13.2 %

5) Convert the below fraction to a percentage.

$$\frac{1}{100}$$

A) 1 %   B) 0.01 %

C) 0.11 %   D) 31 %

6) Convert the below fraction to a percentage.

$$\frac{1}{111}$$

A) 1.111 %   B) 0.$\overline{900}$ %

C) 9.$\overline{009}$ %   D) 0.$\overline{009}$ %

7) Convert the below decimal to a percentage.

0.02

A) 11 %   B) 0.02 %

C) 1 %   D) 2 %

8) Convert the below decimal to a percentage.

0.0024

A) 30.24 %   B) 3.125 %

C) 0.24 %   D) 0.0024 %

©All rights reserved-Math-Knots LLC., VA-USA   www.math-knots.com | www.a4ace.com

Grade 7
Volume 1
Week 11

9) 158 hours is what percent of 456.4 hours?

A) 1.08%   B) 137.2%

C) 34.6%   D) 1.37%

10) 48% of what is 322 inches?

A) 670.8 inches
B) 15456 inches
C) 175.4 inches
D) 154.6 inches

11) What is 35% of 447.2 cm?

A) 3040 cm   B) 47.5 cm

C) 156.5 cm   D) 1277.7 cm

12) 284 miles is what percent of 428.4 miles?

A) 0.66%   B) 66.3%

C) 150.8%   D) 1.51%

13) What is 86% of 174 hours?

A) 9331 hours   B) 149.6 hours

C) 504.7 hours   D) 202.3 hours

14) 90% of what is $140?

A) $126   B) $12600

C) $3892.50   D) $155.56

15) 436 ft is 85% of what?

A) 370.6 ft   B) 37060 ft

C) 512.9 ft   D) 305.9 ft

16) What percent of 480 minutes is 256 minutes?

A) 1.88%   B) 53.3%

C) 187.5%   D) 109.6%

17) 210% of what is 472.8 cm?

   A) 99288 cm   B) 146.2 cm

   C) 14620 cm   D) 225.1 cm

18) 65 miles is what percent of 90 miles?

   A) 139.6%   B) 138.5%

   C) 72.2%   D) 1.4%

19) What percent of 218 cm is 308.7 cm?

   A) 141.6%   B) 0.36%

   C) 275.6%   D) 2.8%

20) 54 minutes is what percent of 398 minutes?

   A) 1.07%   B) 13.6%

   C) 1.29%   D) 106.7%

21) Find the percentage change from $88 to $128. Round to the nearest tenth.

   A) 31.3% decrease
   B) 145.5% increase
   C) 45.5% increase
   D) 31.3% increase

22) Find the percentage change from 146 km to 80 km. Round to the nearest tenth.

   A) 82.5% decrease
   B) 45.2% decrease
   C) 82.5% increase
   D) 6% decrease

23) Find the percentage change from $14 to $68. Round to the nearest tenth.

   A) 385.7% increase
   B) 485.7% increase
   C) 54% increase
   D) 179.4% increase

24) Find the percentage change from 50 miles to 96 miles. Round to the nearest tenth.

   A) 46% increase
   B) 192% increase
   C) 92% decrease
   D) 92% increase

Grade 7
Volume 1
Week 11

25) Find the percentage change from 96 inches to 24 inches. Round to the nearest tenth.

A) 75% decrease

B) 400% decrease

C) 300% increase

D) 79.6% increase

26) Find the percentage change from 68 m to 64 m. Round to the nearest tenth.

A) 4% increase

B) 94.1% decrease

C) 5.9% decrease

D) 79.9% increase

27) The Original price of a sweater is $5.50. The store offers a discount of 15%. Find the selling price.

A) $5.22   B) $0.82

C) $4.67   D) $4.68

28) The Original price of a shoes is $99.95. The store offers a discount of 30%. Find the selling price.

A) $129.94   B) $29.98

C) $94.95   D) $69.97

29) The Original price of a book is $7.95. The store offers a discount of 40%. Find the selling price.

A) $4.77   B) $11.13

C) $7.55   D) $6.36

30) The Original price of a sweater is $5,500. The store charges a tax of 2%. Find the selling price.

A) $5,225.00   B) $5,610.00

C) $5,390.00   D) $4,675.00

31) The Original price of a parrot is $159.99. The store offers a discount of 36%. Find the selling price.

A) $143.99   B) $217.59

C) $102.39   D) $57.60

32) The Original price of a sweater is $99.50. The store offers a discount of 50%. Find the selling price.

A) $49.75   B) $149.25

C) $79.60   D) $109.45

©All rights reserved-Math-Knots LLC., VA-USA   www.math-knots.com | www.a4ace.com

Grade 7
Volume 1
Week 11

33) The Original price of a book is $32.95. The store charges a tax of 1%. Find the selling price.

A) $37.89   B) $33.28

C) $0.33   D) $39.54

34) The Original price of a sweater is $9.99. The store charges a tax of 4%. Find the selling price.

A) $10.39   B) $11.99

C) $9.59   D) $0.40

35) The Original price of a car is $26,800. The store charges a tax of 1%. Find the selling price.

A) $27,068.00   B) $268.00

C) $26,532.00   D) $25,460.00

36) The Original price of a microphone is $30. The store marked up the price by 50%. Find the selling price.

A) $15.00   B) $33.00

C) $34.50   D) $45.00

37) The Original price of a CD is $21.50. The store marked up the price by 40%. Find the selling price.

A) $25.80   B) $12.90

C) $8.60   D) $30.10

38) The Original price of a sweater is $49.99. The store marked up the price by 50%. Find the selling price.

A) $25.00   B) $57.49

C) $74.99   D) $74.98

39) The Original price of a sweater is $39.95. The store charges a tax of 1%. Find the selling price.

A) $40.35   B) $33.96

C) $45.94   D) $0.40

40) The Original price of a calendar is $10.50. The store marked up the price by 70%. Find the selling price.

A) $12.07   B) $17.85

C) $7.35   D) $3.15

©All rights reserved-Math-Knots LLC., VA-USA   www.math-knots.com | www.a4ace.com

41) The Original price of a pen is $2.95. The store marked up the price by 90%. Find the selling price.

A) $2.66   B) $5.61

C) $3.10   D) $0.29

45) Write the below number in scientific notation.
59000000

A) 0.59×10^7 B) 0.59×10^{-1}

C) 5.9×10^7 D) 0.59×10^{-2}

42) 24% of what is $401?

A) $49.53   B) $63.80

C) $9624   D) $1670.83

46) Write the below number in scientific notation.
5400000

A) 5.4×10^{-6} B) 54×10^6

C) 5.4×10^6 D) 54×10^{-6}

43) 92% of 56 cm is what?

A) 5152 cm   B) 236.1 cm

C) 34960 cm   D) 51.5 cm

47) 78% of what is $343?

A) $439.74   B) $59.25

C) $4997.30   D) $49.97

44) An investment of $51,000 at simple interest of 7% for 5 years is made at a bank. Find the ending balance.

A) $17,850.00   B) $71,530.14

C) $68,850.00   D) $54,570.00

48) An investment of $10,000 at simple interest of 12% for 2 years is made at a bank. Find the ending balance.

A) $2,400.00   B) $12,544.00

C) $1,200.00   D) $12,400.00

49) Write the below number in scientific notation.

6400000

A) 6.4×10^2 B) 64×10^2

C) 6.4×10^{-6} D) 6.4×10^6

50) Write the below number in scientific notation.

56800000

A) 5.68×10^3 B) 5.68×10^{-3}

C) 5.68×10^7 D) 5.68×10^{-5}

51) Write the below number in standard notation.

8.1×10^{-3}

A) 0.081   B) 0.0081

C) 8.1   D) 0.81

52) Write the below number in standard notation.

5.3×10^0

A) 0.053   B) 5.3

C) 53   D) 0.53

53) An investment of $335 at simple interest of 11% for 2 years is made at a bank. Find the ending balance.

A) $408.70   B) $371.85

C) $412.75   D) $73.70

54) An investment of $53,100 at simple interest of 4% for 3 years is made at a bank. Find the ending balance.

A) $59,730.28   B) $6,372.00

C) $59,472.00   D) $55,224.00

55) Write the below number in standard notation.

7.98×10^{-4}

A) 0.0798   B) 798000000

C) 0.00798   D) 0.000798

56) Write the below number in standard notation.

5.09×10^5

A) 5090   B) 5090000

C) 509000   D) 0.00509

57) An investment of $315 at simple interest of 5% for 2 years is made at a bank. Find the ending balance.

A) $330.75 B) $347.29

C) $31.50 D) $346.50

58) An investment of $50,400 at simple interest of 3% for 4 years is made at a bank. Find the ending balance.

A) $56,725.64 B) $51,912.00

C) $56,448.00 D) $6,325.64

59) Pick the greater rational number.

A) $\left(\dfrac{-4}{7}\right)$ B) 4

60) State True or False:

$\left(\dfrac{-15}{7}\right)$ is less than $\left(\dfrac{15}{7}\right)$

61) State True or False:

$\left(\dfrac{21}{6}\right)$ lies to the right of $\left(\dfrac{21}{-6}\right)$ on the number line.

62) Compare the following rational numbers using the correct symbol

$\left(\dfrac{6}{-7}\right)$ _____ 0

63) State True or False:

$\left(\dfrac{-35}{-91}\right)$ lies to the left of 0 on the number line.

64) Arrange the following rational numbers in descending order:

$\left(\dfrac{11}{-25}\right)$, $\left(\dfrac{-8}{75}\right)$, $\left(\dfrac{9}{-10}\right)$, $\left(\dfrac{-7}{15}\right)$

Grade 7
Volume 1
Week 11

65) Evaluate

$$\left(\frac{3}{7}\right)+\left(\frac{7}{3}\right)+\left(\frac{-11}{7}\right)+\left(\frac{-2}{3}\right)$$

66) Evaluate

$$\left(\frac{2}{-9}\right)+\left(\frac{-7}{8}\right)+\left(\frac{5}{3}\right)+\left(\frac{-5}{16}\right)$$

67) Fill in the blank with appropriate symbol

$$\left(\frac{5}{13}\right) \underline{\qquad} \left(\frac{-8}{13}\right)$$

68) Fill in the blank with appropriate symbol

$$\left(\frac{-5}{5}\right) \underline{\qquad} \left(\frac{5}{5}\right)$$

69) Evaluate:

$$\left(\frac{11}{18}\right)+\left(\frac{-5}{18}\right)$$

70) Find the sum of:

$$\frac{19}{6}+\frac{3}{7}+\left(\frac{-1}{8}\right)$$

71) Which of the following rational numbers lies between $\left(\frac{2}{5}\right)$ and $\left(\frac{4}{5}\right)$?

A) $\left(\frac{7}{5}\right)$

B) $\left(\frac{1}{5}\right)$

C) $\left(\frac{3}{5}\right)$

D) $\left(\frac{9}{5}\right)$

©All rights reserved-Math-Knots LLC., VA-USA   www.math-knots.com | www.a4ace.com

72) Find the difference:

$$\left(\frac{-5}{21}\right) - \left(\frac{7}{18}\right)$$

73) What rational number should be added to $\left(\frac{-4}{7}\right)$ so as to get $\left(\frac{-6}{35}\right)$?

74) Simplify

$$\left(\frac{-8}{15}\right) + \left(\frac{2}{-5}\right)$$

75) Add

$$\left(\frac{-3}{8}\right) + \left(\frac{-5}{12}\right) + \left(\frac{-1}{6}\right)$$

76) The sum of two rational numbers is $\left(\frac{-17}{8}\right)$. If one of the numbers is $\left(\frac{-15}{4}\right)$ find the other.

77) What number should be added to $\left(\frac{-3}{5}\right)$ so as to get $\left(\frac{-8}{9}\right)$?

78) Daniel used $\frac{11}{2}$ ft green color cloth, $\frac{12}{5}$ ft orange color cloth and $\frac{17}{4}$ ft white color cloth to stitch a doll house. How much cloth did he use in all to make the doll house?

Grade 7
Volume 1
Week 11

79) The weight of a full jam bottle is $\frac{4}{3}$ pounds and the weight of an empty jam bottle is $\frac{1}{5}$ pounds. What is the weight/quantity of jam in the bottle?

80) The sum of two rational numbers is $\left(\frac{11}{3}\right)$. If one of the numbers is $\left(\frac{-11}{15}\right)$, find the other number.

81) Find the additive inverse of $\left(\frac{16}{-21}\right)$

82) Subtract $\left(\frac{4}{7}\right)$ from $\left(\frac{3}{4}\right)$

83) Find the multiplicative inverse of $\frac{15}{17}$

84) Multiply $\frac{4}{5}$ by $\frac{7}{3}$

85) Fill in the blank

$$\frac{84}{(-147)} = \left(\frac{4}{\_}\right)$$

86) Simplify :

$$\left[\left(\frac{36}{7}\right) \times \left(\frac{14}{-9}\right)\right] + \left[\left(\frac{26}{-125}\right) \times \left(\frac{100}{39}\right)\right]$$

87) Convert $\dfrac{3}{12}$ into a decimal number.

88) Represent 0.583333....... as a rational number.

89) Find the denominator of the rational number 0

90) State True or False:

Every rational number can be expressed as a repeating decimal.

91) Fill in the blank

$$\left(\dfrac{-12}{13}\right) = \left(\dfrac{}{52}\right)$$

92) By what rational number should $\dfrac{25}{8}$ be divided to get $\dfrac{5}{2}$?

93) State True or False:

$$\dfrac{5}{13} \div \dfrac{25}{26} = \dfrac{25}{26} \div \dfrac{5}{13}$$

94) Fill in the blank

$$\dfrac{49}{4} \div \underline{} = \dfrac{7}{2}$$

95) Which of the following rational numbers is in its standard form?

A) $\left(\dfrac{-3}{28}\right)$ B) $\left(\dfrac{-6}{14}\right)$

C) $\left(\dfrac{8}{36}\right)$ D) $\left(\dfrac{9}{42}\right)$

96) Which of the following rational numbers is in its standard form?

A) $\left(\dfrac{-2}{12}\right)$ B) $\left(\dfrac{9}{19}\right)$

C) $\left(\dfrac{6}{16}\right)$ D) $\left(\dfrac{-5}{15}\right)$

Grade 7
Volume 1
Week 12

Grade 7
Volume 1
Week 12

1) Evaluate the below expression

$15 - 12 + 6 + 17 - 8$

A) 18   B) 32

C) 23   D) 30

2) Evaluate the below expression

$19 - 15 \div (12 - (17 - 8))$

A) 2   B) 5

C) 1   D) 14

3) Evaluate the below expression

$(9)(16) - (15 + 2^3)$

A) 129   B) 121

C) 109   D) 134

4) Evaluate the below expression

$(((5)(2))(2)) \div 2$

A) 1   B) 22

C) 10   D) 2

5) Evaluate the below expression

$16 - 2 + 7 - (8 + 3)$

A) 17   B) 10

C) 22   D) 7

6) Evaluate the below expression

$(28 - 10) \div 2 + (7)(4)$

A) 40   B) 49

C) 38   D) 37

7) Evaluate the below expression

$((26)(2)) \div (1 + 15 - 12)$

A) 8   B) 13

C) 25   D) 15

8) Evaluate the below expression

$(2)(16) - 22 \div 2 + 12$

A) 34   B) 13

C) 23   D) 33

9) Evaluate the below expression

$$(11)(14-10)-(19-11)$$

A) 28 B) 36

C) 37 D) 50

10) Evaluate the below expression

$$((14+16-15) \div 5)(16)$$

A) 51 B) 50

C) 48 D) 34

11) Evaluate the below expression

$$(37+4+1) \div (5+9)$$

A) 22 B) 3

C) 20 D) 7

12) Evaluate the below expression

$$(24-11) \div (6+12-5)$$

A) 10 B) 1

C) 7 D) 17

13) Evaluate the below expression

$$13-(1+(5)(2))+16$$

A) 18 B) 10

C) 22 D) 2

14) Evaluate the below expression

$$14+((14)(2)) \div 4 - 3$$

A) 15 B) 10

C) 18 D) 31

15) Evaluate the below expression

$$((35)(2)) \div 5$$

A) 29 B) 3

C) 16 D) 14

16) Evaluate the below expression

$$8^2 - (36 \div (19-17) - 17)$$

A) 51 B) 69

C) 63 D) 46

Grade 7
Volume 1
Week 12

17) Evaluate the below expression

$(18) (((36) (2)) \div (16 + 3 - 7))$

A) 104   B) 97

C) 107   D) 108

18) Evaluate the below expression

$16 + 19 - 54 \div 18 + 6 - 1$

A) 37   B) 33

C) 38   D) 43

19) Evaluate the below expression

$((2 + 1) (2) + 32 - 12) \div 2$

A) 6   B) 13

C) 25   D) 9

20) Evaluate the below expression

$(18 - (3) (3)) \div (20 - (7 + 10))$

A) 3   B) 6

C) 8   D) 20

21) Evaluate the below expression

$(2 + 34) \div (8 + 14 - 19 - 1)$

A) 13   B) 7

C) 18   D) 32

22) Evaluate the below expression

$(9 - 7)^2 + 4 - 4 + 8$

A) 15   B) 12

C) 29   D) 24

23) Evaluate the below expression

$(41 + 25) \div (6 + (14) (6 - 6))$

A) 11   B) 16

C) 23   D) 21

24) Evaluate the below expression

$(22 - 19 + 1 + 1 + 8) \div 13$

A) 1   B) 6

C) 15   D) 21

25) Evaluate the below expression

$$16b + 2 + 6 + 24b$$

A) $40b + 8$ B) $36b + 7$

C) $9b + 7$ D) $5b + 7$

26) Evaluate the below expression

$$-29b - 7b$$

A) $68b$ B) $-36b$

C) $33b$ D) $27b$

27) Evaluate the below expression

$$k + 50 + 44k + 44$$

A) $45k + 94$ B) $65k + 94$

C) $16k + 36$ D) $29k + 94$

28) Evaluate the below expression

$$35v + 11 + 45v - 26$$

A) $4v$ B) $v - 21$

C) $80v - 15$ D) $-15v$

29) Evaluate the below expression

$$5n + 22n$$

A) $-72n$ B) $-44n$

C) $27n$ D) $-70n$

30) Evaluate the below expression

$$48n + 16n$$

A) $-26n - 9$ B) $44n - 4$

C) $96n$ D) $64n$

31) Evaluate the below expression

$$15n + 43 + n - 48$$

A) $16n - 5$ B) $-54 - 48n$

C) $-19n$ D) $-54 - 35n$

32) Evaluate the below expression

$$50k + 4 - 22 - 26k$$

A) $24k - 18$ B) $-10k - 18$

C) $-22k$ D) $-50k$

33) Evaluate the below expression

$$19 + 42k - 16$$

A) $3 + 42k$ B) $-4k + 6$

C) $53k - 11$ D) $23k - 11$

34) Evaluate the below expression

$$-6n + 11n$$

A) $35 + 33n$ B) $-29n$

C) $5n$ D) $n - 47$

35) Evaluate the below expression

$$-49n + 9 + 15$$

A) $31n + 20$ B) $-55n + 24$

C) $-49n + 24$ D) $-15n + 20$

36) Evaluate the below expression

$$8x + 12x$$

A) $20x$ B) $47x$

C) $41x$ D) $86x$

37) Evaluate the below expression

$$x - 35 + 40x$$

A) $-15 + 50x$ B) $41x - 35$

C) $-15 + 7x$ D) $13 - 13x$

38) Evaluate the below expression

$$22 + 21r + r + 21$$

A) $43 + 22r$ B) $36r$

C) $18r$ D) $20r$

39) Evaluate the below expression

$$-44v - 26v$$

A) $-9v$ B) $33v$

C) $-48v$ D) $-70v$

40) Evaluate the below expression

$$-11n + 33n$$

A) $22n$ B) $21n$

C) $41n + 27$ D) $33n$

Grade 7
Volume 1
Week 12

41) Evaluate the below expression

$$-34r + r$$

A) $3r + 27$ B) $33r + 27$

C) $-33r$ D) $-8r + 27$

42) Evaluate the below expression

$$-30p + 34p$$

A) $-10p - 59$ B) $p + 28$

C) $23p$ D) $4p$

43) Evaluate the below expression

$$15p - 32p$$

A) $-17p$ B) $2p$

C) $15p + 3$ D) $51p + 3$

44) Evaluate the below expression

$$-39n + 25n$$

A) $-19n$ B) $10n$

C) $-14n$ D) $-31n$

45) Evaluate the below expression

$$47 - 48m + 47m + 12$$

A) $59 + 61m$ B) $59 + 48m$

C) m D) $59 - m$

46) Evaluate the below expression

$$-2a - 30a$$

A) $50a$ B) $65a$

C) $27a$ D) $-32a$

47) Evaluate the below expression

$$-31m - 20m$$

A) $19m + 13$ B) $26m + 13$

C) $48m + 13$ D) $-51m$

48) Evaluate the below expression

$$m + 40 + m - 17$$

A) $20m - 36$ B) $49 + 47m$

C) $-14m - 36$ D) $2m + 23$

©All rights reserved-Math-Knots LLC., VA-USA   www.math-knots.com | www.a4ace.com

Grade 7
Volume 1
Week 12

49) Evaluate the below expression

$$-49a + 29a$$

A) $23a$ B) $16a$

C) $-20a$ D) $62a$

50) Evaluate the below expression

$$\frac{99}{14}b + \frac{43}{8}b$$

A) $-\frac{3}{5}b - \frac{145}{88}$ B) $\frac{94}{11}b - \frac{4}{3}$

C) $\frac{697}{56}b$ D) $2 + \frac{235}{42}b$

51) Evaluate the below expression

$$-13m + \frac{4}{3} + \frac{7}{6}m$$

A) $\frac{45}{7}m - 2$ B) $-\frac{23}{2}m + \frac{4}{3}$

C) $-\frac{71}{6}m + \frac{4}{3}$ D) $-\frac{27}{2}m + \frac{4}{3}$

52) Evaluate the below expression

$$\frac{43}{6}b + \frac{2}{3} - 3\frac{2}{3}$$

A) $\frac{43}{6}b - 3$ B) $-3 + \frac{26}{3}b$

C) $-3 + \frac{151}{24}b$ D) $-\frac{3}{5}b - \frac{145}{88}$

53) Evaluate the below expression

$$-\frac{3}{13}a + \frac{38}{5} + \frac{27}{10}a$$

A) $\frac{433}{40}a$ B) $\frac{66}{5}a$

C) $\frac{321}{130}a + \frac{38}{5}$ D) $\frac{1097}{260}a + \frac{38}{5}$

54) Evaluate the below expression

$$-2m - \frac{5}{3} - \frac{23}{12}m - 2$$

A) $-\frac{23}{12}m - \frac{11}{3}$

B) $-\frac{47}{12}m - \frac{11}{3}$

C) $\frac{1187}{180}m - \frac{11}{3}$

D) $-\frac{31}{60}m - \frac{11}{3}$

Grade 7
Volume 1
Week 12

55) Evaluate the below expression

$$\frac{5}{4}n - \frac{17}{10}n$$

A) $-\frac{9}{20}n$ B) $\frac{183}{44}n + \frac{13}{6}$

C) $-\frac{15}{2}n$ D) $-\frac{7}{4}n + \frac{13}{6}$

58) Evaluate the below expression

$$n + \frac{4}{7} + \frac{3}{4}n + \frac{5}{3}$$

A) $\frac{2}{3} - \frac{146}{77}n$ B) $\frac{59}{9} + \frac{299}{42}n$

C) $\frac{7}{4}n + \frac{47}{21}$ D) $\frac{59}{9} + \frac{25}{14}n$

56) Evaluate the below expression

$$\frac{22}{3}n + 1 - \frac{17}{5}n + \frac{16}{13}$$

A) $\frac{29}{13} + \frac{499}{165}n$ B) $-\frac{106}{9}n$

C) $\frac{29}{13} + \frac{59}{15}n$ D) $\frac{29}{13} + \frac{1831}{660}n$

59) Evaluate the below expression

$$\frac{1}{6}n + \frac{5}{8} - \frac{1}{6}n$$

A) $\frac{5}{8}$ B) $\frac{109}{14}n - \frac{19}{12}$

C) $2n + \frac{5}{8}$ D) $n - \frac{19}{12}$

57) Evaluate the below expression

$$\frac{4}{5}x + \frac{2}{3}x$$

A) $\frac{97}{10}x$ B) $\frac{451}{195}x$

C) $\frac{22}{15}x$ D) $\frac{2986}{195}x$

60) Evaluate the below expression

$$7x - 3x$$

A) $\frac{247}{56}x$ B) $\frac{59}{5}x$

C) $\frac{29}{10} + \frac{129}{52}x$ D) $4x$

61) Evaluate the below expression

$$-\frac{1}{2}x + 1 - \frac{3}{2}$$

A) $\frac{7}{5}x - \frac{4}{5}$ B) $\frac{29}{8}x - \frac{1}{2}$

C) $-\frac{1}{2} + 4x$ D) $-\frac{1}{2} - \frac{1}{2}x$

62) Evaluate the below expression

$$\frac{16}{9}n - \frac{4}{7} - \frac{55}{14}n + \frac{2}{5}$$

A) $\frac{516}{65}n + \frac{103}{30}$ B) $\frac{667}{84}n - \frac{20}{9}$

C) $-\frac{271}{126}n - \frac{6}{35}$ D) $\frac{927}{110}n$

63) Evaluate the below expression

$$x - \frac{3}{4} - x$$

A) $\frac{15}{4}x$ B) $-\frac{3}{4}$

C) $\frac{13}{2}x$ D) $\frac{51}{8}x$

64) Evaluate the below expression

$$\frac{4}{3}n - 2n$$

A) $-\frac{2}{3}n$ B) $\frac{8}{13}n - \frac{34}{9}$

C) $\frac{1}{12}n + \frac{109}{56}$ D) $-\frac{1}{2}n + \frac{109}{56}$

65) Evaluate the below expression

$$-\frac{17}{9}a + a$$

A) $-\frac{134}{9}a$ B) $-\frac{122}{9}a$

C) $-\frac{8}{9}a$ D) $-\frac{1039}{72}a$

66) Evaluate the below expression

$$\frac{49}{11}n + \frac{25}{6} + \frac{10}{11}n$$

A) $\frac{59}{11}n + \frac{25}{6}$ B) $\frac{1029}{143}n$

C) $\frac{283}{35}n$ D) $\frac{577}{44}n + \frac{25}{6}$

67) Evaluate the below expression

$$\frac{1}{3}x + \frac{51}{4} - \frac{1}{2}x$$

A) $-\frac{1}{6}x + \frac{51}{4}$ B) $\frac{611}{42}x + \frac{51}{4}$

C) $\frac{23}{42}x + \frac{51}{4}$ D) $\frac{82}{7}x + \frac{51}{4}$

68) Evaluate the below expression

$$\frac{5}{6}x + \frac{22}{3} + \frac{7}{13}x + \frac{5}{3}$$

A) $\frac{5}{36}x$ B) $9 + \frac{268}{39}x$

C) $-\frac{1}{9}x + \frac{71}{12}$ D) $\frac{107}{78}x + 9$

69) Evaluate the below expression
$$5.2n + 7.6 - 4.2n$$

A) $6.8n + 3.6$

B) $4.648n + 7.6$

C) $n + 7.6$

D) $-5.434n + 7.6$

70) Evaluate the below expression

$$-\frac{1}{3}v - \frac{14}{9}v$$

A) $-\frac{166}{65}v$ B) $-\frac{353}{520}v$

C) $\frac{232}{63}v$ D) $-\frac{17}{9}v$

71) Evaluate the below expression
$$3a - 10.7a$$

A) $1.73a$ B) $-7.7a$

C) $4.63a$ D) $9.5a$

72) Evaluate the below expression
$$2.4p - 11.4p$$

A) $-9p$ B) $7 - 9.8p$

C) $-3.92p$ D) $6.9p$

73) Evaluate the below expression

$-0.1n + 1.9n$

A) $1.8n$ B) $-0.5 - 1.1n$

C) $12.1n$ D) $3.9n$

74) Evaluate the below expression

$5.5a + 0.2a$

A) $a - 4$ B) $5.7a$

C) $-8.4a - 4$ D) $7a$

75) Evaluate the below expression

$-2.4n - 9.2 - 8.9n + 2.9$

A) $13.241n - 10.5$

B) $21.13n - 10.5$

C) $11.13n - 10.5$

D) $-11.3n - 6.3$

76) Evaluate the below expression

$7v + 3.6v$

A) $22.53v + 0.2$ B) $10.6v$

C) $13.1v + 0.2$ D) $17.4v$

77) Evaluate the below expression

$1.3x + 11x$

A) $8.2x - 4.5$ B) $-11.9x$

C) $12.3x$ D) $x - 4.5$

78) Evaluate the below expression

$-0.6b - 10.5b$

A) $-3.8 + 7.5b$ B) $-11.1b$

C) $-0.7b + 9.4$ D) $0.5b$

79) Evaluate the below expression

$1 - 8.6p + 11.4p$

A) $1.7 - 5.23p$ B) $1.7 - 2.1p$

C) $1.7 + 4.47p$ D) $1 + 2.8p$

80) Evaluate the below expression

$-10.9k + 9.8k$

A) $-1.1k$ B) $-12.3k$

C) $-3.1k$ D) $-13.1k$

81) Evaluate the below expression

$$-9.1p - 6.3p$$

A) $-20.1\,p$ B) $-15.4\,p$

C) $-11.7\,p$ D) $-9.2\,p$

82) Evaluate the below expression

$$4.45r + 4.8r$$

A) $9.25r$ B) $14.71r$

C) $0.75r$ D) $-12.9r$

83) Evaluate the below expression

$$10.68x + 10.9x$$

A) $2.3x - 12.2$

B) $6.8x - 12.2$

C) $21.58x$

D) $13.8x - 12.2$

84) Evaluate the below expression

$$6.4 + 6.48x + 11.6x + 11.9$$

A) $3.2 - 12.1x$

B) $18.3 + 6.78x$

C) $18.3 + 18.08x$

D) $-10.3x$

85) Evaluate the below expression

$$b + 7.3 + b - 11.8$$

A) $2b - 4.5$ B) b

C) $0.956b$ D) $6.8b$

86) Evaluate the below expression

$$7.9b + 7.97b$$

A) $20.1b$ B) $15.87b$

C) $12.5b$ D) $23.7b$

87) Evaluate the below expression

$$6.77 - 1.9p - 8.4p$$

A) $-15.2\,p$ B) $-7.2\,p$

C) $0.5\,p$ D) $6.77 - 10.3\,p$

88) Evaluate the below expression

$$11.3a + 0.7a$$

A) $-3.4a$

B) $10.08a + 6.9$

C) $12.83a + 6.8$

D) $12a$

89) Evaluate the below expression

$5x + 11.3x$

A) $-9.18x$ B) $16.3x$

C) $-5.3x$ D) $-11.58x$

93) Evaluate the below expression.

$(44 \div 11)(22 + 48 \div 12)$

A) 98   B) 117

C) 104   D) 113

90) Evaluate the below expression.

$(6)(30) + 28 + (9)(4)$

A) 254   B) 252

C) 244   D) 222

94) Evaluate the below expression.

$28 - (27 + (5)(3)) \div 21$

A) 6   B) 22

C) 26   D) 42

91) Evaluate the below expression.

$((72)(2)) \div (23 + 4 - 9)$

A) 8   B) 2

C) 24   D) 25

95) Evaluate the below expression.

$27 + 11 - (17 - 9 + 17)$

A) 37   B) 13

C) 27   D) 28

92) Simplify the below expression

$-21p - 50p$

A) $-71p$ B) $27 - 39p$

C) $14p - 7$ D) $-12p - 7$

96) Simplify the below expression

$-35x - 17x$

A) $-19x$ B) $47 + 14x$

C) $-52x$ D) $32 + 14x$

97) Simplify the below expression

$28v - 50v$

A) $24v$ B) $-22v$

C) $81 + 57v$ D) $-61v$

98) Simplify the below expression

$20m - 18 + m - 20$

A) $21m - 38$ B) $-43m - 35$

C) $-63m$ D) $86m$

99) Simplify the below expression

$1 + 45a - 48$

A) $-26a - 79$ B) $a + 80$

C) $-47 + 45a$ D) $23a - 79$

100) Simplify the below expression

$-4n + 11n$

A) $7n$ B) $88n$

C) $46n$ D) $62n$

101) Simplify the below expression

$24b - 39b$

A) $-5b$ B) $-30 + 20b$

C) $-15b$ D) $-30 - 17b$

102) Simplify the below expression

$1 - 12n - 22n$

A) $54n$ B) $93n$

C) $1 - 34n$ D) $22n$

103) Simplify the below expression

$11k - 22k$

A) $28k$ B) $-9k$

C) $-11k$ D) $39k$

104) Simplify the below expression

$8a - 26 + 38a - 8$

A) $46a - 34$ B) $51a$

C) $47a$ D) $20a$

105) Simplify the below expression

$-3x - 27x$

A) $-30x$ B) $-44x$

C) $-69x$ D) $-85x$

106) Simplify the below expression

$-39v - 41 + 25$

A) $-99v - 16$ B) $-39v - 16$

C) $-60v - 16$ D) $-72v - 16$

107) Simplify the below expression

$37n - 28n$

A) 0 B) $9n$

C) $-25n$ D) $8n - 23$

108) Simplify the below expression

$\frac{41}{10}x + \frac{33}{10}x$

A) $\frac{37}{5}x$ B) $\frac{47}{5}x$

C) $\frac{1391}{90}x$ D) $\frac{139}{10}x$

109) Simplify the below expression

$42r - 30r$

A) $22r$ B) $35r$

C) $12r$ D) 0

110) Simplify the below expression

$-26p - 3p$

A) $54p$ B) $93p$

C) $-29p$ D) $-70p$

111) Simplify the below expression

$\frac{27}{10}n + \frac{17}{6} + \frac{2}{3}$

A) $\frac{27}{10}n + \frac{7}{2}$ B) $\frac{319}{180}n$

C) $\frac{9}{5}n$ D) $\frac{91}{45}n$

112) Simplify the below expression

$-\frac{13}{5}v + \frac{5}{2}v$

A) $\frac{3}{10}v$ B) $-\frac{1}{10}v$

C) $\frac{12}{5}v$ D) $-\frac{1}{5}v$

113) Simplify the below expression

$$\frac{5}{4}a + \frac{25}{6}a$$

A) $\frac{35}{8}a$ B) $\frac{25}{84}a$

C) $\frac{65}{12}a$ D) $-\frac{99}{28}a$

114) Simplify the below expression

$$-2x + \frac{3}{4}x$$

A) $\frac{199}{90}x - \frac{91}{36}$ B) $-\frac{5}{4}x$

C) $\frac{11}{18}x - \frac{91}{36}$ D) $-\frac{97}{40}x + \frac{11}{2}$

115) Simplify the below expression

$$-\frac{1}{4}n + \frac{3}{4} + \frac{14}{3}n + \frac{1}{3}$$

A) $-\frac{8}{21}n$ B) $-\frac{1}{2}n$

C) $-\frac{19}{8}n$ D) $\frac{53}{12}n + \frac{13}{12}$

116) Simplify the below expression

$$\frac{17}{6}a + \frac{7}{3}a$$

A) $\frac{31}{6}a$ B) $\frac{11}{8}a$

C) $\frac{25}{24}a$ D) $\frac{103}{24}a$

117) Simplify the below expression

$$-x + \frac{1}{8}x$$

A) $-\frac{97}{56}x$ B) $-7x - \frac{59}{36}$

C) $-\frac{7}{8}x$ D) $1 - \frac{11}{9}x$

118) Simplify the below expression

$$-\frac{10}{7}k + \frac{21}{4}k$$

A) $\frac{107}{28}k$ B) $\frac{41}{84}k$

C) $\frac{23}{28}k$ D) $\frac{51}{28}k$

119) Simplify the below expression

$$\frac{11}{4}x - \frac{12}{7}x$$

A) $\frac{759}{140}x$ B) $\frac{29}{28}x$

C) $x - \frac{7}{10}$ D) $\frac{859}{140}x$

120) Evaluate the below expression

$$((17)(11)) \div (18 + 17 - 18)$$

A) 20   B) 11

C) 6   D) 4

Grade 7 Volume 1 Week 13

Grade 7
Volume 1
Week 13

1) Simplify the below expression

$$-9(1-6x)$$

A) $-9+54x$ B) $-11+54x$

C) $8+54x$ D) $3+54x$

2) Simplify the below expression

$$-9(1-3p)$$

A) $-7+27p$ B) $4+27p$

C) $-20+27p$ D) $-9+27p$

3) Simplify the below expression

$$-6(-11r-10)$$

A) $4r-39$ B) $63r-18$

C) $66r+60$ D) $4r-32$

4) Simplify the below expression

$$7(-n+13)$$

A) $-7n+91$ B) $-7n+110$

C) $-7n+101$ D) $-11n+110$

5) Simplify the below expression

$$14(1-10k)$$

A) $-k-9$ B) $9-140k$

C) $14-140k$ D) $3-140k$

6) Simplify the below expression

$$-6(5n-7)$$

A) $23+24n$ B) $-12n+60$

C) $32+24n$ D) $-30n+42$

7) Simplify the below expression

$$9(14a+10)$$

A) $24a+31$ B) $24a+33$

C) $24a+30$ D) $126a+90$

8) Simplify the below expression

$$-3(10v-13)$$

A) $-30v+39$ B) $60v-30$

C) $91-39v$ D) $60v-16$

**Grade 7
Volume 1
Week 13**

9) Simplify the below expression

$$8(1+8p)$$

A) $20p - 18$

B) $8 + 64p$

C) $-117p + 126$

D) $-182p - 56$

10) Simplify the below expression

$$-14(9r - 1)$$

A) $10r + 112$ B) $-126r + 14$

C) $10r + 120$ D) $50 - 60r$

11) Simplify the below expression

$$-4(11v + 11)$$

A) $-44v - 44$ B) $21 + 50v$

C) $26 + 117v$ D) $21 + 36v$

12) Simplify the below expression

$$10(8 + 12x)$$

A) $-168 + 42x$ B) $80 + 120x$

C) $91 - 182x$ D) $82 + 120x$

13) Simplify the below expression

$$-6(9 + 13p)$$

A) $-55 - 80p$ B) $-56 - 80p$

C) $-49 - 80p$ D) $-54 - 78p$

14) Simplify the below expression

$$-7(14 + 13a)$$

A) $-98 - 91a$ B) $-a + 1$

C) $5a + 1$ D) $78a - 30$

15) Simplify the below expression

$$12(-4 - 6r)$$

A) $102 - 100r$ B) $-48 - 72r$

C) $98 - 112r$ D) $102 - 112r$

16) Simplify the below expression

$$-(3 - 4a)$$

A) $-3 + 4a$ B) $10 - 20a$

C) $21 - 22a$ D) $21 - 20a$

17) Simplify the below expression

$$-6(5-12r)$$

A) $-50+72r$ B) $-41+72r$

C) $-30+72r$ D) $-45+72r$

18) Simplify the below expression

$$9(4-7k)$$

A) $-16k-34$ B) $-7k-28$

C) $-16k-28$ D) $36-63k$

19) Simplify the below expression

$$-(n-2)$$

A) $-88-48n$ B) $-n+2$

C) $-94-48n$ D) $-99-48n$

20) Simplify the below expression

$$6(1+14m)$$

A) $9+84m$ B) $6+84m$

C) $-4+84m$ D) $3+84m$

21) Simplify the below expression

$$-(13n+13)$$

A) $-8n-40$ B) $-13n-13$

C) $132n+144$ D) $-99n+63$

22) Simplify the below expression

$$-8(-12+r)$$

A) $-7+44r$ B) $-30-110r$

C) $-16+44r$ D) $96-8r$

23) Simplify the below expression

$$-14(2-14x)$$

A) $-28+196x$ B) $-40x-70$

C) $-40x-60$ D) $-28+192x$

24) Simplify the below expression

$$9(2x+1)$$

A) $27x+9$ B) $-12x-12$

C) $30x+9$ D) $18x+9$

25) Simplify the below expression

$$-2(12m-12)$$

A) $14m+86$ B) $14m+91$

C) $-24m+24$ D) $15m+86$

26) Simplify the below expression

$$6(8m+4)$$

A) $48m+24$ B) $43m+8$

C) $48m+21$ D) $43m+21$

27) Simplify the below expression

$$-11(1+6n)$$

A) $50n-110$ B) $-11-60n$

C) $-11-66n$ D) $50n-97$

28) Simplify the below expression

$$16(14x+10)+4x$$

A) $228x+160$ B) $228x+123$

C) $228x+141$ D) $44-72x$

29) Simplify the below expression

$$-(13v+9)$$

A) $-9-99v$ B) $-13v-9$

C) $-25-99v$ D) $-18-99v$

30) Simplify the below expression

$$-8(1+12x)$$

A) $5-96x$ B) $-8-96x$

C) $-8+32x$ D) $1-96x$

31) Simplify the below expression

$$-(6x+13)$$

A) $-6x-22$ B) $-70x-35$

C) $-6x-13$ D) $-6x-17$

32) Simplify the below expression

$$-(2p+5)-4p$$

A) $-6p-5$ B) $114+42p$

C) $-67-90p$ D) $-6p-25$

33) Simplify the below expression

$$-10k + 16(16 + 6k)$$

A) $17 - 36k$ B) $-253k - 120$

C) $86k + 256$ D) $65k + 91$

34) Simplify the below expression

$$2(12n + 2) - 2n$$

A) $28n + 4$ B) $30n + 4$

C) $22n + 4$ D) $30n - 16$

35) Simplify the below expression

$$15 - 16(8 + 14p)$$

A) $-113 - 224p$

B) $-22p + 23$

C) $-22p + 14$

D) $133p - 238$

36) Simplify the below expression

$$12(-6 + 19n) + 6n$$

A) $33n + 64$ B) $-21n - 28$

C) $28n + 64$ D) $-72 + 234n$

37) Simplify the below expression

$$-10(-11 + 13n) - 10$$

A) $111 - 130n$ B) $11n - 6$

C) $38n + 126$ D) $100 - 130n$

38) Simplify the below expression

$$-18n - 5(2 - 6n)$$

A) $-15 - 39n$ B) $-80 + 79n$

C) $12n - 10$ D) $-12 - 39n$

39) Simplify the below expression

$$-4x + (13 + 2x) \cdot 16$$

A) $-153x + 15$ B) $28x + 208$

C) $-12x + 134$ D) $-12x + 125$

40) Simplify the below expression

$$19p - 6(5 - 14p)$$

A) $103p - 33$ B) $103p - 11$

C) $103p - 23$ D) $103p - 30$

Grade 7
Volume 1
Week 13

41) Simplify the below expression

$$10(-19b+6)-8$$

A) $-190b+52$ B) $-60b+12$

C) $-75b+12$ D) $-182b+52$

42) Simplify the below expression

$$6-2(1+4n)$$

A) $4-8n$ B) $5-8n$

C) $7-8n$ D) $-8-8n$

43) Simplify the below expression

$$-15+(1-x)\cdot 5$$

A) $-28x-10$ B) $-19x+40$

C) $-10-5x$ D) $-2-5x$

44) Simplify the below expression

$$(1+4n)\cdot -5-5n$$

A) $51n+189$ B) $-5-23n$

C) $-5-25n$ D) $-5-14n$

45) Simplify the below expression

$$9(3p+13)+7p$$

A) $2-63p$ B) $-24-63p$

C) $34p+117$ D) $-4-63p$

46) Simplify the below expression

$$-6+7(n+15)$$

A) $-3-21n$ B) $97+7n$

C) $99+7n$ D) $18n+238$

47) Simplify the below expression

$$-16(16x-8)-19x$$

A) $-275x+128$

B) $-96+101x$

C) $-108+101x$

D) $-24+85x$

48) Simplify the below expression

$$4x+7(10x+14)$$

A) $74x+98$ B) $13+77x$

C) $152x+118$ D) $152x+119$

Grade 7
Volume 1
Week 13

49) Simplify the below expression

$$-9 + 10(1 - 9x)$$

A) $10 - 90x$ B) $28 - 94x$

C) $1 - 90x$ D) $10 - 94x$

50) Simplify the below expression

$$-13r + 9(18r + 2)$$

A) $149r + 18$ B) $5 - 8r$

C) $-51 + 165r$ D) $149r + 32$

51) Simplify the below expression

$$(-10 + 13x) \cdot 10 + 9x$$

A) $-100 + 139x$

B) $261 + 17x$

C) $277 + 17x$

D) $-151 + 70x$

52) Simplify the below expression

$$-14(-16a + 17) - 19$$

A) $224a - 257$ B) $121 + 56a$

C) $13 - 18a$ D) $121 + 46a$

53) Simplify the below expression

$$4(4 - r) + 9r$$

A) $25 + 85r$ B) $16 - 13r$

C) $30 + 85r$ D) $16 + 5r$

54) Simplify the below expression

$$-6(7k + 6) + 2k$$

A) $-40k - 36$ B) $33k + 240$

C) $-1 - 119k$ D) $-23k - 80$

55) Simplify the below expression

$$-19(19v + 15) - 19v$$

A) $-7 + 116v$ B) $3 + 120v$

C) $-380v - 285$ D) $-7 + 120v$

56) Simplify the below expression

$$10 + (n + 14) \cdot -5$$

A) $-60 - 5n$ B) $-14n - 63$

C) $-14n - 61$ D) $70 + 158n$

57) Simplify the below expression

$$(6+6x) \cdot -17 - 9$$

A) $-109 - 102x$

B) $-101 - 102x$

C) $-111 - 102x$

D) $-117 - 102x$

58) Simplify the below expression

$$-6 - 11(n-1)$$

A) $-23 - 96n$ B) $8 - 11n$

C) $-7 - 11n$ D) $5 - 11n$

59) Simplify the below expression

$$2.9(1 - 5.6p)$$

A) $3.1p + 14.9699$

B) $2.9 - 16.24p$

C) $45.56 - 13.6p$

D) $10.9 - 16.24p$

60) Simplify the below expression

$$9.1(1 + 8.7r)$$

A) $14 + 79.17r$

B) $10.74 + 79.17r$

C) $5.2 + 79.17r$

D) $9.1 + 79.17r$

61) Simplify the below expression

$$-20(8x - 8) + 14$$

A) $-157x + 174$

B) $18x + 77$

C) $-160x + 174$

D) $-143 - 15x$

62) Simplify the below expression

$$12(2 - 19x) + 12x$$

A) $24 - 201x$ B) $24 - 216x$

C) $4 - 201x$ D) $14 - 201x$

63) Simplify the below expression

$$2.2(x - 0.7)$$

A) $2.2x + 8.46$

B) $11.4x + 8.46$

C) $2.2x - 1.54$

D) $6.5x + 8.46$

64) Simplify the below expression

$$-6.98(x - 1.3)$$

A) $-6.98x + 9.074$

B) $-0.48x + 19.174$

C) $-6.98x + 9.174$

D) $-0.48x + 9.174$

Grade 7
Volume 1
Week 13

65) Simplify the below expression

$-0.4(9.3-8a)$

A) $-3.72 + 3.2a$

B) $-0.57a - 46.13$

C) $-0.57a - 52.93$

D) $-7.37a - 52.93$

66) Simplify the below expression

$-2.6(2.2x+9.1)$

A) $6.15 + 4.92x$

B) $-96.03x + 48.51$

C) $-5.72x - 15.76$

D) $-5.72x - 23.66$

67) Simplify the below expression

$-1.9(k-0.7)$

A) $-1.9k + 1.33$

B) $35.188k - 43.522$

C) $-26.288 - 6.89k$

D) $-7.1k + 31.95$

68) Simplify the below expression

$n - \dfrac{4}{3} + n - \dfrac{39}{10}$

A) $-\dfrac{2}{9}n - \dfrac{157}{30}$ B) $\dfrac{1}{9}n - \dfrac{157}{30}$

C) $2n - \dfrac{157}{30}$ D) $-\dfrac{8}{9}n - \dfrac{157}{30}$

69) Simplify the below expression

$5.079(0.2x - 9.8)$

A) $1.0158x - 41.8742$

B) $1.0158x - 49.7742$

C) $27x + 47.25$

D) $-0.8342x - 41.8742$

70) Simplify the below expression

$4.4(-7.5b + 8.4)$

A) $3.7 - 27.01b$

B) $-33b + 36.96$

C) $4.2 - 27.01b$

D) $13.4 - 27.01b$

71) Simplify the below expression

$-2.6(4.5x + 1.6)$

A) $37.56x + 4.68$

B) $-11.7x - 4.16$

C) $30.96x + 4.68$

D) $34.66x + 4.68$

72) Simplify the below expression

$\dfrac{35}{6}k + \dfrac{1}{6} + \dfrac{9}{2}k - \dfrac{16}{5}$

A) $\dfrac{31}{3}k - \dfrac{91}{30}$ B) $-\dfrac{5}{12}k$

C) $-6k + \dfrac{35}{18}$ D) $-5k + \dfrac{35}{18}$

73) Simplify the below expression

$$\frac{25}{6}a - \frac{19}{6}a$$

A) $\frac{109}{30}a$ B) $-\frac{73}{18}a$

C) a D) $\frac{79}{30}a$

74) Simplify the below expression

$$-\frac{9}{10}b + 1 - \frac{1}{3}b - \frac{5}{3}$$

A) $2 - \frac{11}{14}b$ B) $-\frac{15}{2}b$

C) $-\frac{2}{3} - \frac{37}{30}b$ D) $-\frac{2}{3} - \frac{17}{30}b$

75) Simplify the below expression

$$-\frac{17}{10}v - \frac{6}{5}v$$

A) $\frac{933}{280}v$ B) $-\frac{29}{10}v$

C) $\frac{67}{70}v$ D) $-\frac{143}{70}v$

76) Simplify the below expression

$$\frac{7}{6}n - \frac{9}{8} - \frac{1}{2}n$$

A) $\frac{17}{12}n - \frac{9}{8}$ B) $-\frac{3}{7}n + \frac{46}{45}$

C) $\frac{61}{70}n + \frac{46}{45}$ D) $\frac{2}{3}n - \frac{9}{8}$

77) Simplify the below expression

$$-\frac{7}{2}n + \frac{1}{4} + n - \frac{1}{3}$$

A) $\frac{47}{7} + \frac{1}{6}n$ B) $-\frac{5}{2}n - \frac{1}{12}$

C) $\frac{47}{7} + \frac{11}{30}n$ D) $\frac{41}{30}n + \frac{47}{7}$

78) Simplify the below expression

$$-\frac{13}{8}b + \frac{13}{7} + \frac{31}{8}b + \frac{23}{4}$$

A) $\frac{178}{21}b$ B) $\frac{113}{18}b$

C) $b - \frac{53}{12}$ D) $\frac{9}{4}b + \frac{213}{28}$

79) Simplify the below expression

$$-\frac{7}{2}x - \frac{39}{10}x$$

A) $-\frac{37}{5}x$ B) $-\frac{55}{18}x$

C) $\frac{109}{30}x$ D) $-\frac{14}{9}x$

80) Simplify the below expression

$$\frac{4}{5}p + \frac{1}{2} - 3\frac{1}{6}$$

A) $\frac{29}{20}p$ B) $\frac{209}{30}p + 2$

C) $\frac{4}{5}p - \frac{8}{3}$ D) $\frac{69}{20}p$

Grade 7
Volume 1
Week 14

Grade 7
Volume 1
Week 14

1) Subtract (x + y) from (4y - 5x)

2) Subtract (3a - 3b + 4c) from the sum of (a + 3b - 4c), (4a - b + 9c) and (- 2b + 3c - a)

3) Simplify: (5x - 9y) + (-7x + y)

4) Add: $-2x^2$ - 11x + 3, $-9x^2$ + 10x - 3 and $11x^2$ - x -1

5) Add :

$$\left(2x^2 - \frac{1}{3}x + \frac{5}{7}\right) + \left(\frac{-3}{4}x^2 + \frac{2}{3}x - \frac{1}{6}\right)$$
$$+ \left(-5x^2 + \frac{1}{5}x + 7\right)$$

6) Subtract :

$$\left(-4y^2 + \frac{1}{3}y - 7\right) \text{ from } (-15y^2 - 4y + 11)$$

7) Simplify: $(4x^2 - 2x - 3) + (-5x^2 - 2x + 7)$

8) Simplify: $(-2x^2 + 5x + 1) - (3x^2 + 2x - 6)$

9) The perimeter of a triangle is given by P = 27x + 12y and two of its sides have lengths of x + 2y and 6x + 4y. What is the length of the third side?

10) Add: $(11 - 19b^2 - b) + (b^2 - 3b^3 + b) + (3b^2 + 21)$

11) Add: $(7x^2 - 19x^3 - 4) - (3x^2 - 3x^3 - x)$

12) Subtract: -43ab from -28ab.

13) Simplify the following : (3x + 4y -1) + (-2x + y - 5) - (2x + 5y + 4)

14) Simplify the following : (3x + 2y + 4z) + (2x - y + 3z) - (5x + y - 3z)

15) Simplify : (a + 3b + 4c) - (2b + 5c) + (4a + b - c)

**Grade 7
Volume 1
Week 14**

16) Add: $(x^2 + y^2 + z^2) + (x^2 - y^2 + z^2)$

17) Subtract: $(2a + b)$ from the sum of $(5a - 2b + 3)$ and $(2a + b - 1)$

18) What should be added to $(3a + 5b - c)$ to get $(a - 2b + 5c)$?

19) Add: $(3x^2 - 5x + 2)$, $(4x + 5)$ and $(2x^2 - 4x + 1)$

20) What should be added to $(4p + 6p - 7)$ to get $(5p^2 - 5p + 2)$?

21) Simplify: $(4x^2 + 10x + 25) + (2x^2 - 4x - 7) - (7x^2 - 11x + 13) - (-3x^2 - 4)$

22) Subtract $(11m^2 - 3m - 1)$ from the sum of $(6m^2 + 5m - 7)$ and $(-3m^2 + 12m - 5)$

23) Subtract the sum of $(12x + 7)$ and $(6x^2 - 4x + 1)$ from $(8x^2 - 3x + 5)$

24) The sum of two algebraic expressions is ($-3a^2 - 23a + 11$). If one of the expressions is ($2a^2 - 12a + 4$), find the other expression.

25) Add : ($6a + 3b - 4c$), ($4b + c - 3a$), ($a - 6c - 3b$)

26) Add : ($4x^2 + 5x - 4$), ($2x + 3 - x^2$), ($8 - 4x + 7x^2$)

27) Add :
$$\left[2x^2 - \left(\frac{1}{5}\right)x + \frac{7}{3}\right], \left[2x^2 + \left(\frac{1}{3}\right)x + \frac{1}{6}\right], \left[-x^2 + \left(\frac{2}{3}\right)x + 1\right]$$

28) Add : ($5x^2y + 8xy^2$), ($11x^2y - 13xy^2$), ($7xy^2 - 5x^2y$)

29) Add : ($2x^2 - 4y^2$), ($5x^2 - y^2$), ($-x^2 + 3y^2$)

30) Subtract : ($3a + 4b - 2c$) from ($6a - 3b + 5c$)

31) Subtract : ($3x^2 - 6x - 3$) from ($7 + 5x + 4x^2$)

Grade 7
Volume 1
Week 14

32) Subtract : (3x - y + 5z) from (7x + 4y - 6z)

33) Subtract : ($a^2 + b^2$ - 2ab) from ($a^2 + b^2$ + 2ab)

34) Simplify : (5x - 9y) + (-5x + 4y)

35) Simplify : (2x - 8y) + (-7x + 4y)

36) Simplify : (2x - 8y) - (-7x + 4y)

37) Subtract : ($5a^2 + 5b^2$ + 10ab) from ($5a^2$ - 5b)

38) Subtract : ($a^3 + 3a - a^2 - 6$) from ($2a^2 + a - 2a^3 + 3$)

39) What must be added to (7a - 9b + 12c) to get (5a + b - 2c) ?

**Grade 7
Volume 1
Week 14**

40) What must be added to (m + n + 6) to get (6m + 3n - 8) ?

41) Subtract the sum of $(8a - 6a^2 + 9)$ and $(7a - 8a^2$ from $(3 - 6a^2)$.

42) Subtract the sum of (2a - 4b + 4c) and (5a - 3b + 8c) from the sum of (2a - 5b - 2c) and (7a + 2b + c).

43) Subtract $(2p^2 - pq + 4q^2)$ from $(4p^2 - 2pq + 3q^2)$.

44) If a = $(x^3 - 3y^3)$, b = $(7x^3 + 2y^3)$, find a + b

45) If x = (2a + 3b - 5c), y = (7a + b + 3c), z = (2a - 5b - 7c), find (x + y + z).

46) Subtract $(6x^3 - 5x^2 + 4x - 3)$ from the sum of $(x + 3x^2 - 3x^3)$ and $(4 - 7x - x^2 - x^3)$

47) What should be subtracted from $(2 + 6x + x^2)$ to get $(2x + x^2)$?

48) What must be added to $(1 + 4x - 3x^2)$ to get $(x^2 + x + 5)$?

49) Add $(3a^2 + 6ab - 8b^2)$ and $(4a^2 - 2ab + 4b^2)$

50) Add $(6xy - 3yz + 8xz)$ and $(-2xy + 5yz + 2xz)$

51) Subtract $3a^2 - 5a + 4$ from $-5a^2 + 4a - 3$

52) Subtract $4p^2 + 10pq$ from $6p^2 - 3pq + pr$

53) Four - fifths of a number is greater than two thirds of the number by 6. Find the number.

54) The length of a rectangular plot exceeds its width by 5 yard. If the perimeter of plot is 146 yard, find the dimensions of the plot.

55) After 12 years Sam will be three times as old as he was 4 years ago. Find his present age.

56) Calvin wanted to find three consecutive whole numbers that add up to 58. He wrote the equation as
(n - 1) + n + (n + 1) = 58
What does the variable 'n' represent?

A) The middle whole number.
B) The difference between the least and greatest of the three whole numbers.
C) The least of the three whole numbers.
D) The greatest of the three whole numbers

57) A rectangular garden has a perimeter of 46 feet. The length of the garden is 15 feet and the width is w. Which equation can be used to determine the width of the garden?

A) 2(w) + 15 = 46
B) w + 15 = 46
C) 2(w + 15) = 46
D) w - 12 = 46

58) For the animal shelter, Brant purchased eight dog leashes for $10.56 each, eight bags of cat food for $15.42 each, and eight dog crates for 'x' dollars each. Which equation could be used to determine Brant's total cost, C?

A) ($10.56 + $15.42) / 8 = C
B) ($10.56 + $15.42 + x) × 8 = C
C) ($10.56 + $15.42 + x) + 8 = C
D) $10.56 + $15.42 + 8 + 8 = C

59) The difference of a number and $\frac{1}{4}$ is $-\frac{4}{5}$. Find the number.

60) A store received $928 from the sale of 5 tape recorders and 7 radios. If the receipts from the tape recorders exceeded the receipts from the radios by $112, what is the sale price of a tape recorder?

61) Two supplementary angles differ by 48°. Find the angles.

62) Find three consecutive positive even integers whose sum is 96.

63) The numerator of a fraction is 5 less than its denominator. If 1 is added to both its numerator and denominator, the fraction becomes 1/2. Find the fraction.

64) Two - fifths of a number is 4. What is the number?

65) Thrice a number decreased by 7 equals 11. What is the number?

66) A number exceeds one fourth of itself by 30. What is the number?

67) In a class of 80 students, the number of girls is one - third of the number of boys. Find the number of boys.

68) Ben donated one - fourth of his salary to a blind school. If the amount donated to the blind school is $160, how much is the Ben's salary?

69) The length of a rectangular park is double its width. If the perimeter of the park is 120 yards, find the length and width of the rectangular park.

70) One number exceeds the other number by 6. If the sum of the two numbers is 40, find the two numbers.

71) One fourth of a number increased by 7 equals double the number. Find the number.

72) 10 is added to 3 times a certain number. The result is 4 less than 5 times the same number. Find the number.

73) The sum of two consecutive even numbers is 86. Find the numbers.

74) The ratio of two numbers is 3:5 and their sum is 32. Find the two numbers.

75) One number exceeds the other number by 2. Four times the smaller number added to the larger number gives 97. Find the numbers.

76) The denominator of a fraction is 3 more than the numerator. If 2 is added to the numerator and 3 is added to the denominator, the fraction becomes $\frac{1}{2}$. Find the fraction.

77) The sum of three consecutive odd numbers is 69. Find the numbers.

78) The sum of two consecutive odd numbers is 92. Find the numbers.

79) The sides of a triangle are in the ratio 3:5:4. If the perimeter of the triangle is 48 inches, find the measure of its sides.

80) A bag contains 96 red and blue marbles. If the ratio of red and blue marbles is 7:5, how many blue marbles are in the bag?

81) The ratio of two numbers is 3:7 and their sum is 100. Find the two numbers.

82) Eight added to twice a number gives 42. Find the number.

83) The sum of two consecutive multiples of 2 is 66. Find the numbers.

84) Five times a number subtracted by 4 equals 11. Find the number.

85) A number subtracted by 7 gives 55. Find the number.

86) Thrice a number increased by 7 equals 49. Find the number.

87) Half of a number is 4 more than one third of the number. What is the number?

88) There are 55 students in a class. The ratio of the number of boys to the number of girls is 2 : 3. Find the number of girls in the class.

89) There are 40 students in a class. The ratio of the number of boys to the number of girls is 7 : 1. Find the number of boys in the class.

90) The measures of the three angles in a triangle are in the ratio 5:6:7. What is the measure of the largest angle?

91) June is now twice as old as Alex. But six years ago June was five times as old as Alex was. How old is June now?

92) In 7 years Henna will be twice as old as she was 8 years ago. How old is Henna now?

93) In a family of three, the father weighed 5 times as the child, and the mother weighed 3/4 as much as the father. If three of them weighed a total of 390 pounds, how much did the child weigh?

94) In a family of three, the father weighed 5 times as the child, and the mother weighed 3/4 as much as the father. If three of them weighed a total of 390 pounds, how much did the mother weigh?

95) A box contains only red, yellow and green beads. The number of red beads is 4/5 the number of green beads, and the number of green beads is 3/4 the number of yellow ones. If there are 470 beads in all, how many of them are yellow?

96) A box contains only red, yellow and green beads. The number of red beads is 4/5 the number of green beads, and the number of green beads is 3/4 the number of yellow ones. If there are 470 beads in all, how many of them are red?

97) The sum of the two consecutive numbers is 87. Find the numbers.

98) Four - fifth of a number is greater than three - fourth of the number by 5. Find the number.

99) Two - third of a number is greater than one - third of the number by 10. Find the number.

100) A number has two digits whose sum is 5. If 27 is added to the number, its digits are reversed. Find the number.

101) The length of a rectangular plot exceeds its width by 6 yards. If the perimeter of the plot is 164 yards, find the length of the plot.

102) The denominator of a fraction is 4 more than the numerator. If 2 is added to the numerator and 1 is added to the denominator, the fraction becomes 1/2. Find the fraction.

103) Two complementary angles differ by 12°. Find the measure of the angles.

104) Two supplementary angles differ by 36°. Find the measure of the angles.

105) The sum of the three consecutive odd numbers is 177. Find the numbers.

106) Bill weighs 40 pounds more than Steve. How much does Steve weigh, if they together weigh 370 pounds?

107) 4 consecutive natural numbers add up to 1850. What is the smallest number?

108) Eight times a number is same as that number plus 84. What is the number?

109) 8 chairs or 12 tables cost $1728. Find the cost of 8 tables and 2 chairs.

110) A number when subtracted from 27, reduces to 15 less than twice the number. Find the number.

111) A man's age is three times that of his daughter's age. After 12 years, the father's age will be double the daughter's age. Find the man's present age.

112) Charles's age is four times that of his daughter. 5 years ago, his age was seven times the age of his daughter. Find their present ages.

113) The length of a rectangular field is twice its width. If the perimeter of the field is 204 yards, find its dimensions.

114) The sum of digits of a 2-digit number is 9. If 27 is added to the number, the digits are reversed. Find the number.

115) The length of a rectangular field is three times its width. If the perimeter of the field is 720 yards, find its dimensions.

116) Myranda's age and her mother's age together is 72 years. If the mother's age is 3 times that of Myranda's age, what are their ages?

117) The sum of two numbers is 89. If one number is 13 more than the other, find the numbers.

Grade 7
Volume 1
Week 14

118) Find three consecutive even numbers whose sum is 114.

119) The sum of three consecutive numbers is 36. Find the numbers.

120) If 10 is subtracted from thrice a number, the result is 35. Find the number.

121) A number when divided by 7 gives the quotient 5. Find the number.

122) The product of 7 and a number is 56. Find the number.

123) If 21 is subtracted from a number, the result is 81. What is the number?

124) The sum of a number and 13 is 46. Find the number.

125) Alex is 7 years older than Max. The sum of their ages is 27. What are their ages?

126) After buying a notebook, I am left with $15 from my $50 pocket money. What is the cost of the notebook?

127) How much pure alcohol should be added to 400 mL of a 15% solution to make its strength 32%?

128) The total cost of 3 chairs and 2 tables is $745. If a chair costs $40 more than the table, find the price of each.

129) The sum of digits of a 2 - digit number is 8. If 18 is added to the number, its digits are reversed. Find the number.

130) The numerator of a fraction is 4 less than the denominator. If 1 is added to both its numerator and denominator, the fraction becomes 1/2. Find the fraction.

131) 5 years ago, a man was 7 times as old as his daughter. After 5 years, the father would be 3 times as old as his daughter. Find their present ages.

132) Twice a certain number decreased by 7 gives the result 45. Find the number.

133) Thrice a number when increased by 5 gives the result 44. Find the number.

134) Twice a number when increased by 4 gives (26/5). Find the number.

135) A number when added to its half yields 72. Find the number.

136) A number when added to its two - thirds equals 55. Find the number.

137) A number multiplied by 4, exceeds itself by 45. What is the number?

138) A number is as much greater than 21 as it is less than 71. What is the number?

139) Two - thirds of a number is less than the original number by 20. What is the number?

140) A number is (2/5) times another number. Their sum is 70. What are the numbers?

141) Two - thirds of a number is greater than one -third of the number by 3. What is the number?

142) The 5th part of a number when increased by 5 is equal to its 4th part decreased by 5. What is the number?

143) The sum of two consecutive natural numbers is 63. Find the numbers.

143) The sum of two consecutive positive odd integers is 76. Find the numbers.

145) The sum of three consecutive positive even integers is 90. Find the numbers.

146) In a school, grade 7 has 49 students. The number of girls in this grade is 1/6 times the number of boys. Find the number of boys and girls in grade 7.

147) Find the measure of two complementary angles which differ by 8°.

148) Find the measure of two supplementary angles that differ by 44°.

149) Divide 184 in two parts such that (1/3) of one part exceeds (1/7) of other part by 8.

150) The length of a rectangle is twice its width. If the perimeter of the rectangle is 150 yards, find its dimensions.

151) In an isosceles triangle, each of the two equal sides are 5 in. less than twice the third side. If the perimeter of the triangle is 55 in., find the length of its sides.

152) In an examination, the total points to pass are 40%. If Charles gets 185 points and fails by 15, find the total points.

153) The ages of Alex and Tammy are in the ratio 7:5. After 10 years, the ratio of their ages will be 9:7. Find their present ages.

154) A father is 30 years older than his daughter. After 12 years, the man will be three times as old as his daughter. Find their present ages.

155) Sam is 19 years younger than his cousin. 5 years hence, their ages will be in the ratio 2:3. Find their present ages.

156) Thrice a number when increased by 9 gives 45. Find the number.

157) Find two consecutive multiples of 3 whose sum is 69.

158) The sum of digits of a 2 - digit number is 9. If 27 is added to the number, the digits are reversed. Find the number.

159) A dealer sold an article for $714 and earned a profit of 5%. Find the cost price of the article.

160) A dealer earned a profit of 10% by selling an article for $495. Find the cost price of the article.

161) 50 kg of an alloy of lead and tin contains 60% tin. How much tin must be melted into it to make the alloy contain 75% tin?

162) State True or False : The absolute value of a rational number is never less than the actual rational number

163) Find | x - y |, when x = (- 5) and y = $\left(\dfrac{-11}{5}\right)$

164) Evaluate:

$$\left|\left(\dfrac{-9}{7}\right)+\left(\dfrac{-11}{3}\right)\right|$$

165) State True or False : The absolute value of a rational number is never negative.

166) State True or False : For any rational number x and y, | x + y| = |x| + |y|.

167) State True or False : For any rational numbers x and y, |x × y| = |x| × |y |

168) State True or False : There is always a rational number between any two distinct rational numbers.|

169) Find the value of : $\left|\dfrac{-21}{10}\right|$

170) Find the value of :

$$\left|-\left\{\dfrac{-19}{3}\right\}\right|$$

171) Find the value of :

$$\left|\dfrac{18}{-11}\right|$$

172) Evaluate :

$$\left|\left(\dfrac{7}{-12}\right)+\left(\dfrac{-11}{8}\right)\right|$$

173) Evaluate :

$$\left|\left(\dfrac{-3}{4}\right)-\left(\dfrac{10}{7}\right)\right| \div \left|\left(\dfrac{2}{9}\right)-\left(\dfrac{5}{6}\right)\right|$$

**Grade 7
Volume 1
Week 14**

174) Evaluate :

$$\left| \left(\frac{-8}{19}\right) \times \left(\frac{11}{-6}\right) \times \left(\frac{-57}{88}\right) \right|$$

175) Evaluate :

$$\left| (-21) - \left(\frac{-3}{5}\right) \right|$$

176) Find | x + y | , when x = $\left(\frac{-15}{16}\right)$ and y = $\left(\frac{3}{4}\right)$

177) Find | x + y | , when x = $\frac{1}{6}$ and y = $\frac{-9}{8}$

178) Find | x - y | , when x = $\left(\frac{-3}{7}\right)$ and y = $\left(\frac{-1}{3}\right)$

179) Find | x - y | , when x = $\left(\frac{-15}{8}\right)$ and y = $\left(\frac{5}{6}\right)$

180) Find the value of | 2x - 3 |, when x = (-2).

181) Find the value of | (4) + 3 - (-6) + 1 |

399

182) Evaluate :

$$\left|\left(\frac{-7}{8}\right)+\left(\frac{3}{4}\right)+\left(\frac{-2}{3}\right)-\left(\frac{-1}{2}\right)\right|$$

183) Evaluate :

$$|(-3)-(-1)|$$

184) Find the value of $|x^2 + 2x - 5|$, when $x = (-3)$.

185) Find the value of $|3x - 2y|$ when $x = \left(\frac{-5}{6}\right)$ and $y = \left(\frac{-3}{4}\right)$

186) Evaluate :

$$\left|-(9)\left(\frac{-1}{2}\right)+\left(\frac{-1}{5}\right)\right|$$

187) Evaluate :

$$\left|\left(\frac{3}{4}\right)-\left(\frac{2}{5}\right)\right|$$

188) Evaluate :

$$\left|\left(\frac{-3}{4}\right)\times\left(\frac{7}{11}\right)\right|$$

189) Evaluate :

$$\left|\left(\frac{5}{7}\right)\times\left(\frac{7}{-2}\right)\right|$$

**Grade 7
Volume 1
Week 14**

190) Evaluate :

$$\left| (-3) - \left(\frac{-2}{7} \right) \right|$$

191) Find the absolute value of $\left(\frac{7}{-12} \right)$

192) State True or False:

$$\left| -\left\{ \frac{-7}{19} \right\} \right| = \frac{7}{19}$$

193) Fill in the blank

$$\left| \left(\frac{-13}{2} \right) + \left(\frac{15}{2} \right) \right| = \underline{\qquad}$$

194) State True or False:

$$|0| = 0$$

195) Find | x - y | , when x = $\left(\frac{5}{8} \right)$ and y = $\left(\frac{5}{2} \right)$

196) Find | x - y | , when x = $\left(\frac{9}{7} \right)$ and y = $\left(\frac{-3}{2} \right)$

197) Compare the following with suitable symbol:

$$\left| \left(\frac{-5}{6} \right) + \left(\frac{3}{4} \right) \right| \underline{\qquad} \left(\frac{1}{12} \right)$$

**Grade 7
Volume 1
Week 14**

198) Compare the following with suitable symbol:

$$\left|\left(\frac{-6}{7}\right)\right| \underline{\quad} (-1)$$

199) Compare the following with suitable symbol:

$$\left|\left(\frac{-3}{5}\right)+\left(\frac{7}{10}\right)\right| \underline{\quad} \left|\left(\frac{-3}{5}\right)\right| + \left|\frac{7}{10}\right|$$

200) Compare the following with suitable symbol:

$$\left|\left(\frac{-3}{5}\right)-\left(\frac{2}{3}\right)\right| \underline{\quad} \left|\left(\frac{-3}{5}\right)\right| - \left|\frac{2}{3}\right|$$

201) Compare the following with suitable symbol:

$$\left|\left(\frac{-4}{5}\right)\times\left(\frac{7}{4}\right)\right| \underline{\quad} \left(-1\frac{2}{5}\right)$$

202) Evaluate:

$$\left|\left(\frac{-5}{8}\right)-\left(\frac{3}{4}\right)\right|$$

203) Evaluate:

$$\left|\left(\frac{2}{3}\right)-\left(\frac{8}{11}\right)\right|$$

204) State True or False:

$$\left|(-2) \div \left(\frac{2}{3}\right)\right| < 3$$

Grade 7 Volume 1 Week 15

Grade 7
Volume 1
Week 15

1) Find the area of the figure given below.

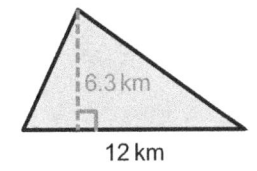

6.3 km
12 km

A) 75.6 km²   B) 18.9 km²

C) 46.9 km²   D) 37.8 km²

2) Find the area of the figure given below.

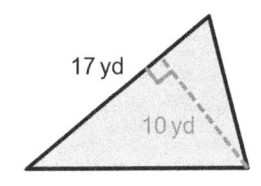

17 yd
10 yd

A) 83.7 yd²   B) 85 yd²

C) 42.5 yd²   D) 170 yd²

3) Find the area of the figure given below.

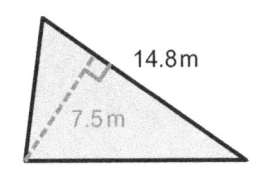

14.8 m
7.5 m

A) 27.8 m²   B) 47.4 m²

C) 111 m²   D) 55.5 m²

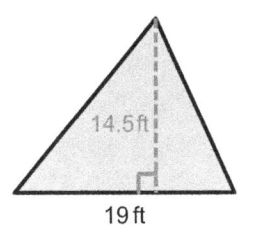

4) Find the area of the figure given below.

14.5 ft
19 ft

A) 275.5 ft²   B) 137.75 ft²

C) 136.55 ft² D) 68.9 ft²

5) Find the area of the figure given below.

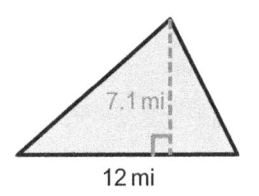

7.1 mi
12 mi

A) 85.2 mi²   B) 48.4 mi²

C) 42.6 mi²   D) 21.3 mi²

6) Find the area of the figure given below.

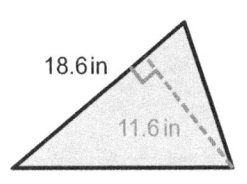

18.6 in
11.6 in

A) 109.68 in²   B) 215.76 in²

C) 107.88 in²   D) 53.9 in²

Grade 7
Volume 1
Week 15

7) Find the area of the figure given below.

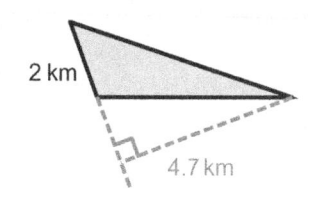

2 km
4.7 km

A) 4.7 km²   B) 1.8 km²

C) 9.4 km²   D) 0.9 km²

10) Find the area of the figure given below.

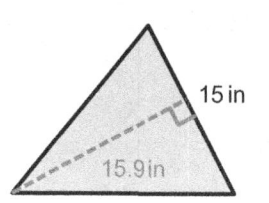

15 in
15.9 in

A) 126.55 in²   B) 238.5 in²

C) 119.25 in²   D) 59.6 in²

8) Find the area of the figure given below.

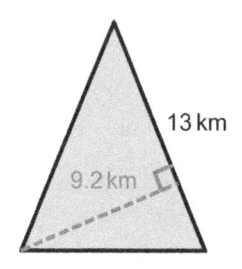

13 km
9.2 km

A) 29.9 km²   B) 119.6 km²

C) 68.7 km²   D) 59.8 km²

11) Find the area of the figure given below.

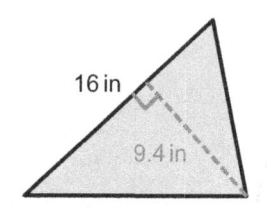

16 in
9.4 in

A) 75.2 in²   B) 82 in²

C) 150.4 in²   D) 37.6 in²

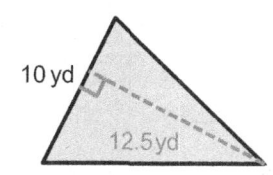

9) Find the area of the figure given below.

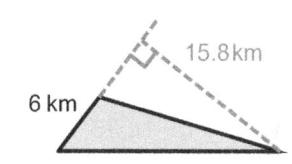

15.8 km
6 km

A) 28.1 km²   B) 23.7 km²

C) 56.1 km²   D) 47.4 km²

12) Find the area of the figure given below.

10 yd
12.5 yd

A) 125 yd²   B) 62.5 yd²

C) 67.7 yd²   D) 31.3 yd²

Grade 7
Volume 1
Week 15

13) Find the area of the figure given below.

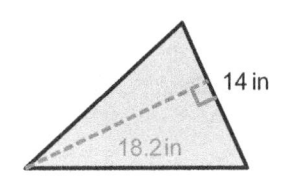

14 in
18.2 in

A) 254.8 in²   B) 127.4 in²

C) 67.5 in²   D) 134.9 in²

16) Find the area of the figure given below.

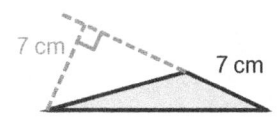

7 cm   7 cm

A) 24.5 cm²   B) 30.9 cm²

C) 12.3 cm²   D) 49 cm²

14) Find the area of the figure given below.

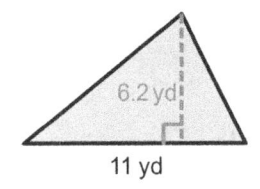

6.2 yd
11 yd

A) 68.2 yd²   B) 31.6 yd²

C) 34.1 yd²   D) 17.1 yd²

17) Find the area of the figure given below.

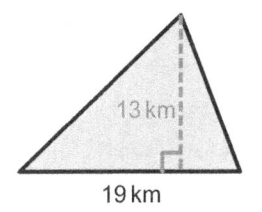

13 km
19 km

A) 247 km²   B) 61.8 km²

C) 123.5 km²   D) 126.1 km²

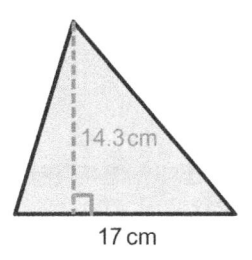

15) Find the area of the figure given below.

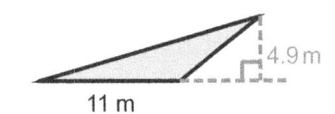

4.9 m
11 m

A) 53.9 m²   B) 23.85 m²

C) 13.5 m²   D) 26.95 m²

18) Find the area of the figure given below.

14.3 cm
17 cm

A) 60.8 cm²   B) 130.05 cm²

C) 121.55 cm²   D) 243.1 cm²

Grade 7
Volume 1
Week 15

19) Find the area of the figure given below.

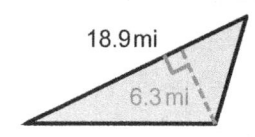

18.9 mi
6.3 mi

A) 29.8 mi² B) 50.335 mi²

C) 100.67 mi² D) 59.535 mi²

20) Find the missing measurement from the figure given below. Round the answer to the nearest tenth.

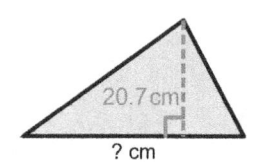

20.7 cm
? cm
Area = 420.2 cm²

A) 40.6 cm B) 25.5 cm

C) 29.4 cm D) 47.5 cm

21) Find the missing measurement from the figure given below. Round the answer to the nearest tenth.

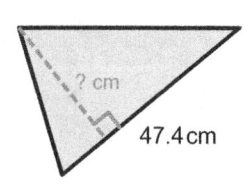

? cm
47.4 cm
Area = 684.9 cm²

A) 28.9 cm B) 29.4 cm

C) 38.8 cm D) 25.7 cm

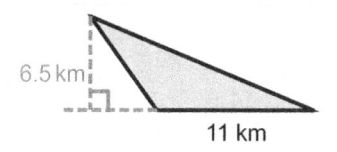

22) Find the area of the figure given below.

6.5 km
11 km

A) 17.9 km² B) 35.75 km²

C) 28.85 km² D) 71.5 km²

23) Find the missing measurement from the figure given below. Round the answer to the nearest tenth.

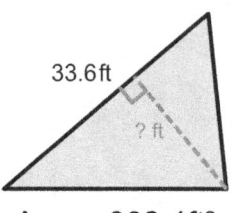

33.6 ft
? ft
Area = 302.4 ft²

A) 19 ft B) 18 ft

C) 23.6 ft D) 15.6 ft

24) Find the missing measurement from the figure given below. Round the answer to the nearest tenth.

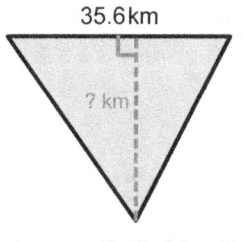

35.6 km
? km
Area = 510.9 km²

A) 30.1 km B) 19.1 km

C) 23.3 km D) 28.7 km

**Grade 7
Volume 1
Week 15**

25) Find the missing measurement from the figure given below. Round the answer to the nearest tenth.

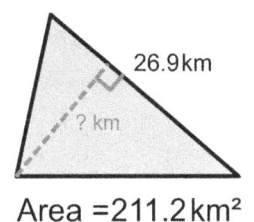

26.9 km
? km
Area = 211.2 km²

A) 11.9 km   B) 15.7 km
C) 10.4 km   D) 20 km

26) Find the missing measurement from the figure given below. Round the answer to the nearest tenth.

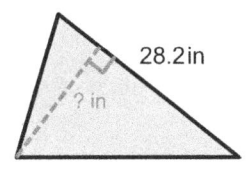

28.2 in
? in
Area = 236.9 in²

A) 15.7 in   B) 11.1 in
C) 16.8 in   D) 21.8 in

27) Find the missing measurement from the figure given below. Round the answer to the nearest tenth.

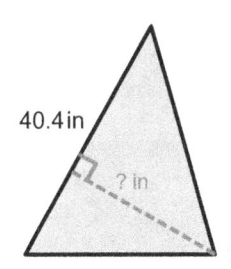

40.4 in
? in
Area = 527.2 in²

A) 26.1 in   B) 29.7 in
C) 26.2 in   D) 31.5 in

28) Find the missing measurement from the figure given below. Round the answer to the nearest tenth.

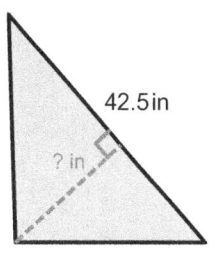

42.5 in
? in
Area = 433.5 in²

A) 17.5 in   B) 20.4 in
C) 26.4 in   D) 17.8 in

29) Find the missing measurement from the figure given below. Round the answer to the nearest tenth.

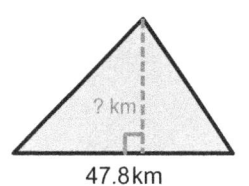

? km
47.8 km
Area = 674 km²

A) 28.2 km   B) 18 km
C) 35.3 km   D) 32.5 km

30) Find the missing measurement from the figure given below. Round the answer to the nearest tenth.

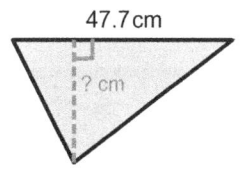

47.7 cm
? cm
Area = 612.9 cm²

A) 19.1 cm   B) 18.4 cm
C) 25.7 cm   D) 24.9 cm

31) Find the missing measurement from the figure given below. Round the answer to the nearest tenth.

50 cm
? cm

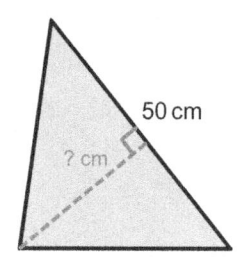

Area = 717.5 cm²

A) 18.9 cm   B) 35.7 cm
C) 22.6 cm   D) 28.7 cm

32) Find the missing measurement from the figure given below. Round the answer to the nearest tenth.

47.7 mi
? mi

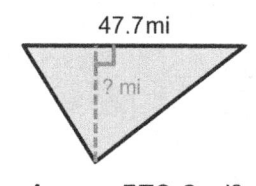

Area = 579.6 mi²

A) 24.3 mi   B) 23.8 mi
C) 24.6 mi   D) 29.3 mi

33) Find the missing measurement from the figure given below. Round the answer to the nearest tenth.

22.5 mi
? mi

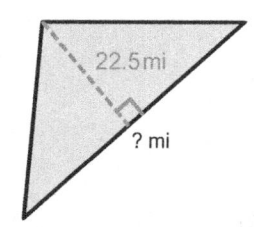

Area = 558 mi²

A) 34.3 mi   B) 62.6 mi
C) 69.4 mi   D) 49.6 mi

34) Find the missing measurement from the figure given below. Round the answer to the nearest tenth.

? ft
45.3 ft

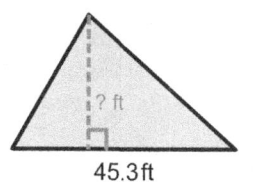

Area = 602.5 ft²

A) 26.6 ft   B) 35.8 ft
C) 31 ft   D) 24.4 ft

35) Find the missing measurement from the figure given below. Round the answer to the nearest tenth.

? mi
25.3 mi

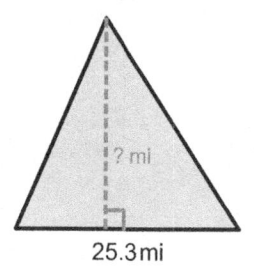

Area = 296 mi²

A) 31.1 mi   B) 23.4 mi
C) 15 mi   D) 20.2 mi

36) Find the missing measurement from the figure given below. Round the answer to the nearest tenth.

18.5 cm   ? cm

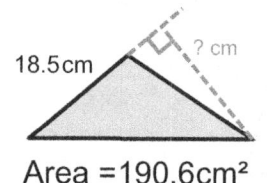

Area = 190.6 cm²

A) 18.1 cm   B) 25.9 cm
C) 13.8 cm   D) 20.6 cm

37) Find the missing measurement from the figure given below. Round the answer to the nearest tenth.

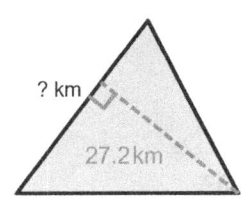

Area = 451.5 km²

A) 41 km   B) 35.5 km
C) 33.2 km D) 37.2 km

38) Find the missing measurement from the figure given below. Round the answer to the nearest tenth.

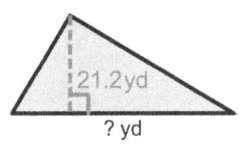

Area = 524.7 yd²

A) 65.9 yd   B) 66.9 yd
C) 49.5 yd   D) 32.8 yd

39) Find the missing measurement from the figure given below. Round the answer to the nearest tenth.

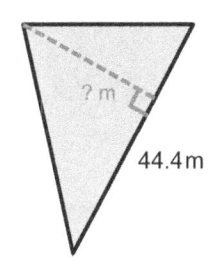

Area = 586.1 m²

A) 23 m   B) 22.9 m
C) 24 m   D) 26.4 m

40) Find the missing measurement from the figure given below. Round the answer to the nearest tenth.

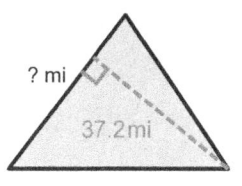

Area = 757 mi²

A) 32.4 mi   B) 36.1 mi
C) 53.6 mi   D) 40.7 mi

41) Find the missing measurement from the figure given below. Round the answer to the nearest tenth.

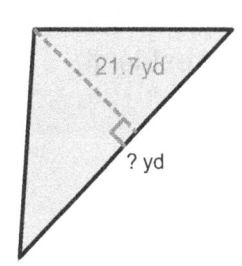

Area = 505.6 yd²

A) 43.1 yd   B) 57.8 yd
C) 36.8 yd   D) 46.6 yd

42) Find the missing measurement from the figure given below. Round the answer to the nearest tenth.

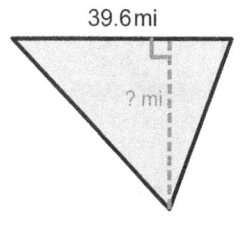

Area = 586.1 mi²

A) 30.6 mi   B) 27.4 mi
C) 29.6 mi   D) 23 mi

43) Find the missing measurement from the figure given below. Round the answer to the nearest tenth.

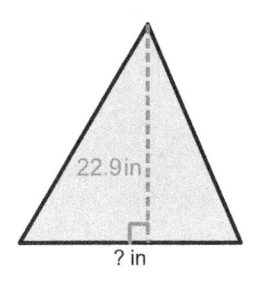

22.9 in
? in
Area = 273.7 in²

A) 31.2 in B) 29.1 in
C) 23.3 in D) 23.9 in

44) Find the missing measurement from the figure given below. Round the answer to the nearest tenth.

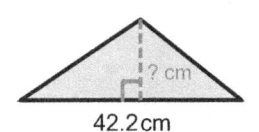

? cm
42.2 cm
Area = 314.4 cm²

A) 18.3 cm B) 18.8 cm
C) 17.5 cm D) 14.9 cm

45) Find the area of the figure given below.

33 m
3300 cm

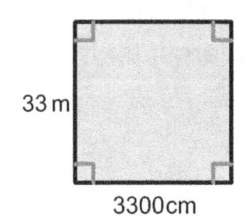

A) 544.5 m² B) 2178 m²
C) 1089 m² D) 1094.8 m²

46) Find the missing measurement from the figure given below. Round the answer to the nearest tenth.

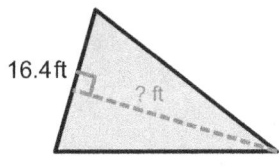

16.4 ft
? ft
Area = 199.3 ft²

A) 24.3 ft B) 25.1 ft
C) 27.4 ft D) 29.5 ft

47) Find the area of the figure given below.

43.7 ft
283.2 in

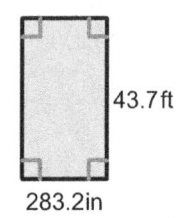

A) 1034.52 ft² B) 515.7 ft²
C) 1031.32 ft² D) 2062.64 ft²

48) Find the area of the figure given below.

45.6 cm
456 mm

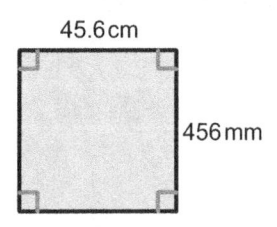

A) 2078.46 cm²
B) 4156.92 cm²
C) 2079.36 cm²
D) 2072.86 cm²

49) Find the area of the figure given below.

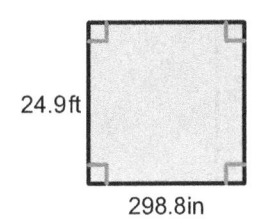

24.9 ft
298.8 in

A) 310 ft² B) 1240.02 ft²
C) 611.41 ft² D) 620.01 ft²

50) Find the area of the figure given below.

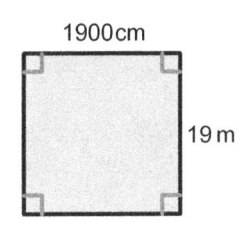

1900 cm
19 m

A) 354.7 m² B) 180.5 m²
C) 361 m² D) 722 m²

51) Find the area of the figure given below.

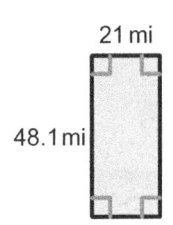

21 mi
48.1 mi

A) 1010.1 mi² B) 505.1 mi²
C) 2020.2 mi² D) 1014.7 mi²

52) Find the area of the figure given below.

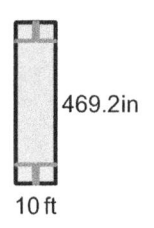

469.2 in
10 ft

A) 782 ft² B) 391 ft²
C) 381 ft² D) 195.5 ft²

53) Find the area of the figure given below.

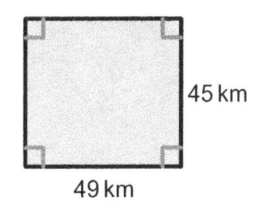

45 km
49 km

A) 2203.1 km² B) 2205 km²
C) 1102.5 km² D) 4410 km²

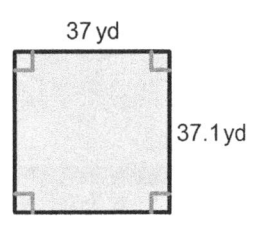

54) Find the area of the figure given below.

37 yd
37.1 yd

A) 1376.2 yd² B) 2745.4 yd²
C) 686.4 yd² D) 1372.7 yd²

55) Find the area of the figure given below.

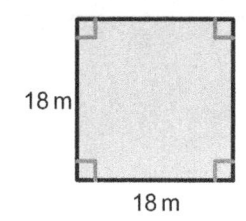

18 m
18 m

A) 324 m²   B) 648 m²

C) 651.8 m²   D) 325.9 m²

58) Find the area of the figure given below.

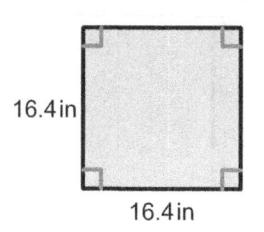

16.4 in
16.4 in

A) 537.92 in²   B) 134.5 in²

C) 268.96 in²   D) 272.86 in²

56) Find the area of the figure given below.

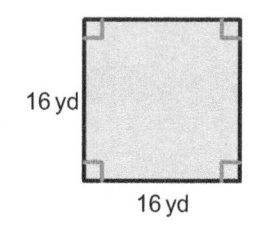

16 yd
16 yd

A) 512 yd²   B) 256 yd²

C) 128 yd²   D) 248.6 yd²

59) Find the area of the figure given below.

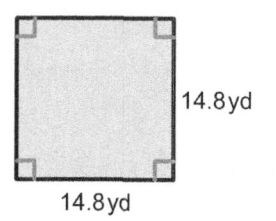

14.8 yd
14.8 yd

A) 109.5 yd²   B) 217.84 yd²

C) 224.94 yd²   D) 219.04 yd²

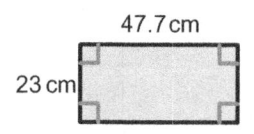

57) Find the area of the figure given below.

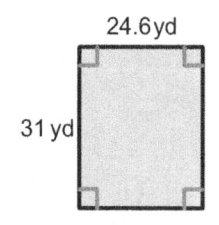

24.6 yd
31 yd

A) 762.6 yd²   B) 758.4 yd²

C) 381.3 yd²   D) 1525.2 yd²

60) Find the area of the figure given below.

47.7 cm
23 cm

A) 1097.1 cm²   B) 548.6 cm²

C) 1090.7 cm²   D) 2194.2 cm²

Grade 7
Volume 1
Week 15

61) Find the area of the figure given below.

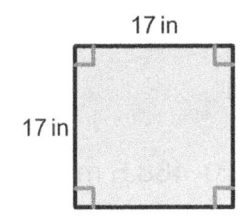

17 in
17 in

A) 294.7 in²   B) 144.5 in²

C) 578 in²   D) 289 in²

62) Find the area of the figure given below.

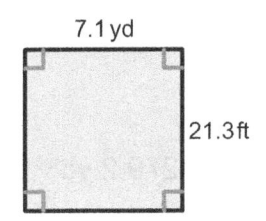

7.1 yd
21.3 ft

A) 50.41 yd²   B) 53.31 yd²

C) 100.82 yd²   D) 25.2 yd²

63) Find the area of the figure given below.

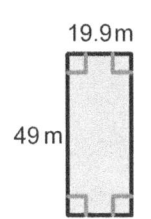

19.9 m
49 m

A) 969.3 m²   B) 487.6 m²

C) 975.1 m²   D) 1950.2 m²

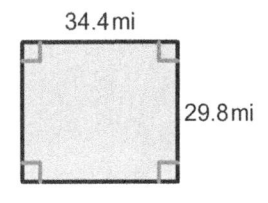

64) Find the area of the figure given below.

34.4 mi
29.8 mi

A) 2050.24 mi²

B) 512.6 mi²

C) 1033.22 mi²

D) 1025.12 mi²

65) Find the area of the figure given below.

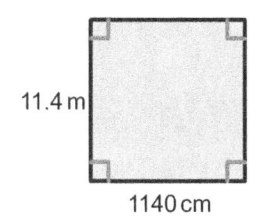

11.4 m
1140 cm

A) 123.36 m²   B) 65 m²

C) 129.96 m²   D) 259.92 m²

66) Find the area of the figure given below.

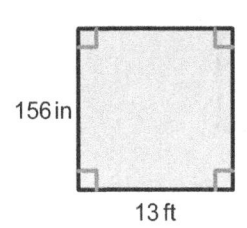

156 in
13 ft

A) 169 ft²   B) 84.5 ft²

C) 338 ft²   D) 159.5 ft²

67) Find the area of the figure given below.

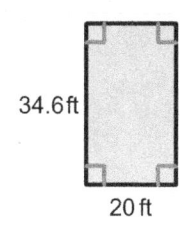

34.6 ft
20 ft

A) 694.6 ft²   B) 346 ft²

C) 1384 ft²   D) 692 ft²

68) Find the area of the figure given below.

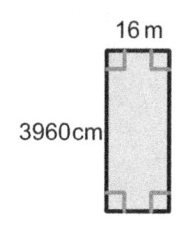

16 m
3960 cm

A) 639.3 m²   B) 640.4 m²

C) 633.6 m²   D) 316.8 m²

69) Find the area of the figure given below.

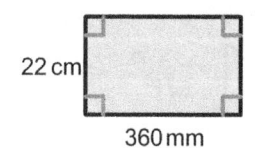

22 cm
360 mm

A) 1584 cm²   B) 784 cm²

C) 792 cm²   D) 396 cm²

70) Find the area of the figure given below.

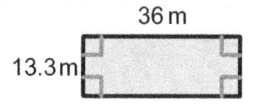

36 m
13.3 m

A) 239.4 m²   B) 488.6 m²

C) 957.6 m²   D) 478.8 m²

71) Find the area of the figure given below.

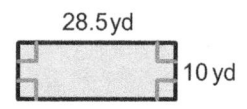

28.5 yd
10 yd

A) 142.5 yd²   B) 279.2 yd²

C) 570 yd²   D) 285 yd²

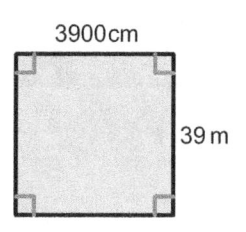

72) Find the area of the figure given below.

3900 cm
39 m

A) 1521 m²   B) 1517.7 m²

C) 760.5 m²   D) 3042 m²

73) Find the area of the figure given below.

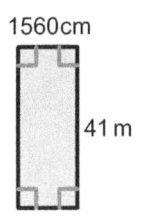

1560 cm
41 m

A) 319.8 m²   B) 642 m²

C) 643.2 m²   D) 639.6 m²

74) Find the area of the figure given below.

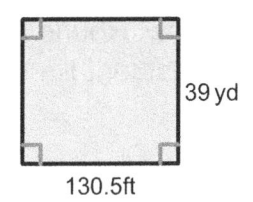

39 yd
130.5 ft

A) 1690.3 yd²   B) 848.3 yd²

C) 1696.5 yd²   D) 3393 yd²

75) Find the missing measurement from the figure given below. Round the answer to the nearest tenth.

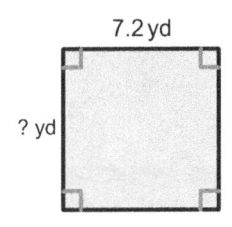

7.2 yd
? yd
Area = 51.8 yd²

A) 8.2 yd   B) 7.2 yd

C) 5.6 yd   D) 7.7 yd

76) Find the area of the figure given below.

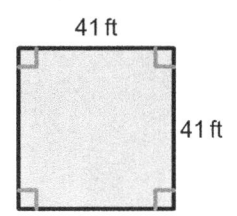

41 ft
41 ft

A) 1688.7 ft²   B) 3362 ft²

C) 840.5 ft²   D) 1681 ft²

77) Find the missing measurement from the figure given below. Round the answer to the nearest tenth.

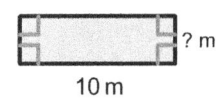

? m
10 m
Area = 30 m²

A) 3 m   B) 2.4 m

C) 2.1 m   D) 1.8 m

78) Find the missing measurement from the figure given below. Round the answer to the nearest tenth.

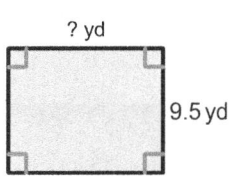

? yd
9.5 yd
Area = 116.9 yd²

A) 16.8 yd   B) 10.4 yd

C) 12.3 yd   D) 15.1 yd

Grade 7
Volume 1
Week 15

79) Find the missing measurement from the figure given below. Round the answer to the nearest tenth.

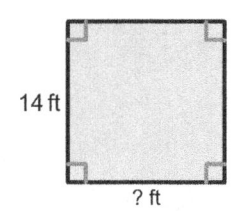

14 ft
? ft
Area = 196 ft²

A) 8.5 ft   B) 19.2 ft
C) 14 ft   D) 11.7 ft

80) Find the missing measurement from the figure given below. Round the answer to the nearest tenth.

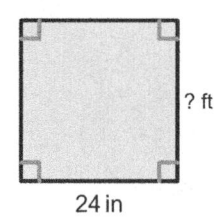

? ft
24 in
Area = 4 ft²

A) 2.7 ft   B) 2.1 ft
C) 2 ft   D) 1.2 ft

81) Find the missing measurement from the figure given below. Round the answer to the nearest tenth.

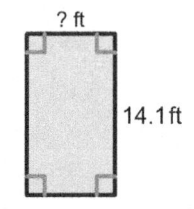

? ft
14.1 ft
Area = 112.8 ft²

A) 10.7 ft   B) 6.9 ft
C) 8 ft   D) 10 ft

82) Find the missing measurement from the figure given below. Round the answer to the nearest tenth.

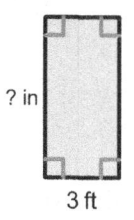

? in
3 ft
Area = 18.9 ft²

A) 75.6 in   B) 94.2 in
C) 6.3 in   D) 100.6 in

83) Find the missing measurement from the figure given below. Round the answer to the nearest tenth.

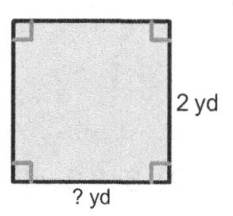

2 yd
? yd
Area = 4 yd²

A) 2.6 yd   B) 1.8 yd
C) 2 yd   D) 1.7 yd

84) Find the missing measurement from the figure given below. Round the answer to the nearest tenth.

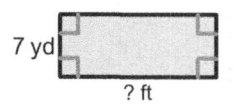

7 yd
? ft
Area = 126 yd²

A) 52.3 ft   B) 18 ft
C) 54.3 ft   D) 54 ft

©All rights reserved-Math-Knots LLC., VA-USA
www.math-knots.com | www.a4ace.com

85) Find the missing measurement from the figure given below. Round the answer to the nearest tenth.

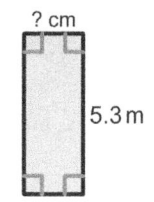

? cm
5.3 m
Area = 10.6 m²

A) 276.2 cm   B) 2 cm

C) 126.3 cm   D) 200 cm

86) Find the missing measurement from the figure given below. Round the answer to the nearest tenth.

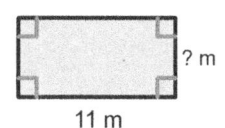

? m
11 m
Area = 59.4 m²

A) 5.4 m   B) 7.3 m

C) 4.9 m   D) 5.7 m

87) Find the missing measurement from the figure given below. Round the answer to the nearest tenth.

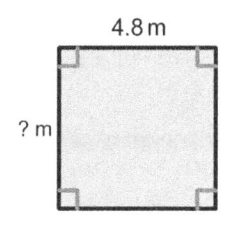

4.8 m
? m
Area = 23 m²

A) 3.1 m   B) 4.8 m

C) 3.8 m   D) 4.5 m

88) Find the missing measurement from the figure given below. Round the answer to the nearest tenth.

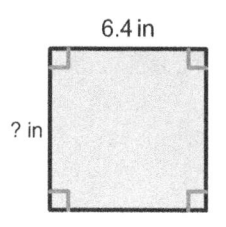

6.4 in
? in
Area = 41 in²

A) 8.5 in   B) 7.7 in

C) 6.4 in   D) 6.8 in

89) Find the missing measurement from the figure given below. Round the answer to the nearest tenth.

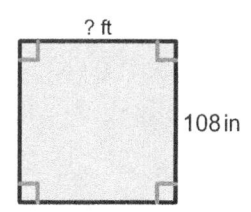

? ft
108 in
Area = 81 ft²

A) 9 ft   B) 0.8 ft

C) 6.8 ft   D) 10.4 ft

90) Find the missing measurement from the figure given below. Round the answer to the nearest tenth.

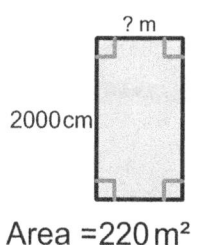

? m
2000 cm
Area = 220 m²

A) 11 m   B) 6.9 m

C) 7.2 m   D) 12.7 m

Grade 7
Volume 1
Week 15

91) Find the missing measurement from the figure given below. Round the answer to the nearest tenth.

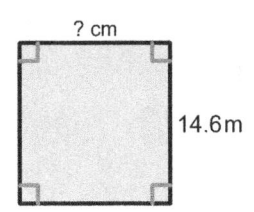

? cm
14.6 m
Area = 205.9 m²

A) 14.1 cm
B) 963.9 cm
C) 970 cm
D) 1410.3 cm

92) Find the missing measurement from the figure given below. Round the answer to the nearest tenth.

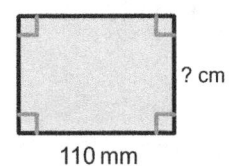

? cm
110 mm
Area = 88 cm²

A) 0.8 cm
B) 7.7 cm
C) 8 cm
D) 11.1 cm

93) Find the missing measurement from the figure given below. Round the answer to the nearest tenth.

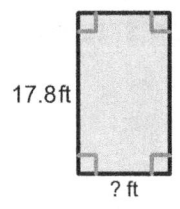

17.8 ft
? ft
Area = 186.9 ft²

A) 13.3 ft
B) 6.8 ft
C) 7.7 ft
D) 10.5 ft

94) Find the missing measurement from the figure given below. Round the answer to the nearest tenth.

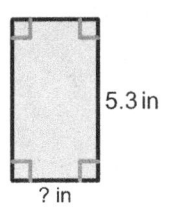

5.3 in
? in
Area = 15.4 in²

A) 2.5 in
B) 3.5 in
C) 2.8 in
D) 2.9 in

95) Find the missing measurement from the figure given below. Round the answer to the nearest tenth.

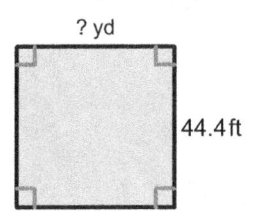

? yd
44.4 ft
Area = 219 yd²

A) 14.3 yd
B) 4.9 yd
C) 14.8 yd
D) 17.9 yd

96) Find the missing measurement from the figure given below. Round the answer to the nearest tenth.

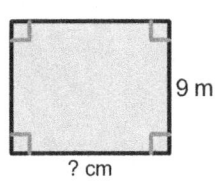

9 m
? cm
Area = 99 m²

A) 819.7 cm
B) 949.9 cm
C) 11 cm
D) 1100 cm

Grade 7
Volume 1
Week 15

97) Find the missing measurement from the figure given below. Round the answer to the nearest tenth.

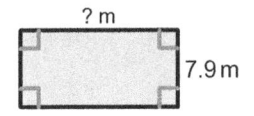

? m
7.9 m
Area = 130.4 m²

A) 17.3 m   B) 19 m

C) 20.5 m   D) 16.5 m

98) Find the missing measurement from the figure given below. Round the answer to the nearest tenth.

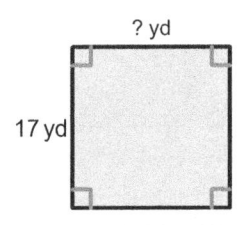

? yd
17 yd
Area = 289 yd²

A) 22.7 yd   B) 19.9 yd

C) 18.4 yd   D) 17 yd

99) Find the area of the figure given below.

5.8 ft
11.7 ft
16 ft

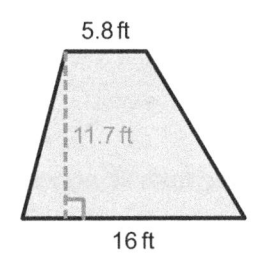

A) 127.53 ft²   B) 255.06 ft²

C) 134.53 ft²   D) 63.8 ft²

100) Find the missing measurement from the figure given below. Round the answer to the nearest tenth.

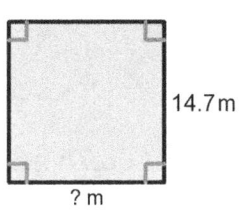

14.7 m
? m
Area = 216.1 m²

A) 20.1 m   B) 20.2 m

C) 17 m   D) 14.7 m

101) Find the missing measurement from the figure given below. Round the answer to the nearest tenth.

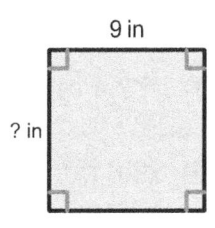

9 in
? in
Area = 81 in²

A) 9 in   B) 8.6 in

C) 12.5 in   D) 7.7 in

102) Find the area of the figure given below.

960 cm
9 m
3000 cm

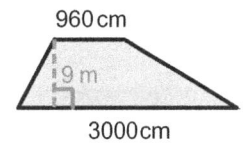

A) 178.2 m²   B) 356.4 m²

C) 89.1 m²   D) 172.8 m²

Grade 7
Volume 1
Week 15

103) Find the area of the figure given below.

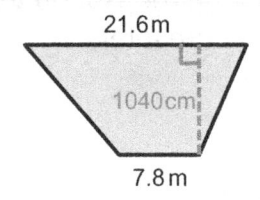

21.6 m
1040 cm
7.8 m

A) 152.88 m²   B) 156.78 m²

C) 76.4 m²   D) 305.76 m²

104) Find the area of the figure given below.

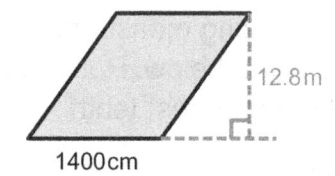

12.8 m
1400 cm

A) 179.2 m²   B) 172.4 m²

C) 358.4 m²   D) 89.6 m²

105) Find the area of the figure given below.

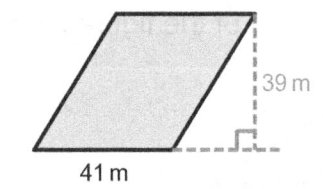

39 m
41 m

A) 3198 m²   B) 1596.4 m²

C) 1599 m²   D) 799.5 m²

106) Find the area of the figure given below.

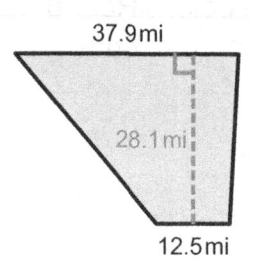

37.9 mi
28.1 mi
12.5 mi

A) 1416.24 mi²   B) 708.12 mi²

C) 354.1 mi²   D) 714.02 mi²

107) Find the area of the figure given below.

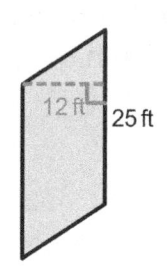

12 ft   25 ft

A) 303.5 ft²   B) 300 ft²

C) 150 ft²   D) 600 ft²

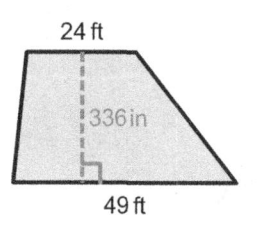

108) Find the area of the figure given below.

24 ft
336 in
49 ft

A) 511 ft²   B) 1029.1 ft²

C) 1022 ft²   D) 2044 ft²

Grade 7
Volume 1
Week 15

109) Find the area of the figure given below.

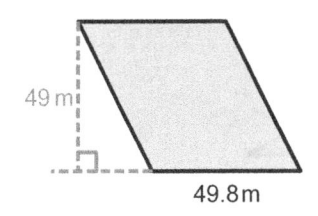

49 m
49.8 m

A) 1220.1 m²   B) 4880.4 m²

C) 2448.3 m²   D) 2440.2 m²

110) Find the area of the figure given below.

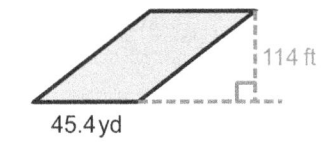

114 ft
45.4 yd

A) 1719.5 yd²   B) 3450.4 yd²

C) 1725.2 yd²   D) 862.6 yd²

111) Find the area of the figure given below.

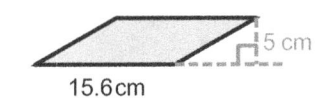

5 cm
15.6 cm

A) 76.6 cm²   B) 156 cm²

C) 39 cm²   D) 78 cm²

112) Find the area of the figure given below.

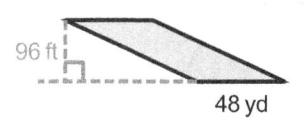

96 ft
48 yd

A) 768 yd²   B) 3072 yd²

C) 1535 yd²   D) 1536 yd²

113) Find the area of the figure given below.

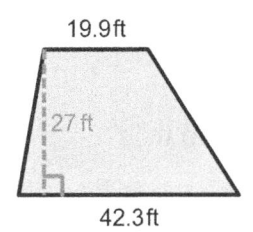

19.9 ft
27 ft
42.3 ft

A) 419.9 ft²   B) 839.8 ft²

C) 839.7 ft²   D) 837.9 ft²

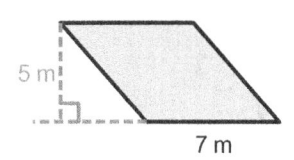

114) Find the area of the figure given below.

5 m
7 m

A) 35 m²   B) 29.3 m²

C) 39.2 m²   D) 70 m²

©All rights reserved-Math-Knots LLC., VA-USA   www.math-knots.com | www.a4ace.com

**Grade 7
Volume 1
Week 15**

115) Find the area of the figure given below.

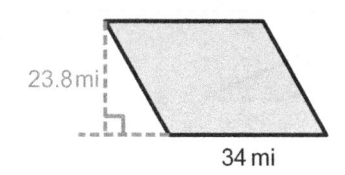

23.8 mi
34 mi

A) 799.2 mi²   B) 404.6 mi²

C) 809.2 mi²   D) 1618.4 mi²

118) Find the area of the figure given below.

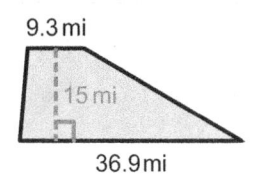

9.3 mi
15 mi
36.9 mi

A) 342.8 mi²   B) 173.3 mi²

C) 341.5 mi²   D) 346.5 mi²

116) Find the area of the figure given below.

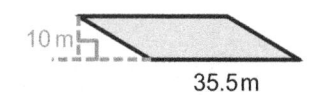

10 m
35.5 m

A) 710 m²   B) 355 m²

C) 354 m²   D) 177.5 m²

119) Find the area of the figure given below.

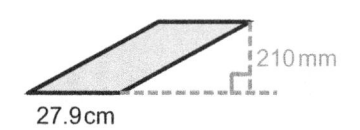

210 mm
27.9 cm

A) 585.9 cm²   B) 586 cm²

C) 589.4 cm²   D) 293 cm²

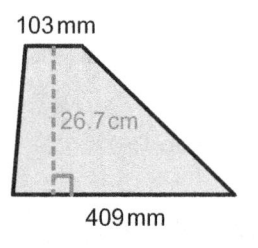

117) Find the area of the figure given below.

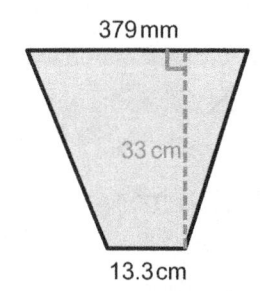

379 mm
33 cm
13.3 cm

A) 422.4 cm²   B) 844.8 cm²

C) 841.4 cm²   D) 1689.6 cm²

120) Find the area of the figure given below.

103 mm
26.7 cm
409 mm

A) 683.52 cm²   B) 1367.04 cm²

C) 677.22 cm²   D) 341.8 cm²

©All rights reserved-Math-Knots LLC., VA-USA

121) Find the area of the figure given below.

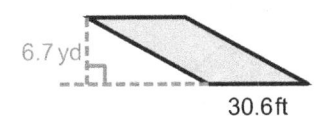

6.7 yd
30.6 ft

A) 136.68 yd²   B) 34.2 yd²

C) 68.34 yd²   D) 76.74 yd²

124) Find the area of the figure given below.

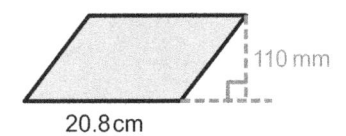

110 mm
20.8 cm

A) 457.6 cm²   B) 462.8 cm²

C) 228.8 cm²   D) 231.4 cm²

122) Find the area of the figure given below.

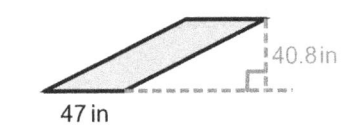

40.8 in
47 in

A) 958.8 in²   B) 1917.6 in²

C) 1912.8 in²   D) 3835.2 in²

125) Find the area of the figure given below.

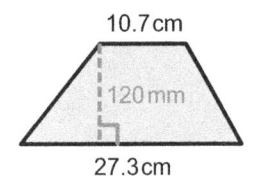

10.7 cm
120 mm
27.3 cm

A) 228 cm²   B) 221.6 cm²

C) 110.8 cm²   D) 114 cm²

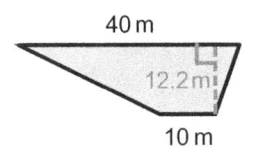

123) Find the area of the figure given below.

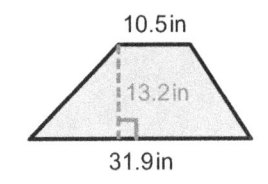

10.5 in
13.2 in
31.9 in

A) 139.9 in²   B) 279.84 in²

C) 287.84 in²   D) 559.68 in²

126) Find the area of the figure given below.

40 m
12.2 m
10 m

A) 152.5 m²   B) 610 m²

C) 313 m²   D) 305 m²

**Grade 7
Volume 1
Week 15**

127) Find the area of the figure given below.

30 ft
456 in

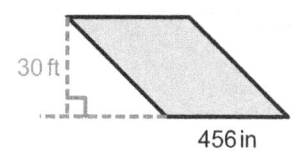

A) 570 ft² B) 1140 ft²

C) 1145.2 ft² D) 2280 ft²

130) Find the area of the figure given below.

35.1 km
48.2 km

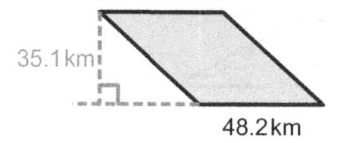

A) 1684.42 km²

B) 3383.64 km²

C) 1691.82 km²

D) 1683.62 km²

128) Find the area of the figure given below.

7.9 mi
12.5 mi
15.7 mi

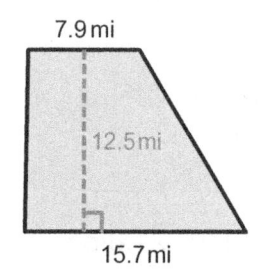

A) 141.1 mi² B) 147.5 mi²

C) 73.8 mi² D) 295 mi²

131) Find the missing measurement from the figure given below. Round the answer to the nearest tenth.

4.5 yd
7.5 yd
10.9 yd

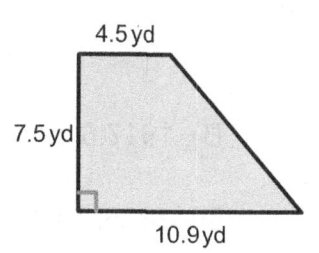

A) 57.75 yd² B) 48.45 yd²

C) 28.9 yd² D) 115.5 yd²

129) Find the missing measurement from the figure given below. Round the answer to the nearest tenth.

? yd
12 yd
17.3 yd

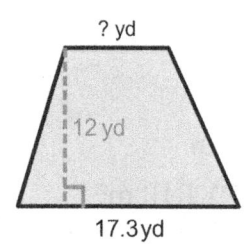

Area = 152.4 yd²

A) 9.7 yd B) 11.3 yd

C) 8.1 yd D) 5 yd

132) Find the missing measurement from the figure given below. Round the answer to the nearest tenth.

18.1 km
19 km
? km

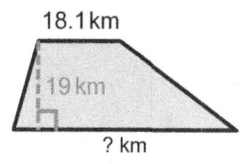

Area = 636.5 km²

A) 48.9 km B) 38.3 km

C) 55.2 km D) 57.1 km

133) Find the missing measurement from the figure given below. Round the answer to the nearest tenth.

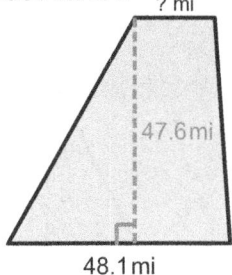

? mi
47.6 mi
48.1 mi
Area =1556.5mi²

A) 20 mi   B) 24.2 mi
C) 17.3 mi   D) 22.4 mi

134) Find the missing measurement from the figure given below. Round the answer to the nearest tenth.

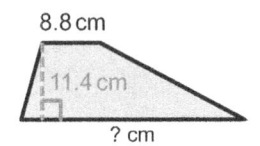

8.8 cm
11.4 cm
? cm
Area =244 cm²

A) 20.8 cm   B) 38.1 cm
C) 24.5 cm   D) 34 cm

135) Find the missing measurement from the figure given below. Round the answer to the nearest tenth.

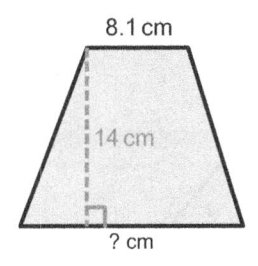

8.1 cm
14 cm
? cm
Area =182 cm²

A) 21.6 cm   B) 24.8 cm
C) 17.9 cm   D) 14 cm

136) Find the missing measurement from the figure given below. Round the answer to the nearest tenth.

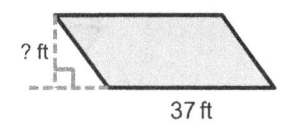

? ft
37 ft
Area =577.2ft²

A) 13.5 ft   B) 15.6 ft
C) 10.2 ft   D) 16.6 ft

137) Find the missing measurement from the figure given below. Round the answer to the nearest tenth.

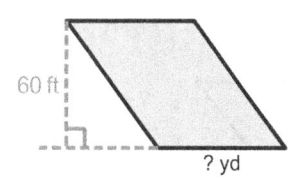

60 ft
? yd
Area =418 yd²

A) 7 yd   B) 21.4 yd
C) 20.9 yd   D) 26.1 yd

138) Find the missing measurement from the figure given below. Round the answer to the nearest tenth.

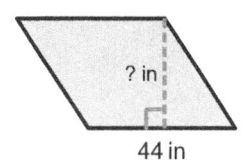

? in
44 in
Area =1381.6in²

A) 31.4 in   B) 32.1 in
C) 34.6 in   D) 25.8 in

139) Find the missing measurement from the figure given below. Round the answer to the nearest tenth.

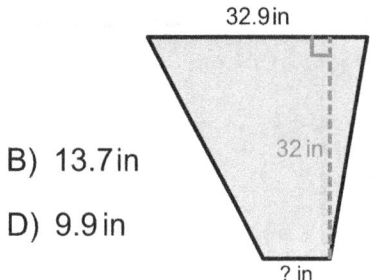

32.9 in
32 in
? in
Area = 684.8 in²

A) 13.2 in   B) 13.7 in
C) 6.2 in   D) 9.9 in

140) Find the missing measurement from the figure given below. Round the answer to the nearest tenth.

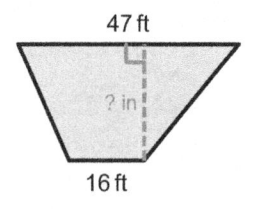

47 ft
? in
16 ft
Area = 756 ft²

A) 247.6 in   B) 288 in
C) 24 in   D) 369.4 in

141) Find the missing measurement from the figure given below. Round the answer to the nearest tenth.

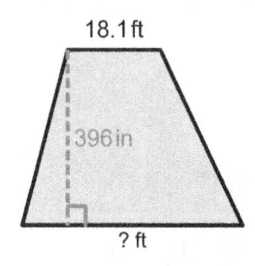

18.1 ft
396 in
? ft
Area = 1006.5 ft²

A) 54.8 ft   B) 33.9 ft
C) 3.6 ft   D) 42.9 ft

142) Find the missing measurement from the figure given below. Round the answer to the nearest tenth.

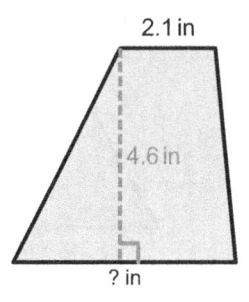

2.1 in
4.6 in
? in
Area = 16.1 in²

A) 5 in   B) 4.2 in
C) 5.6 in   D) 4.9 in

143) Find the missing measurement from the figure given below. Round the answer to the nearest tenth.

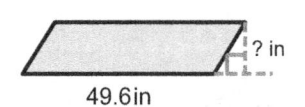

? in
49.6 in
Area = 644.8 in²

A) 8.5 in   B) 10.6 in
C) 18.2 in   D) 13 in

144) Find the missing measurement from the figure given below. Round the answer to the nearest tenth.

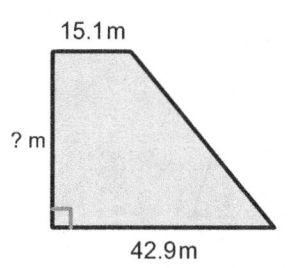

15.1 m
? m
42.9 m
Area = 962.8 m²

A) 42 m   B) 33.2 m
C) 30 m   D) 21 m

Grade 7
Volume 1
Week 15

145) Find the missing measurement from the figure given below. Round the answer to the nearest tenth.

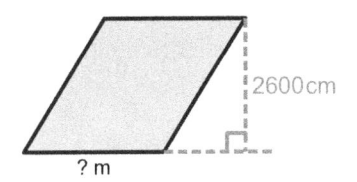

2600 cm
? m
Area = 730.6 m²

A) 0.3 m
B) 17.4 m
C) 28.1 m
D) 27.5 m

146) Find the missing measurement from the figure given below. Round the answer to the nearest tenth.

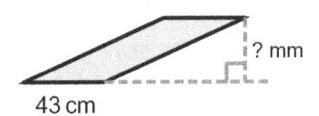

? mm
43 cm
Area = 1419 cm²

A) 321.4 mm
B) 331.9 mm
C) 330 mm
D) 33 mm

147) Find the missing measurement from the figure given below. Round the answer to the nearest tenth.

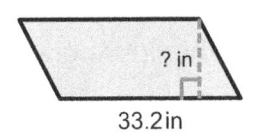

? in
33.2 in
Area = 468.1 in²

A) 10.8 in
B) 14.1 in
C) 11.3 in
D) 14.2 in

148) Find the missing measurement from the figure given below. Round the answer to the nearest tenth.

? in
35 ft
Area = 791 ft²

A) 22.6 in
B) 251 in
C) 255.4 in
D) 271.2 in

149) Find the missing measurement from the figure given below. Round the answer to the nearest tenth.

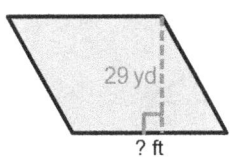

29 yd
? ft
Area = 1160 yd²

A) 40 ft
B) 98.5 ft
C) 119.9 ft
D) 120 ft

150) Find the missing measurement from the figure given below. Round the answer to the nearest tenth.

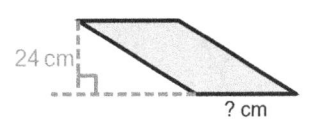

24 cm
? cm
Area = 820.8 cm²

A) 22.2 cm
B) 31.1 cm
C) 34.2 cm
D) 25.6 cm

151) Find the missing measurement from the figure given below. Round the answer to the nearest tenth.

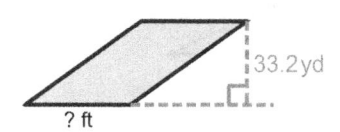

33.2 yd
? ft
Area = 1427.6 yd²

A) 108.4 ft B) 43 ft
C) 129 ft D) 160.5 ft

152) Find the missing measurement from the figure given below. Round the answer to the nearest tenth.

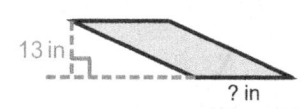

13 in
? in
Area = 299 in²

A) 15.9 in B) 30.7 in
C) 22.1 in D) 23 in

153) Find the missing measurement from the figure given below. Round the answer to the nearest tenth.

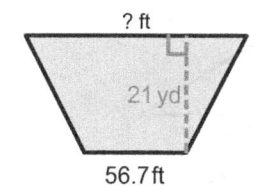

? ft
21 yd
56.7 ft
Area = 627.9 yd²

A) 122.7 ft B) 40.9 ft
C) 154.7 ft D) 140.3 ft

154) Find the missing measurement from the figure given below. Round the answer to the nearest tenth.

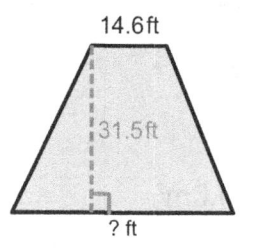

14.6 ft
31.5 ft
? ft
Area = 907.2 ft²

A) 43 ft B) 60.2 ft
C) 45 ft D) 56.2 ft

155) Find the missing measurement from the figure given below. Round the answer to the nearest tenth.

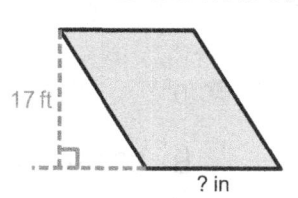

17 ft
? in
Area = 289 ft²

A) 17 in B) 204 in
C) 168.7 in D) 238.3 in

156) Find the area of the figure given below.

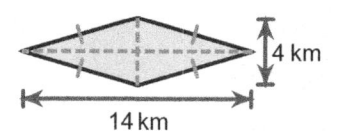

4 km
14 km

A) 18.4 km² B) 28 km²
C) 56 km² D) 14 km²

157) Find the area of the figure given below.

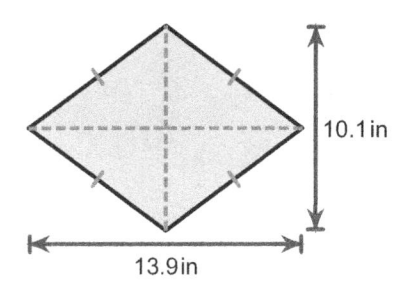

10.1 in
13.9 in

A) 66.695 in²   B) 35.1 in²

C) 70.195 in²   D) 140.39 in²

158) Find the area of the figure given below.

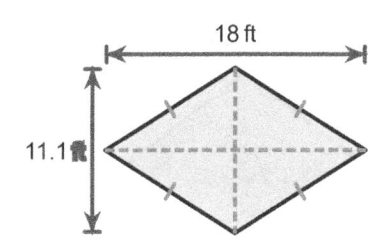

18 ft
11.1 ft

A) 99.9 ft²   B) 97.3 ft²

C) 199.8 ft²   D) 50 ft²

159) Find the area of the figure given below.

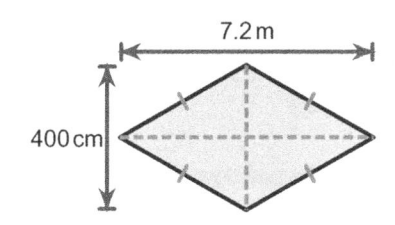

7.2 m
400 cm

A) 28.8 m²   B) 7.2 m²

C) 14.4 m²   D) 17.1 m²

160) Find the area of the figure given below.

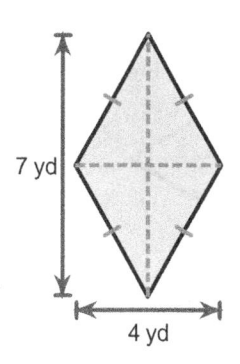

7 yd
4 yd

A) 14 yd²   B) 7 yd²

C) 28 yd²   D) 6.1 yd²

161) Find the area of the figure given below.

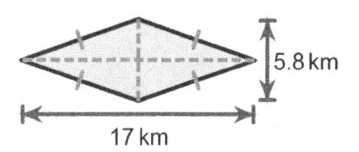

5.8 km
17 km

A) 49.3 km²   B) 56.9 km²

C) 24.7 km²   D) 98.6 km²

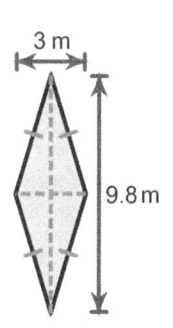

162) Find the area of the figure given below.

3 m
9.8 m

A) 29.4 m²   B) 14.7 m²

C) 7.4 m²   D) 5.1 m²

163) Find the area of the figure given below.

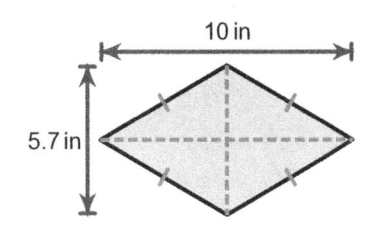

10 in
5.7 in

A) 28.5 in²   B) 57 in²

C) 30.1 in²   D) 14.3 in²

164) Find the area of the figure given below.

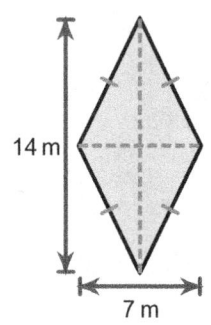

14 m
7 m

A) 49 m²   B) 24.5 m²

C) 43.5 m²   D) 98 m²

165) Find the area of the figure given below.

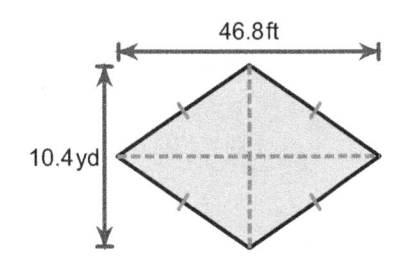

46.8 ft
10.4 yd

A) 81.12 yd²   B) 78.22 yd²

C) 162.24 yd²   D) 40.6 yd²

166) Find the area of the figure given below.

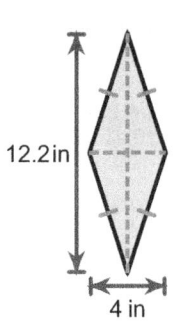

12.2 in
4 in

A) 29.8 in²   B) 48.8 in²

C) 24.4 in²   D) 12.2 in²

167) Find the area of the figure given below.

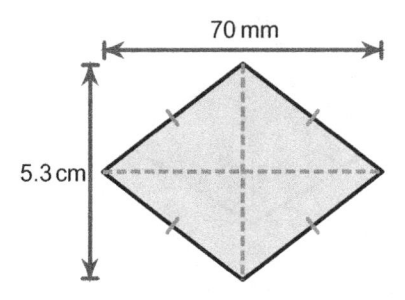

70 mm
5.3 cm

A) 37.1 cm²   B) 18.55 cm²

C) 9.3 cm²   D) 17.05 cm²

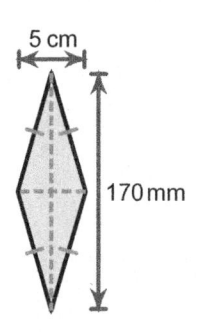

168) Find the area of the figure given below.

5 cm
170 mm

A) 21.3 cm²   B) 42.5 cm²

C) 39.8 cm²   D) 19.9 cm²

169) Find the area of the figure given below.

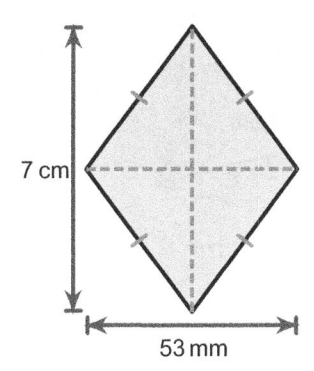

7 cm
53 mm

A) 18.55 cm²   B) 7.8 cm²

C) 15.65 cm²   D) 31.3 cm²

170) Find the area of the figure given below.

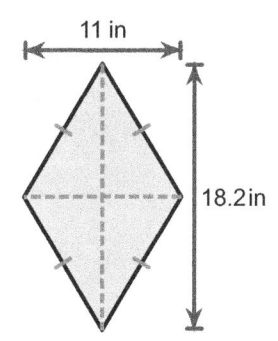

11 in
18.2 in

A) 50.1 in²   B) 200.2 in²

C) 100.1 in²   D) 109.9 in²

171) Find the area of the figure given below.

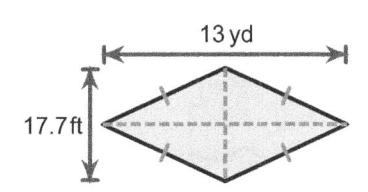

13 yd
17.7 ft

A) 38.35 yd²   B) 31.75 yd²

C) 19.2 yd²   D) 76.7 yd²

172) Find the area of the figure given below.

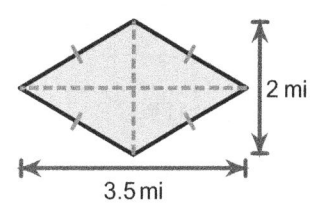

2 mi
3.5 mi

A) 1.8 mi²   B) 3.5 mi²

C) 7.2 mi²   D) 10.3 mi²

173) Find the area of the figure given below.

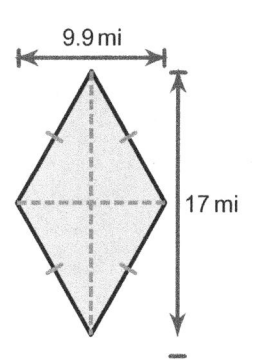

9.9 mi
17 mi

A) 42.1 mi²   B) 168.3 mi²

C) 75.35 mi² D) 84.15 mi²

174) Find the area of the figure given below.

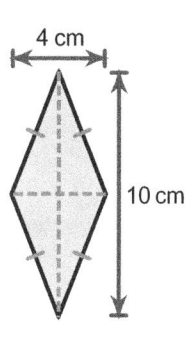

4 cm
10 cm

A) 10 cm²   B) 40 cm²

C) 14.2 cm² D) 20 cm²

175) Find the area of the figure given below.

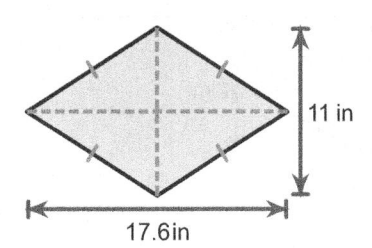

A) 114.4 in² B) 105.5 in²

C) 193.6 in² D) 96.8 in²

176) Find the missing measurement from the figure given below. Round the answer to the nearest tenth.

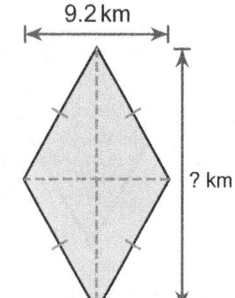

Area =73.6 km²

A) 12.6 B) 16

C) 9.9 D) 20.6

177) Find the missing measurement from the figure given below. Round the answer to the nearest tenth.

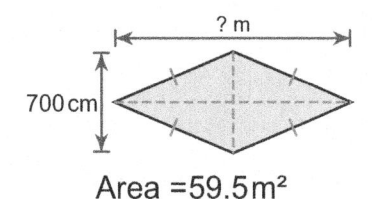

Area =59.5 m²

A) 23.6 B) 23.1

C) 13 D) 17

178) Find the missing measurement from the figure given below. Round the answer to the nearest tenth.

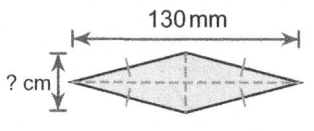

Area =21.45 cm²

A) 4 B) 0.3

C) 3.9 D) 3.3

179) Find the missing measurement from the figure given below. Round the answer to the nearest tenth.

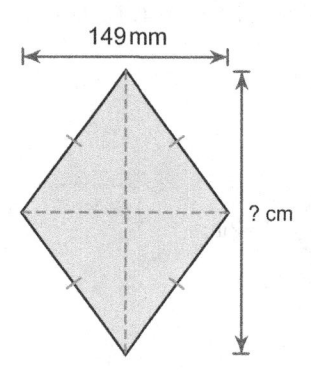

Area =149 cm²

A) 13.9 B) 2

C) 14.9 D) 20

180) Find the missing measurement from the figure given below. Round the answer to the nearest tenth.

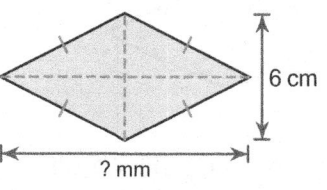

Area =36 cm²

A) 103.3 B) 120

C) 12 D) 159.2

181) Find the missing measurement from the figure given below. Round the answer to the nearest tenth.

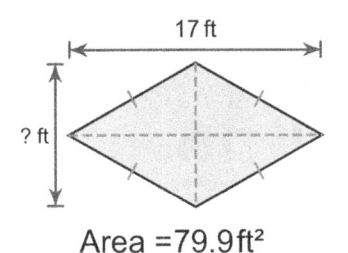

17 ft
? ft
Area =79.9 ft²

A) 7.2   B) 11.1
C) 6.7   D) 9.4

182) Find the missing measurement from the figure given below. Round the answer to the nearest tenth.

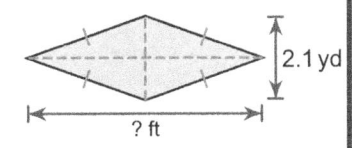

2.1 yd
? ft
Area =6.3 yd²

A) 6   B) 18.1
C) 18   D) 22.8

183) Find the missing measurement from the figure given below. Round the answer to the nearest tenth.

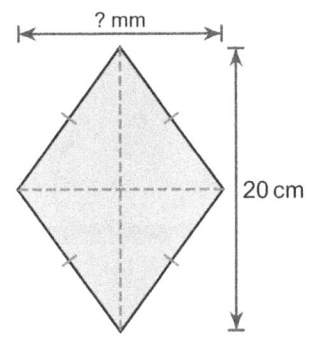

? mm
20 cm
Area =149 cm²

A) 206.5   B) 185.5
C) 14.9   D) 149

184) Find the missing measurement from the figure given below. Round the answer to the nearest tenth.

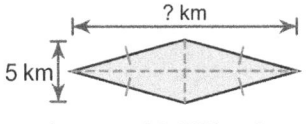

? km
5 km
Area =46.75 km²

A) 16.7   B) 18.7
C) 17   D) 22.9

185) Find the missing measurement from the figure given below. Round the answer to the nearest tenth.

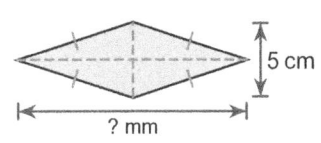

5 cm
? mm
Area =39.25 cm²

A) 179.7   B) 157
C) 211.1   D) 15.7

Grade 7
Volume 1
Week 16

1) The area of a right triangular region is 132.46 ft^2. If one of its sides containing the right angle is 14.8 ft, find the other side.

2) A field is in form of a triangle with area of 3 hectare. If length of its base is 200 m, find its altitude.

3) Find the area of a triangle whose base is 24 ft and height is 15 ft.

4) Find the height of a triangle whose base is 40 inches and area is 300 inches

5) Find the altitude of the triangle whose area is 238 inches^2 and base is 34 inches.

6) Find the base of the triangle whose height is 24 inches and area is 360 inches^2

7) Find the area of an isosceles right triangle of equal sides 22 inches each.

8) The base of an isosceles triangle is 16 inches and its perimeter is 50 inches. Find its area.

9) The legs of a right angle triangle are in the ratio 3:4 and its area is 216 inches^2. Find its hypotenuse.

10) The base and height of a triangular field are 23 yards and 34 yards respectively. Find the cost of leveling the field at the rate of 50 cents per square yard.

11) The base and height of a triangular area are 36 ft and 24 ft respectively. Find the cost of laying tiles at the rate of $1.25 per ft^2

12) The sides of an isosceles triangle measure 10 inch, 10 inch and 12 inch. Find the area of the triangle.

13) Find the altitude of a triangle whose area is 600 inches and base is 30 inches.

14) Find the area of the equilateral triangle whose each side measures 4 inch.
(Take, √3 = 1.732)

15) Find the altitude of a triangular board whose area is 133 ft and base is 14 ft

16) The base of a triangular sign board is 23 ft and its height is 17 ft. Find the cost of painting it at the rate of $2 per ft

17) The cost of painting a triangular shaped sign board at the rate of $3 per square feet is $240. If the base of the sign board is 16 ft, find the height of the sign board.

18) Find the height of a parallelogram, whose one side is 18 inches and area is 153 inches^2

19) The area of rhombus is 119 inches^2 and its perimeter is 56 inches. Find its altitude.

20) The area of rhombus is 24 inches^2 and its height is 4 inches. Find its side.

21) Find the area of a parallelogram whose base is 15 inches and corresponding height is 9 inches.

22) Find the area of rhombus whose diagonals have length 24 inches and 18 inches.

23) Find the area of a parallelogram whose base is 6 inches and the corresponding height is 4 inches.

24) Find the height of the parallelogram whose area is 35 inches^2 and base is 7 inches.

Grade 7
Volume 1
Week 16

25) Find the altitude of a parallelogram whose one of the sides is 9 inches and area is 72 inches^2

26) Find the area of a rhombus whose side is 8 inches and altitude is 5 inches.

27) What is the measure of each side of a rhombus whose area is 280 inches^2 and height is 14 inches?

28) The lengths of the diagonals of a rhombus measure 16 inches and 23 inches respectively. Find the area of the rhombus.

29) The area of a rhombus is 264 inches^2 and its perimeter is 48 inches. Find the length of its altitude.

30) The height of a parallelogram is 2/3 of its base. If the area of the parallelogram is 150 inches^2, find the base and the height of the parallelogram.

31) The base and the corresponding height of a parallelogram are 8 inches and 10 inches respectively. If the other altitude is 5 inches, find the length of each of the other pair of parallel sides.

32) Find the length of each side of the rhombus whose area is 368 ft^2 and altitude is 23 ft.

©All rights reserved-Math-Knots LLC., VA-USA   www.math-knots.com | www.a4ace.com

Grade 7
Volume 1
Week 16

33) Find the area of a parallelogram whose base is 12 ft and the corresponding height is 13 ft.

34) The area of a rhombus is 255 inches^2. If one of its diagonals is of length 34 inches, find the length of the other diagonal.

35) Find the area of a rhombus whose diagonals measure 15 inches and 21 inches.

36) Find the area of a parallelogram, in square inch, whose base is 2 ft and height is 15 inch.

37) In the figure ABCD is a parallelogram. AB & CD measure 8 inches each and AD & BC measure 5 inches each. The perpendicular drawn from A meets DC at M. , If the area of the parallelogram is 24 inches^2, find the measure of DM.

38) In the figure given below, ABCD is a parallelogram. CN $\perp$ AB and BM $\perp$ AD. If AB = 15 cm, AD = 8 cm and CN = 12 cm, find the length of BM.

39) In the figure given below ABCD is a parallelogram. CN $\perp$ AB and BM $\perp$ AD. If AB = 18 cm, CN = 10 cm and BM = 15 cm, find the length of AD.

40) In the figure given below ABCD is a parallelogram. CN $\perp$ AB and BM $\perp$ AD. If AD = 9 cm, CN = 12 cm and BM = 15 cm then find AB.

41) In a parallelogram ABCD (figure given below), AD = 13 cm, AB = 18 cm and the area of the parallelogram is 216 cm^2. If DP$\perp$AB, find the length of AP.

42) In a parallelogram ABCD, DP$\perp$AB, AD = 25 cm and AP = 7 cm. If the area of the parallelogram is 720 cm^2, find the length of AB.

43) Find the height of trapezium, the sum of lengths of whose bases is 16 dm and area is 1.6 m^2

44) Find the sum of lengths of bases of trapezium whose area is 2.025 m^2 and altitude is 9 cm.

45) The parallel sides of a trapezium are 25 inches and 11 inches and non parallel sides are 13 inches and 15 inches. Find its area.

46) The parallel sides of a trapezium are 10 inches and 15 inches in length. The distance between these sides is 8 inches. Find the area of the trapezium.

47) The area of trapezium of height 9 inches is 180 inches^2. If one of the parallel sides of the trapezium is 18 inches, what is the length of the other side?

48) Find the height of the trapezium whose area is 525 inches^2 and whose bases are 23 inches and 19 inches.

49) Find the altitude of the trapezium, whose area is 350 inches^2 and the sum of the lengths of bases, is 56 inches.

50) The area of trapezium is 162 ft^2 and its height is 12 ft. If one of the bases measure double that of the other, find the length of the bases.

51) The area of a trapezium is 245 inches and the altitude is 7 inches. If the bases are in the ratio 2:3, find the length of each base.

52) Find the area of a trapezium whose parallel sides are 23 inches and 26 inches and the distance between them is 16 inches.

53) The parallel sides of a trapezium are 35 inches and 19 inches respectively. Its non parallel sides are equal, each measuring 17 inches. Find the area of the trapezium.

54) The area of a trapezium is 160 ft^2 and its height is 8 ft. If one of the parallel sides is 10 ft longer than the other, find the measure of the two parallel sides.

55) The parallel sides of a trapezium are 15 inches and 9 inches in length. The distance between these sides is 11 inches. Find the area of the trapezium.

56) The area of a trapezium is 168 ft^2 and its height is 8 ft. If the parallel sides are in the ratio 3:4, find the length of each of the parallel sides.

Grade 7
Volume 1
Week 16

57) Find the area of a trapezium whose parallel sides are of length 23 inches and 18 inches and whose height is 15 inches.

58) The area of a trapezium is 351 ft^2. If the parallel sides of the trapezium measure 17 ft and 22 ft respectively, find the height of the trapezium.

59) The area of a trapezium is 180 inches^2 and its height is 12 inches. If one parallel side is half the other side, find the length of the parallel sides.

60) Find the area of a trapezium whose parallel sides are of length 26 inches and 39 inches respectively and the vertical distance between them is 32 inches.

61) Find the area of a trapezium whose parallel sides are 7 inches and 11 inches and the distance between them is 8 inches.

62) Find the area of a trapezium whose parallel sides are 5 inches and 8 inches and the distance between them is 10 inches.

63) Find the area of a trapezium whose parallel sides are 18 inches and 11 inches and the distance between them is 15 inches.

64) Find the area of a trapezium whose parallel sides are 13 inches and 15 inches and the distance between them is 14 inches.

65) Find the area of a trapezium whose parallel sides are 15 inches and 21 inches and the distance between them is 9 inches.

66) The parallel sides of a trapezium are 14 inches and 8 inches. Its non parallel sides are equal and each measure 5 inches. Find the area of the trapezium.

67) The parallel sides of a trapezium are 11 inches and 23 inches. The non parallel sides are equal and measure 10 inches each. Find the area of the trapezium.

68) The parallel sides of a trapezium are 24 inches and 14 inches. The non parallel sides are equal and measure 13 inches each. Find the area of the trapezium.

69) The parallel sides of a trapezium are 13 inches and 25 inches. The non parallel sides are equal and measure 10 inches each. Find the area of the trapezium.

70) The parallel sides of a trapezium are 7 inches and 5.2 inches. The non parallel sides are equal and measure 4.1 inches each. Find the area of the trapezium.

71) The parallel sides of a trapezium are 11 inches and 21 inches. The non parallel sides are equal and measure 13 inches each. Find the area of the trapezium

72) The top surface of a table is in the form of a trapezium. Its parallel sides measure 100 cm and 120 cm respectively and the perpendicular distance between them is 80 cm. Find the area of the top of the table.

Grade 7
Volume 1
Week 16

73) The top surface of a table is in the form of a trapezium. Its parallel sides measure 80 cm and 120 cm respectively and the perpendicular distance between them is 150 cm. Find the area of the top of the table.

74) A field is in the shape of a trapezium. Its parallel sides are 56 ft and 88 ft respectively and the perpendicular distance between them is 80 ft. Find the area of the field.

75) A flower bed is in the shape of a trapezium. Its parallel sides measure 10 m and 15 m respectively and the perpendicular distance between them is 90 cm. Find the area of the flower bed.

76) The parallel sides of a trapezium measure 56 inches and 48 inches and the perpendicular distance between them is 2.5 ft. Find the area of the trapezium.

77) The area of a trapezium is 98 inches^2. The distance between the parallel sides is 7 inches. If one of the parallel sides is 15 inches, find the length of the other side.

78) The area of a trapezium is 203 inches^2. The distance between the parallel sides is 14 inches. If one of the parallel sides is 18 inches, find the length of the other side.

79) The area of a trapezium is 323 inches^2. The distance between the parallel sides is 17 inches. If one of the parallel sides is 21 inches, find the length of the other side.

80) The area of a trapezium is 4 cm^2. The distance between the parallel sides is 1 cm. If one of the parallel sides is 3 cm, find the length of the other side.

Grade 7
Volume 1
Week 16

81) The area of a trapezium is 159.5 inches^2. The distance between the parallel sides is 11 inches. If one of the parallel sides is 7 inches, find the length of the other side.

82) The area of a trapezium is 159.5 inches^2. The parallel sides measure 9 inches and 20 inches respectively. Find the distance between the parallel sides.

83) The area of a trapezium is 34 inches^2. The parallel sides measure 6 inches and 11 inches respectively. Find the distance between the parallel sides.

84) The area of a trapezium is 360 inches^2. The parallel sides measure 19 inches and 11 inches respectively. Find the distance between the parallel sides.

85) The area of a trapezium is 360 inches^2. The parallel sides measure 19 inches and 11 inches respectively. Find the distance between the parallel sides.

86) The area of a trapezium is 98 inches^2. The parallel sides measure 15 inches and 13 inches respectively. Find the distance between the parallel sides.

87) Fill in the blank:
If the sides of a quadrilateral are produced in an order, the sum of the four exterior angles so formed is _____.

88) The angles of a quadrilateral are in the ratio 2 : 3 : 5 : 8. Find the measure of each of the four angles.

89) Fill in the blank:
The sum of the angles of a quadrilateral is _____ right angles.

90) Fill in the blank:
A quadrilateral has _____ diagonals.

91) Fill in the blank:
A quadrilateral has _____ vertices.

92) Fill in the blank:
A quadrilateral has _____ sides.

93) Fill in the blank:
A quadrilateral has _____ angles.

94) The sum of angles of a quadrilateral is _____.

95) The angle A of a quadrilateral ABCD is 120°. The other angles, angle B, C, D are equal. Find the measure of each of these three angles.

96) A quadrilateral has three angles measuring 75°, 55°, 110° respectively. What is the measure of the fourth angle?

97) The angles of a quadrilateral are in the ratio 1 : 2 : 3 : 4. What are the measures of each of the four angles?

98) State True or False:
No three vertices of a quadrilateral are collinear.

99) A quadrilateral has all its four angles of the same measure. Find the measure of each angle.

100) A quadrilateral has three equal acute angles measuring 85° each. Find the measure of the fourth angle.

101) Two angles of a quadrilateral measure 85° and 113° and the other two angles are equal. Find the measure of each of the equal angles.

102) Two equal angles of a quadrilateral measure 88° each. The difference between the other two angles is 12°. Find the measure of each of its remaining angles.

103) Two adjacent angles of a quadrilateral are 80° and 86°. The other two angles are of equal measure. Find the measure of each of the other two angles.

104) The angles of a quadrilateral are in the ratio 3:3:4:5. Find the measure of each of the four angles.

105) The sides of a quadrilateral are in the ratio 2:3:5:7 and its perimeter is 85 inches. Find the length of each side of the quadrilateral.

106) The perimeter of a quadrilateral is 82 inches. Two sides of the quadrilateral are equal in length and measure 12 inches each. The difference between the other two sides of the quadrilateral is 6 inches. Find the lengths of the other two sides of the quadrilateral.

107) All the four sides of a quadrilateral are of equal length. Find the length of each side of the quadrilateral, if the perimeter of the quadrilateral is 48 inches.

108) The three angles of a quadrilateral measure 125°, 60° and 55° respectively. What is the measure of the fourth angle of the quadrilateral?

109) Three angles of a quadrilateral are equal and the fourth angle measures 105°. What is the measure of each of the equal angles?

110) The lengths of the sides of a quadrilateral are in the ratio 2:4:5:7 and the perimeter of the quadrilateral is 144 inches. Find the lengths of sides of the quadrilateral.

111) The angles of a quadrilateral are in the ratio 2:4:5:7. Find the measure of each angle of the quadrilateral.

112) A quadrilateral PQRS has $\angle P = 80°$, $\angle Q = 90°$ and $\angle S = 60°$. Find the measure of $\angle R$.

113) A rectangular field is 75 ft in length and 40 ft in width. Find the length of its diagonal.

114) If the three angles of a quadrilateral measure 30°, 120° and 100°, find the measure of its fourth angle.

115) If all the angles of a quadrilateral are equal, what will be the measure of each angle ?

116) Fill in the blank:
Two angles of a quadrilateral are _____, if they have a common arm.

A) Adjacent

B) Opposite

117) Fill in the blank:
Two angles of a quadrilateral are _____, if they don't have a common arm.

A) Adjacent

B) Opposite

118) Fill in the blank:
Two sides of a quadrilateral are adjacent if they have a common _____.

A) Arm

B) Vertex

119) Fill in the blank:
Two sides of a quadrilateral are _____, if they do not form arms of an angle of the quadrilateral.

A) Opposite

B) Adjacent

120) A convex quadrilateral cannot have an angle greater than _____.

A) 180°

B) 90°

C) 360°

121) A concave quadrilateral will always have an angle which measures greater than _____ .

   A) 180°

   B) 90°

   C) 360°

122) If the line segment joining two points in the interior of a quadrilateral always lies within it, the quadrilateral is _____ .

123) State True or False:
   In a quadrilateral there can be an acute angle.

124) State True or False:
   If the diagonals of a quadrilateral lie outside the quadrilateral region, the quadrilateral is convex.

125) State True or False:
   If the diagonals of a quadrilateral lie outside the quadrilateral region, the quadrilateral is concave.

126) State True or False:
   If all the line segments joining the two points in the interior of the quadrilateral lie within the quadrilateral region, the quadrilateral is convex.

127) State True or False:
   All the angles of a quadrilateral can be obtuse.

128) State True or False:
   All the angles of a quadrilateral can be acute.

129) State True or False:
In a quadrilateral there can be an obtuse angle.

130) State True or False:
In a quadrilateral, all the angles can be right angles.

131) Find the value of 'z' in given figure.

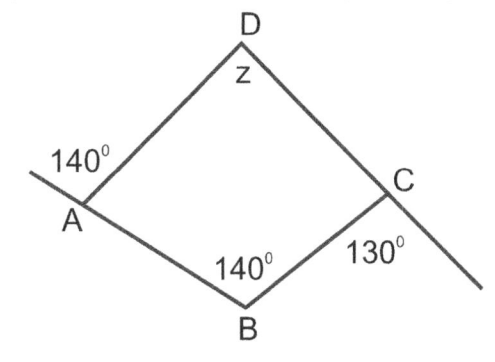

132) Find the value of x in given figure.

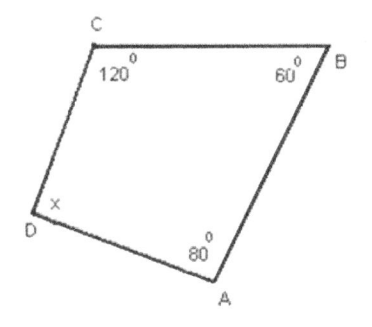

133) In the given figure AB || CD. Find the value of 'x' and 'y'.

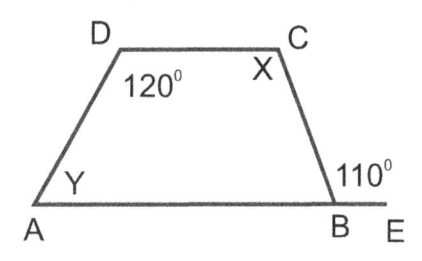

134) In the given figure AD || BC. Find the value of 'x' and 'y'.

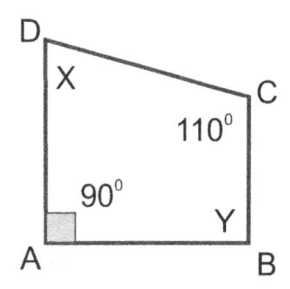

135) The four angles of a quadrilateral are in the ratio 3 : 4 : 5 : 6. Find the angles.

136) In quadrilateral ABCD, name the points in the quadrilateral region.

A) P, X & Y
B) X, Y, Q and Z
C) P, Q & Z

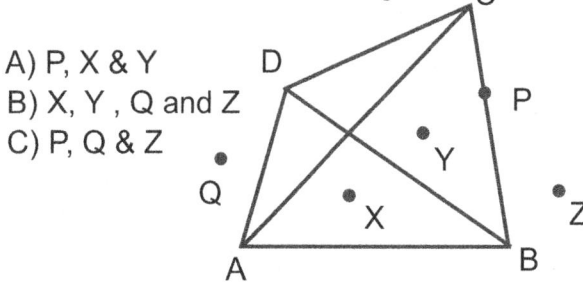

137) In quadrilateral ABCD, pick the point lying on the quadrilateral.

A) X
B) P
C) Q
D) Z

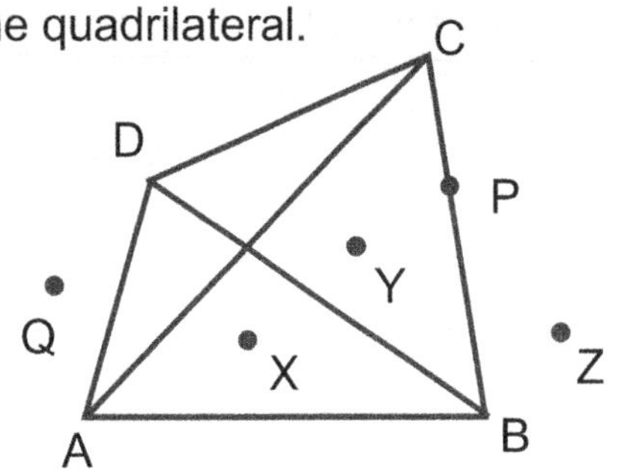

138) In quadrilateral ABCD, name the points in the interior of the quadrilateral.

A) Q & Z
B) X & Y
C) X, Y & Z
D) Q, P & Z

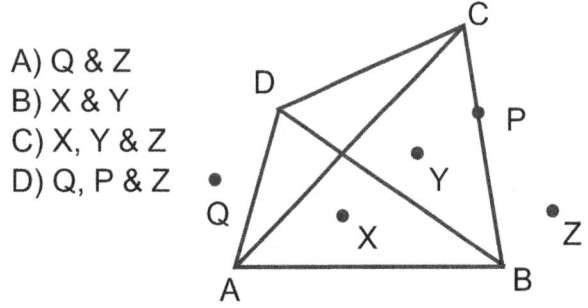

139) In quadrilateral ABCD, name the points in the exterior of the quadrilateral

A) Q & Z
B) X & Y
C) Y & P
D) A, B & Z

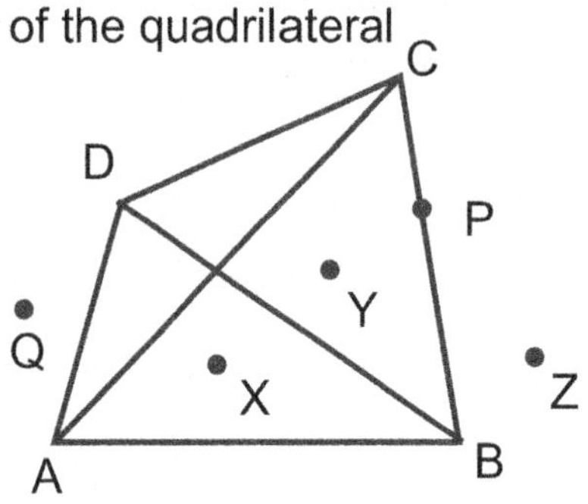

140) In quadrilateral ABCD, name the pair of diagonals
A) AC & BD
B) AB & CD
C) AD & BC

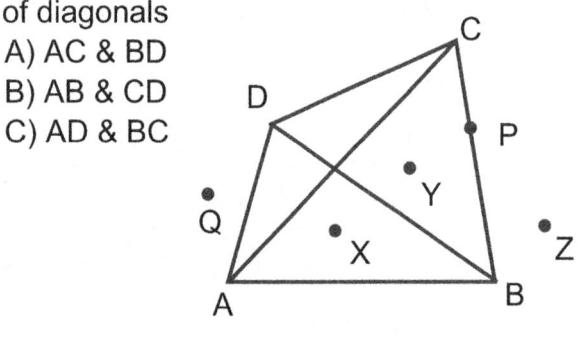

141) In quadrilateral ABCD, pick the pair of opposite sides

A) BC & CD
B) AC & BD
C) BC & AD

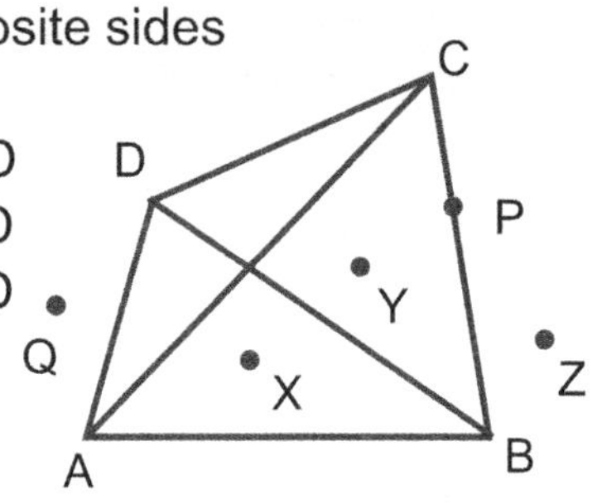

142) In quadrilateral ABCD, pick the pair of opposite sides
A) AB & CD
B) AB & BD
C) AC & CD

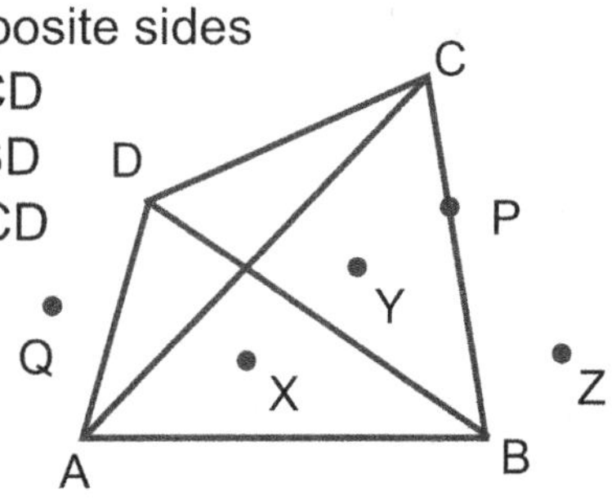

143) The four angles of a quadrilateral are in the ratio 3:5:7:9. Find the angles.

144) Is it possible to have a quadrilateral whose angles are 75°, 65°, 135° and 120°?

145) The three angles of a quadrilateral are in the ratio 1:2:3. Of these, the sum of its greatest angle and the least angle is 180°. Find the measure of each of the four angles.

146) The three angles of a quadrilateral measure 54°, 80° and 116°. Find the measure of fourth angle.

147) An angle of a quadrilateral measures 165° and the remaining three angles are of equal measure. Find the measure of each of the equal angles.

148) The three angles of a quadrilateral measure 55°, 62° and 108°. Find the measure of fourth angle.

149) In the figure below, ABCD is a quadrilateral in which AB || CD. If ∠A = ∠B = 60°, what are the measures of the other two angles?

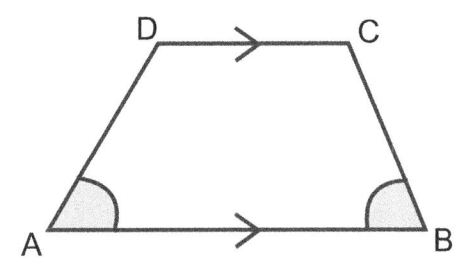

150) In the figure below, ABCD is a quadrilateral in which AB || CD. If ∠A = ∠B = 75°, what are the measures of the other two angles?

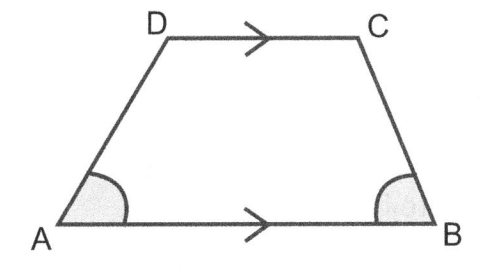

151) In the given figure, the bisectors of ∠P and ∠Q meet at point O in the interior of the quadrilateral PQRS. If ∠R = 50° and ∠S = 100°, find the measure of ∠POQ.

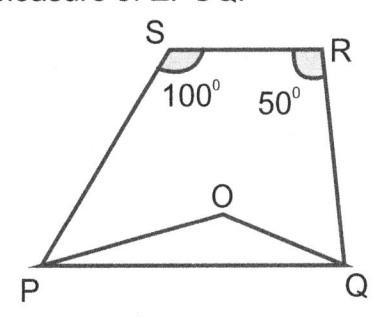

152) How many pairs of opposite angles are there in a quadrilateral?
A) One
B) Two
C) Three
D) Four

153) How many pairs of adjacent sides are there in a quadrilateral?
A) One
B) Two
C) Three
D) Four

154) How many pairs of opposite sides are there in a quadrilateral?
A) One
B) Two
C) Three
D) Four

155) A quadrilateral has three acute angles, each measuring 85°. What is the measure of the fourth angle?

156) A quadrilateral has three equal angles, each measuring 92°. What is the measure of the fourth angle?

157) Two angles of a quadrilateral measure 80° each and the other two angles are equal. What is the measure of each of these angles?

158) Two angles of a quadrilateral measure 65° each and the other two angles are equal. What is the measure of each of these angles?

159) In the figure below, ABCD is a quadrilateral in which AB || CD. If ∠A = ∠B = 40°, what are the measures of the other two angles?

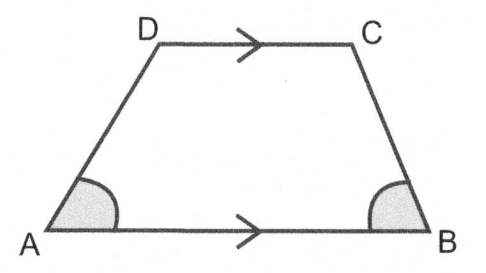

160) In the figure below, Y is a point in the interior of ∠AOB. YZ ⊥ OA and YX ⊥ OB. If ∠AOB = 40°, what is the measure of ∠XYZ?

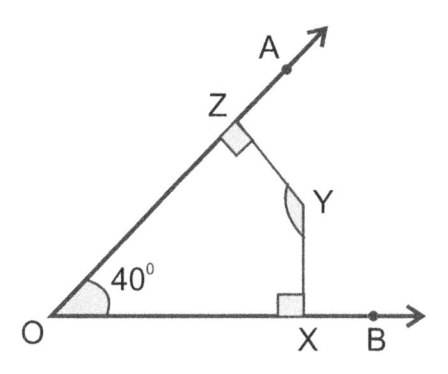

161) In the figure below, Y is a point in the interior of ∠AOB. YZ ⊥ OA and YX ⊥ OB. If ∠XYZ = 105°, what is the measure of ∠AOB?

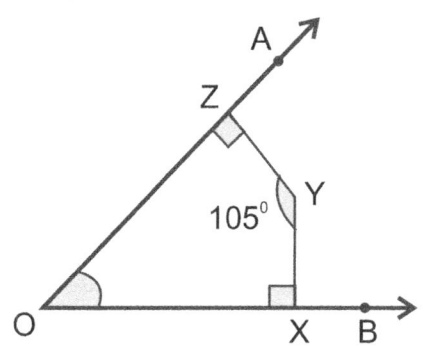

162) In the figure below, ABCD is a quadrilateral in which AB ∥ CD. If ∠A = 45° and ∠B = 70°, what are the measures of the other two angles?

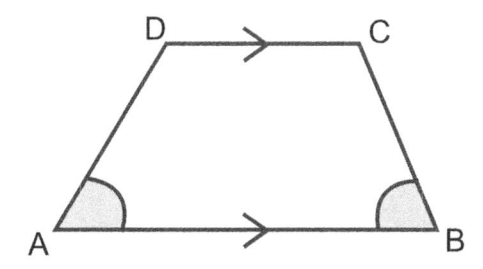

163) If the sum of two angles of a quadrilateral is equal to 180°, what is the sum of the remaining two angles?

164) If the sum of two angles of a quadrilateral is equal to 200°, what is the sum of the remaining two angles?

Grade 7
Volume 1
Week 16

Grade 7
Volume 1
Week 17

Grade 7
Volume 1
Week 17

1) The perimeter of the floor of a room is 30 ft and its height is 3.5 ft. Find the area of the four walls of the room.

2) Find the cost of plastering the four walls of the swimming pool which is 24 ft long, 17 ft wide and 8 ft deep at the rate of $20 per ft^2.

3) The perimeter of the floor of a hall is 80 inches and its height is 58 inches. Find the area of the four walls of the hall.

4) Find the cost of plastering the four walls of the room which are 20 ft long, 30 ft wide and 50 ft high at the rate of $2.5 per square ft.

5) A room is 7 yards long, 9 yards wide and 8.5 yards high. If the room has one door 1.5 yards by 4.5 yards and two windows each of size 1.5 yards by 2 yards, find the remaining area of the four walls of the room.

6) The area of four walls of the room is 180 ft^2. The width and height of the room are 8 ft and 6 ft respectively. Find the length of the room.

7) A room is 15 ft long and 10 ft wide. If the area of the four walls of the room is 275 ft^2, find the height of the room.

8) A hall of length 10 ft, width 8 ft and height 12 ft is to be fixed with a wall paper on all the four walls. How much will it cost for papering the four walls, if the cost of fixing the wall paper is $6 per square ft.

9) Find the area of four walls of a room of length 12 ft, width 8 ft and height 4 ft.

10) A room is 3 ft long, 5 ft wide and 10 ft high. Find the cost of painting the four walls of the room at the cost of $4 per square ft.

11) The area of four walls of a room is 133 ft^2. If the height of the room is 3.5 ft, what is the perimeter of the floor of the room?

12) The area of four walls of a room is 60 ft^2. If the length and width of a room are 6.5 ft and 3.5 ft respectively, find the height of the room.

13) A swimming pool is 40 ft long, 30 ft wide and 8 ft deep. Find the cost of replacing the tiles of the walls at the rate of $3.50 per sq.ft.

14) What is the area of four walls of a room 12 ft long, 8 ft wide and 9 ft high?

15) The area of four walls of a room 18 ft long and 12 ft wide is 420 ft^2. Find the height of the room.

16) A drawing room is 12 ft long, 7 ft wide and 10 ft high. There is a door, measuring 4 ft by 7 ft and two windows each measuring 4 ft by 3 ft for the room. Find the cost of painting the walls of the room at the rate of $2.80 per ft^2.

17) The dimensions of a room are 15 ft by 13 ft by 11 ft. What is the cost of papering its walls by a wall paper at the rate of $1.9 per ft^2.

18) The dimensions of a room are 18 ft by 15 ft by 12 ft. Find the cost of papering the walls of the room by a 2 ft wide wall paper, at the rate of $3 per ft.

19) A swimming pool is 80 ft long, 50 ft wide and 20 ft deep. Find the cost of cementing its floor and walls at the $0.85 per 10 ft^2.

20) The area of four walls of a room 20 ft long and 16 ft wide is 648 ft^2. Find the height of the room.

21) The area of the four walls of a 11 ft high bathroom is 396 ft^2. Find the length of the bathroom, if it is 8 ft wide.

22) A rectangular water tank, whose length, width and height are 8 ft, 6 ft and 5 ft, respectively is to be cemented. Find the cost of cementing the four walls of the tank at the rate of $0.90 per ft^2.

23) Find the area of four walls of a room 23 ft long, 19 ft wide and 11 ft high.

24) The area of four walls of a room of length 17 ft and height 9 ft is 522 ft^2. Find the width of the room.

25) A huge hall is 80 ft long, 30 ft wide and 20 ft high. There are 8 windows each measuring 5 ft by 4 ft in the hall. Find the remaining area of the four walls of the room.

26) A hall of length 15 ft, width 10 ft and height 8 ft is to be fixed with a wall paper on all four walls. If the area of the windows and doors amounts to 28 ft^2, how much wall paper is required for the wall?

27) A room 3 yards high has area of its four walls as 300 yards^2. If the length and width of the room are in the ratio 2:3, find their measure.

28) The length and width of an auditorium are 50 ft and 25 ft respectively. If the cost of painting the walls of the auditorium at the rate of $1.50 per square foot is $2700, find the height of the auditorium.

29) The dimensions of a room are; Length = 8 meters, width = 6 meters and height = 5 meters. Find the cost of white washing its four walls at the rate of $2.75 per square meter.

30) The perimeter of the floor of a hall is 120 ft and the height of the hall is 11 ft. Find the cost of plastering the 4 walls of the hall at the rate of $0.80 per ft^2.

31) Find the area of 4 walls of a room which is 8 m long, 5 m wide and 4 m high.

32) The area of 4 sides of a square box is 576 inches^2. Find the length of each side of the box.

33) The area of 4 walls of a room is 168 ft^2. The room is 8 ft wide and 6 ft high. Find the length of the room.

34) Find the area of 4 walls of a room which is 10 ft long, 7 ft wide and 4 ft high.

35) A room is 13 ft long and 4.5 ft wide. If the area of 4 walls of the room is 91 ft^2, find its height.

36) A room is 7.4 m long, 5.6 m wide and 3.5 m high. There are 2 doors each measuring 1.5 m by 2 m and 2 windows each measuring 1.5 m by 1 m. Find the cost of painting the walls of the room at $15 per sq meter.

37) A room is 8 m long, 6.5 m wide and 4 m high. Find the cost of papering its walls with a 50 cm wide wall paper at the rate of $25 per meter.

38) A lawn is 4.5 m long and 1.2 m wide. It has a 40 cm wide emented border all around it on the inside. Find the cost of recarpeting the cemented border at $2 per cm^2. (Given that 1 m = 100 cm)

39) A square lawn has a 2 ft wide path surrounding it. If the area of the path is 168 ft^2, find the area of the lawn.

40) A rectangular grassy lawn measuring 42 yards by 36 yards has been surrounded externally by a 2.5 yards wide path. Calculate the cost of gravelling the path at the rate of $2.50 per square yard.

41) A rectangular plot of length 120 yards and width 80 yards has two crossroads, each 6 yards wide. These crossroads are running through the middle of the plot, one parallel to the length and the other parallel to the width. Find the total cost for gravelling the roads at $11 per square yard.

42) A rectangular plot of length 120 yards and width 80 yards has two crossroads, each 6 yards wide. These crossroads are running through the middle of the plot, one parallel to the length and the other parallel to the width. Find the total cost for laying the grass in the remaining part at $3 per square yard.

43) A room 10 ft long and 8 ft wide is surrounded by a 2.5 ft wide balcony. Calculate the cost of cementing the floor of the balcony at $12 per square feet.

44) A room is 16 ft long and 9 ft wide. If the area of its 4 walls in 170 ft^2, find the height of the room.

45) A rectangular park has its length and width in the ratio 2 : 1 and its perimeter as 240 yards. A 3 ft wide path runs along the boundary of the park, inside it. Find the cost of paving the path at $3.5 per square yard.

46) A room 14 ft by 12 ft is to be paved with stones, each measuring 0.25 ft by 0.2 ft. Find the number of stones required to pave the room.

47) find the cost of carpeting a room 12 ft by 15 ft with a carpet 2 ft broad at the rate of $9 per ft.

48) A room is 32 ft long and 18 ft wide. Find the height of the room if the area of its four walls is 800 ft^2.

49) A rectangular field is 25 ft by 28 ft. It has two cross roads through its centre running parallel to its sides. The width of the longer and shorter roads are 1.5 ft and 2.2 ft respectively. Find the area of the remaining portion of the field.

50) A lawn is 95 yards long and 72 yard wide. A swimming pool measuring 20 yards by 12 yards is in this lawn. What will be the cost of grassing the remaining part of the field at $2 per square yard.

51) A garden is 275 yards long and 100 yards wide. A path 5 yards wide is to be built outside around it. Find the area of the path.

52) A square park is of side 24 yards. A road 2 yards wide is made all around the park inside it. Find the area of the road.

53) A 4 yards wide path runs along inside the boundary of a square field whose side measures 60 yards. If the field is to be manured except the path, find the cost of manuring the field at $1.25 per square yard.

54) A square lawn of side 50 yards is to be surrounded externally by a path which is 6 ft wide. Find the area of the path.

55) A path 12 yards wide runs inside along the boundary of a square field whose side is 70 yards. Find the area of the path.

56) A path 12 ft wide, runs inside along the boundary of a square park whose side is 120 ft. Find the area of the path in square yards.

57) A rectangular park is 40 ft wide and 55 ft long. A path 5 ft wide is running externally along the boundary of the park. Find the cost of leveling the path at the rate of $0.50 per ft^2.

58) The central hall of a hotel is 50 ft wide and 75 ft long. A carpet is laid on the floor leaving 3 ft space from the walls uncovered. Find the area of the carpet.

59) The side of a square flower bed is 150 yards. It is enlarged by digging a strip 10 yards wide all around it. Find the increase in area of flower bed.

60) A path 2 ft wide is built along the border inside a square garden of side 40 ft. Find the area of the path.

61) A path 2 ft wide is built along the border inside a square garden of side 30 ft. Find the cost of planting grass in the remaining portion of the garden at the rate of $1.25 per ft^2.

62) A room is 20 ft long and 15 ft wide. It is surrounded by a balcony which is 2 ft wide. Find the cost of laying marbles in the balcony at the rate of $1.50 per square feet.

63) A hall is 24 ft long and 18 ft wide. Find the cost of carpeting the hall at $3 per ft^2, after leaving a margin of 2 ft all around it.

64) A sheet of paper is 15 inches long and 11 inches wide. A strip 2 inches wide is cut from all around the paper. What is the area of the cut-out strip?

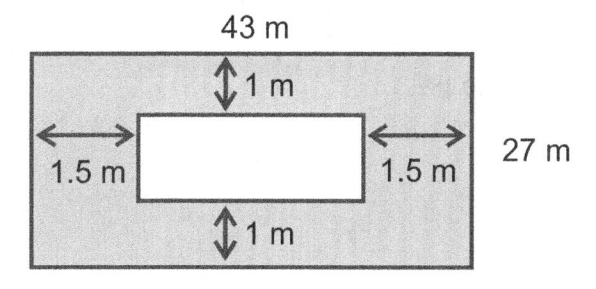

65) A rectangular lawn, 75 ft by 60 ft has two cross roads, each 4 ft wide, running through the middle of the lawn, one parallel to the length and the other parallel to the width of the lawn. Find the area of the roads.

66) Daniel wanted to paint a wall of dimensions 12 ft by 15 ft in bright yellow. He left 0.5 ft margin all around as border and painted the remaining wall in yellow. What is the area of the margin left?

67) A 3 ft wide flower bed was laid inside along the perimeter of a square plot of side 34 ft. What is the area of the flower bed?

68) A 3 ft wide path runs outside along the perimeter of a square park of side 12 yards. Find the area of the path.

69) A color paper of side 25 inches is cut by 2 inches all around its border as the edges were untidy. What is the area of the remaining paper?

70) A square card board of side 32 inches is divided into four equal squares by drawing two cross lines of width two inches in the center of the board. Find the area of the each of the small squares thus formed.

71) A square farm of side 50 yards has a 3 yards wide flower bed all around the inner border of the farm. How much would it cost to manure the flower bed at the rate of $3 per 10 yards^2?

72) Find the area of the shaded region:

73) A square lawn of each side 24 yards, has two roads each 6 ft wide, running though the middle of the lawn and perpendicular to each other. Find the cost of gravelling the roads at $0.70 per yard^2.

74) Ashley is doing a project using a square card-board of side 14 inches. She left a 2 inch wide border all around the card-board and used the remaining area to write her presentation. What is the area of the border?

75) A 3 ft wide track runs outside along the perimeter of a tennis court of side 40 ft. What is the cost of leveling the track at the rate of $1.25 per ft^2?

76) A square plot of side 25 yards has a 2 yards wide path outside along the boundary of the plot. Find the area of the path.

77) A 5 ft wide path runs inside along the perimeter of a square plot of side 50 ft. A house has been built in the remaining area of the plot at the rate of $55 per ft^2. Find the cost of construction of the house.

78) A square shaped drawing room of side 15 ft is to be carpeted leaving a margin of 2 ft from the walls of the room. Find the length of a 5 ft wide carpet required for this purpose.

79) A square park of side 45 ft has two cross paths in the centre of the park at right angles to each other and parallel to its sides. One of the paths is 2 ft wide and the other path is 3 ft wide. Find the cost of leveling the path ways at the cost of $1.20 per ft^2.

80) Linda colored a square paper of side 15 inches in green and yellow. She colored one inch wide border in green all around the paper. The remaining middle portion she colored with yellow. What is the area of the paper colored in yellow?

81) The length and width of a rectangular building are 12 ft and 9 ft respectively. A corridor of 1.5 ft width runs all around the outside of the building. Find the area of the corridor.

82) The length and width of a rectangular room are 15 ft and 12 ft respectively. A carpet is to be laid in the middle of the room leaving 2.5 ft on all sides of the wall. Find the cost of carpeting at $4.50 per square foot.

83) A rectangular garden 100 yards long and 75 yards wide has two roads 6 yards wide, running at right angles to each other midway within it, one parallel to the width and other parallel to the length. Find the cost of cementing the roads at $3.5 per square yard.

84) The length and width of a plot are 60 ft and 50 ft respectively. A path goes all around it on the outside having a width of 4.5 ft. Find the area of the path.

85) The central hall of a school, 30 ft long and 22 ft wide, is decorated for the annual day celebrations. A red carpet 3 ft wide is laid in the center of the hall, parallel to the length of the hall. What is the cost of laying the carpet at $2.25 per ft^2?

86) A painting is 18 inches long and 15 inches wide. It is to be framed with a 1 inch wide frame. What is the cost of framing the painting at the rate of 20¢ per inch^2 ?

87) William colored a 12 inches by 10 inches paper in red, leaving a border of 1 inch wide to be colored in blue. What is the area of the paper to be colored in blue?

88) A play ground is in the shape of a rectangle of dimensions 80 yards by 50 yards. To renovate this play ground, a 5 yards wide flower bed has been laid inside along the perimeter of the park. In the centre of the park a circular water body of 3.5 yard radius is built. Find the area of the rest of the play ground.

Grade 7
Volume 1
Week 17

89) Find the area of the shaded region:

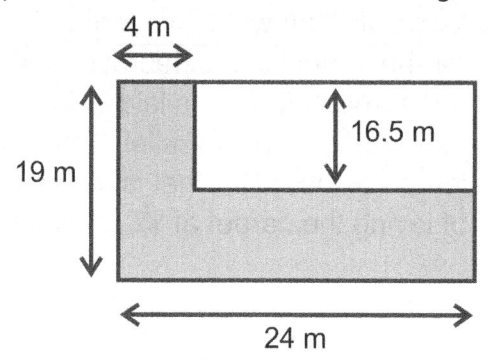

4 m
19 m
16.5 m
24 m

90) Find the area of the shaded region:

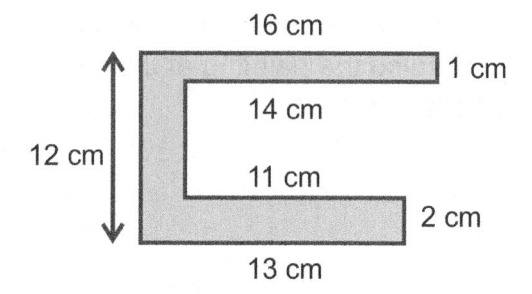

16 cm
1 cm
14 cm
12 cm
11 cm
2 cm
13 cm

91) Find the area of the shaded region:

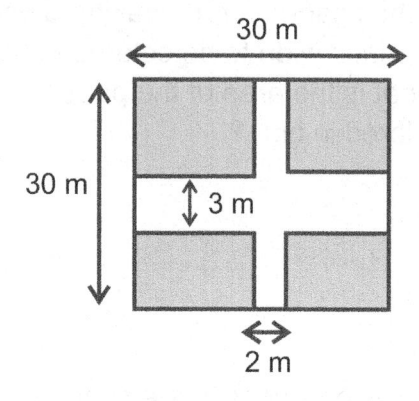

30 m
30 m
3 m
2 m

92) Find the area of the shaded region:

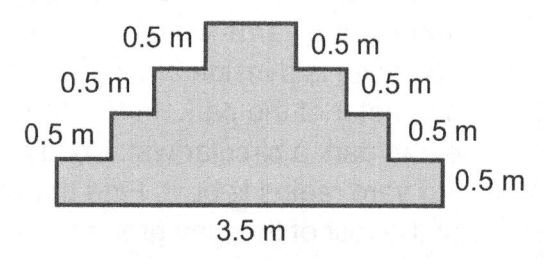

0.5 m 0.5 m
0.5 m 0.5 m
0.5 m 0.5 m
0.5 m
3.5 m

93) Calculate the area of shaded region.

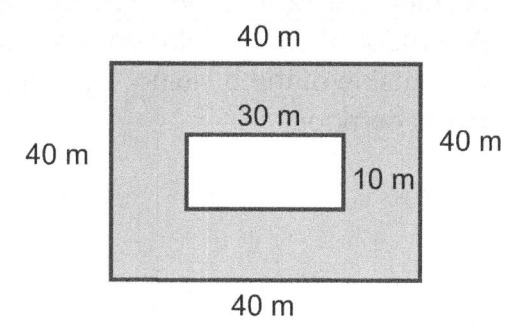

40 m
40 m   30 m   40 m
       10 m
40 m

94) Calculate the area of shaded region.

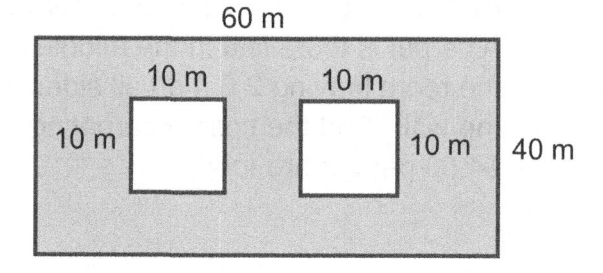

60 m
10 m   10 m
10 m   10 m   40 m

95) Calculate the area of shaded region.

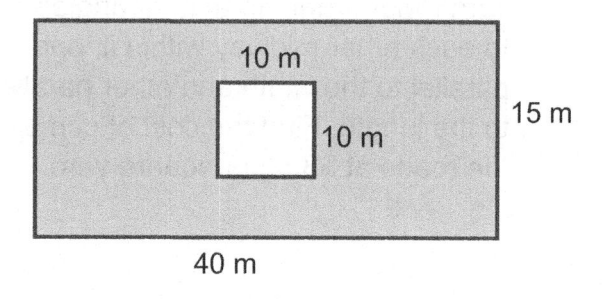

10 m
10 m   15 m
40 m

96) Calculate the area of shaded region.

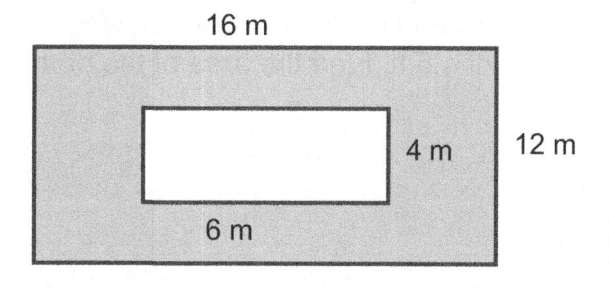

16 m
4 m   12 m
6 m

97) Calculate the area of shaded region.

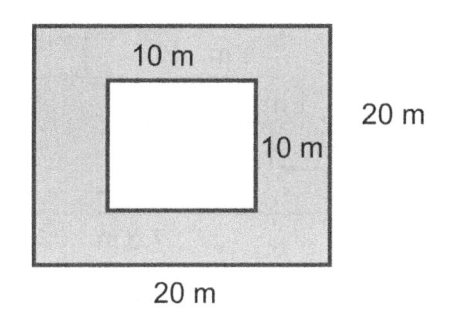

10 m
20 m
10 m
20 m

98) Calculate the area of shaded region.

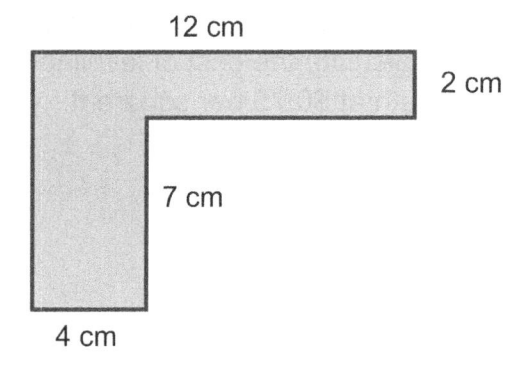

12 cm
2 cm
7 cm
4 cm

99) Find the area of given figure :

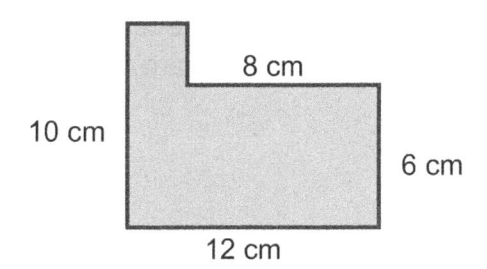

8 cm
10 cm
6 cm
12 cm

100) Find the area of given figure :

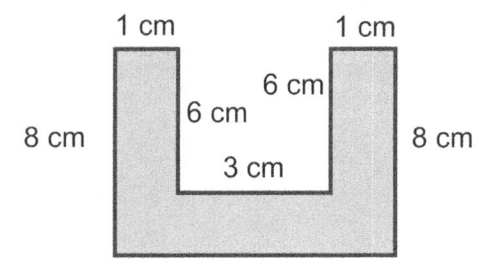

1 cm   1 cm
6 cm
6 cm
8 cm   8 cm
3 cm

101) Find the area of given figure :

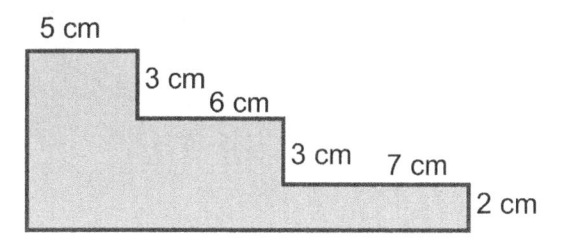

5 cm
3 cm
6 cm
3 cm   7 cm
2 cm

102) A path of uniform width runs around outside a square plot of side 15 m. If the area of the path is 400 m², find its width.

103) A path of uniform width runs around outside a square plot of side 20 m. If the area of the path is 276 m², find its width.

104) A path of uniform width runs around outside a square plot of side 25 m. If the area of the path is 399 m², find its width.

105) A path of uniform width runs around outside a square plot of side 18 m. If the area of the path is 517 m², find its width.

106) A path of uniform width 4 m runs around outside a square plot. If the area of the path is 448 m², find the length/side of the square plot.

107) A path of uniform width 4 m runs around outside a square plot. If the area of the path is 288 m², find the length/side of the square plot.

108) A path of uniform width 2 m runs around outside a square plot of side 22 m. Find the area of the path.

109) Find the area of the shaded region:

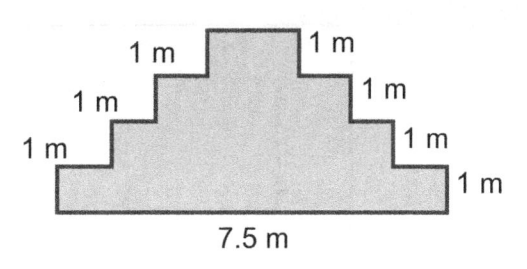

10) A lawn 100 ft long and 90 ft wide is surrounded by a 4 ft wide path. Calculate the cost of leveling the path at $0.75 per square ft.

111) A lawn 5 ft long and 1.5 ft wide is surrounded by a 0.75 ft wide path. Calculate the cost of leveling the path at $0.25 per square ft.

112) A lawn measuring 50 ft by 36 ft is surrounded externally by a path 2 ft wide. Find the area of the path.

113) A lawn measuring 50 ft by 36 ft is surrounded externally by a path 2 ft wide. Calculate the cost of leveling the path at $1.50 per square ft.

114) A hall 9.5 m long and 6 m wide is surrounded by a veranda which is 1.25 m wide. Find the cost of cementing the floor of the veranda at $16.00 per square meter.

115) A canvas cloth is 5 m long and 1.3 m wide. A border of width 25 cm is printed along its sides. Find the cost of printing the border at $1.00 per 100 cm^2.

116) A path 2.5 yard wide runs along inside the boundary of a 75 yard by 60 yard park. Find the cost of graveling the path at $3.60 per square yard.

117) A square flower bed is 2 m 80 cm long. It is extended by digging a strip 30 cm wide all around it. Find the area of the enlarged flower bed.

118) A square flower bed is 2 m 80 cm long. It is extended by digging a strip 30 cm wide all around it. Find the increase in the area of flower bed.

119) A rectangular plot is 95 ft long and 72 ft wide. Inside the plot, a path of uniform width of 3.5 ft is constructed all around. The rest of the plot is to be laid with grass. Find the total cost involved in constructing the path at $6.50 per square ft and laying the grass at $0.75 per square ft.

120) The length and width of a park are in the ratio 2:1. The perimeter of the park is 240 yards. A path 2 yards wide runs inside it, along its boundary. Find the cost of cementing the path at $3.00 per square yard.

**Grade 7
Volume 1
Week 17**

121) The length and width of a park are in the ratio 5:2. A path 2.5 m wide runs outside it, along its boundary and has an area of 305 m². Find the length and width of the park.

122) Find the cost of carpeting a room 13 m by 9 m with a carpet 75 cm broad at $10 per meter.

123) In the figure below, the shaded rectangular path has same width throughout. Find the area of the path.

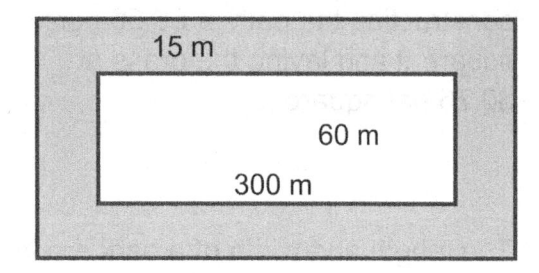

15 m
60 m
300 m

124) A rectangular land for a housing colony has length 450 yards and width 375 yards and three roads 7 yards wide are running along the length of the land, two on either side and one in the middle. Find the total area of the roads.

125) Find the area of the shaded region:

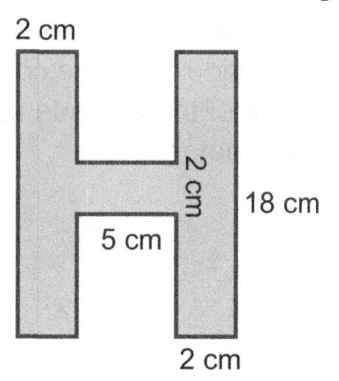

2 cm
2 cm
18 cm
5 cm
2 cm

126) Find the area of the shaded region:

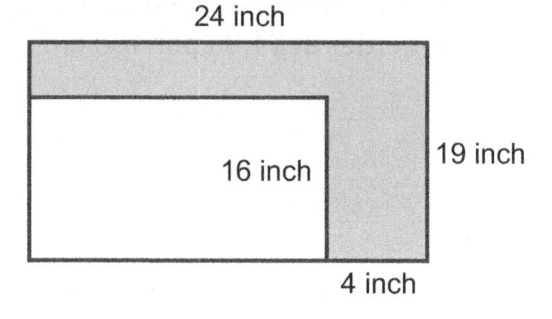

24 inch
16 inch
19 inch
4 inch

127) Find the area of the shaded region:

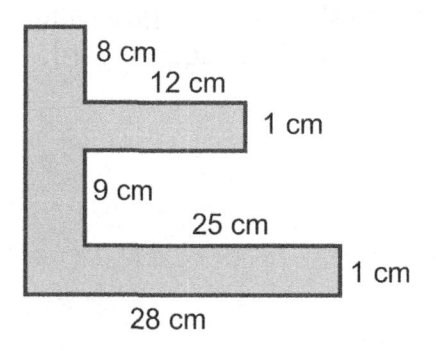

8 cm
12 cm
1 cm
9 cm
25 cm
1 cm
28 cm

**Grade 7
Volume 1
Week 18**

1) Find the missing measurement from the figure given below. Round the answer to the nearest tenth.

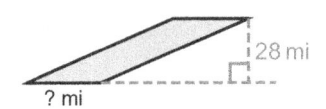

   28 mi
   ? mi
   Area = 943.6 mi²

   A) 41 mi   B) 33.7 mi
   C) 44.7 mi D) 23.6 mi

2) Find the missing measurement from the figure given below. Round the answer to the nearest tenth.

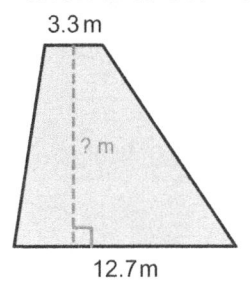

   3.3 m
   ? m
   12.7 m
   Area = 89.6 m²

   A) 7.5 m   B) 14.3 m
   C) 14.4 m   D) 11.2 m

3) Find the missing measurement from the figure given below. Round the answer to the nearest tenth.

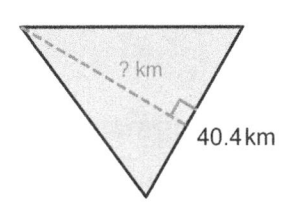

   ? km
   40.4 km
   Area = 818.1 km²

   A) 42.5 km B) 40.5 km
   C) 34.4 km D) 44.8 km

4) Find the missing measurement from the figure given below. Round the answer to the nearest tenth.

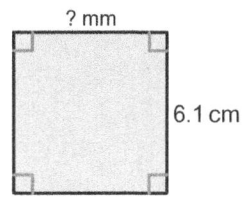

   ? mm
   6.1 cm
   Area = 36.6 cm²

   A) 6 mm   B) 48.7 mm
   C) 58.4 mm D) 60 mm

5) Find the missing measurement from the figure given below. Round the answer to the nearest tenth.

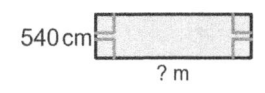

   540 cm
   ? m
   Area = 108 m²

   A) 23 m   B) 20 m
   C) 19.4 m   D) 22.1 m

6) Find the missing measurement from the figure given below. Round the answer to the nearest tenth.

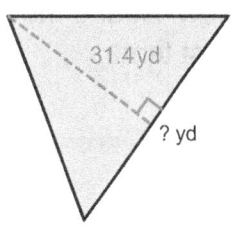

   31.4 yd
   ? yd
   Area = 676.7 yd²

   A) 36 yd   B) 43.1 yd
   C) 34.4 yd D) 33.2 yd

Grade 7
Volume 1
Week 18

7) Find the area of the figure given below.

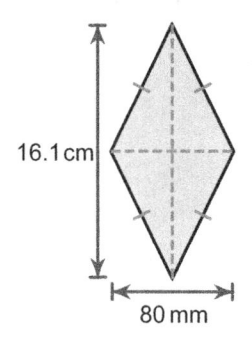

16.1 cm
80 mm

A) 128.8 cm²   B) 64.4 cm²
C) 32.2 cm²   D) 70.5 cm²

8) Find the missing measurement from the figure given below. Round the answer to the nearest tenth.

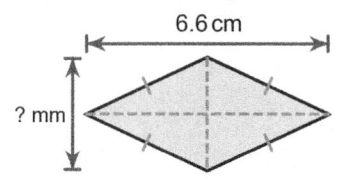

6.6 cm
? mm
Area = 9.9 cm²

A) 3   B) 19.2
C) 31.1 D) 30

9) Find the missing measurement from the figure given below. Round the answer to the nearest tenth.

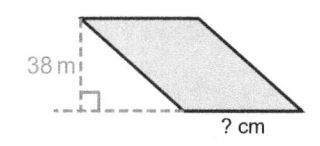

38 m
? cm
Area = 1900 m²

A) 5000 cm   B) 50 cm
C) 999.8 cm D) 999.6 cm

10) Find the area of the figure given below.

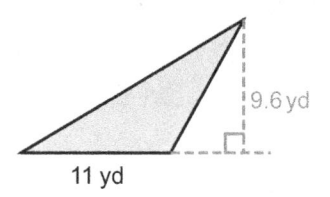

9.6 yd
11 yd

A) 49.3 yd²   B) 105.6 yd²
C) 26.4 yd²   D) 52.8 yd²

11) Find the missing measurement from the figure given below. Round the answer to the nearest tenth.

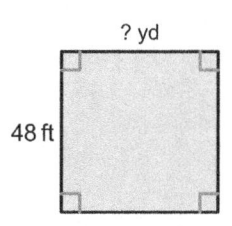

? yd
48 ft
Area = 256 yd²

A) 16 yd   B) 10.8 yd
C) 5.3 yd   D) 17.7 yd

12) Find the area of the figure given below.

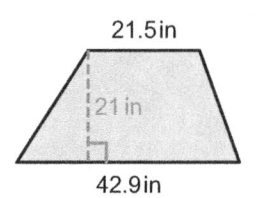

21.5 in
21 in
42.9 in

A) 338.1 in²   B) 1357.8 in²
C) 676.2 in²   D) 678.9 in²

13) Find the missing measurement from the figure given below. Round the answer to the nearest tenth.

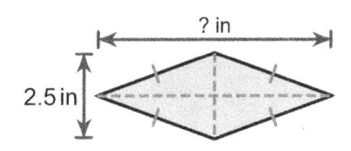

2.5 in
? in
Area = 8.75 in²

A) 7 B) 6.8
C) 6.9 D) 5.7

14) Find the area of the figure given below.

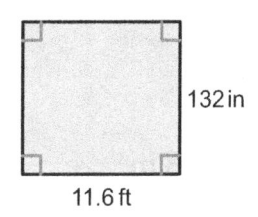

132 in
11.6 ft

A) 127.6 ft² B) 273 ft²
C) 136.5 ft² D) 255.2 ft²

15) Find the area of the figure given below.

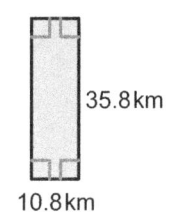

35.8 km
10.8 km

A) 385.54 km² B) 773.28 km²
C) 386.64 km² D) 193.3 km²

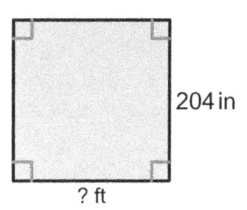

16) Find the missing measurement from the figure given below. Round the answer to the nearest tenth.

204 in
? ft
Area = 289 ft²

A) 18.1 ft B) 17 ft
C) 17.8 ft D) 1.4 ft

17) Find the area of the figure given below.

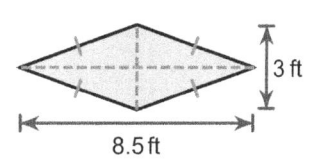

3 ft
8.5 ft

A) 3.45 ft² B) 6.4 ft²
C) 12.75 ft² D) 25.5 ft²

18) Find the area of the figure given below.

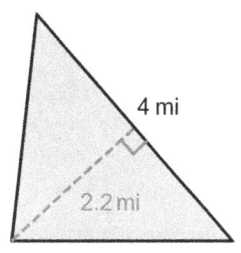

4 mi
2.2 mi

A) 2.2 mi² B) 14.1 mi²
C) 4.4 mi² D) 8.8 mi²

**Grade 7
Volume 1
Week 18**

19) Find the area of the figure given below.

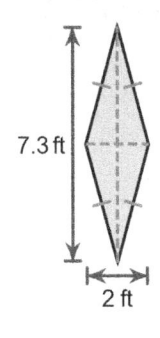

7.3 ft
2 ft

A) 3.7 ft²   B) 14.6 ft²
C) 7.3 ft²   D) 0.1 ft²

20) Find the area of the figure given below.

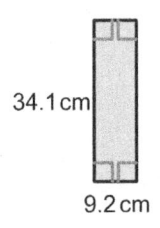

34.1 cm
9.2 cm

A) 156.9 cm²   B) 627.44 cm²
C) 313.72 cm² D) 323.12 cm²

21) Find the area of the figure given below.

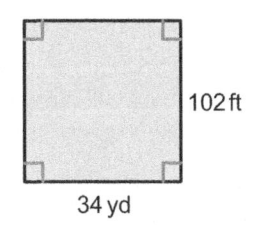

102 ft
34 yd

A) 1156 yd²   B) 1162.3 yd²
C) 578 yd²   D) 2312 yd²

22) Find the area of the figure given below.

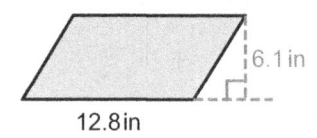

6.1 in
12.8 in

A) 83.38 in²   B) 78.08 in²
C) 80.38 in²   D) 156.16 in²

23) Find the area of the figure given below.

12 mi
3.7 mi

A) 11.1 mi²   B) 12.3 mi²
C) 22.2 mi²   D) 44.4 mi²

24) Find the area of the figure given below.

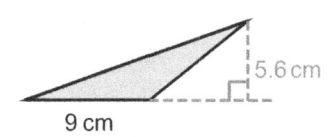

5.6 cm
9 cm

A) 33.3 cm²   B) 25.2 cm²
C) 12.6 cm²   D) 50.4 cm²

©All rights reserved-Math-Knots LLC., VA-USA   www.math-knots.com | www.a4ace.com

Grade 7
Volume 1
Week 18

25) Find the area of the figure given below.

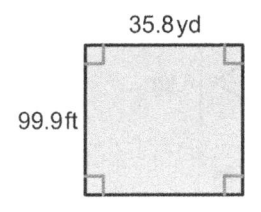

35.8 yd
99.9 ft

A) 2384.28 yd²
B) 596.1 yd²
C) 1183.64 yd²
D) 1192.14 yd²

26) Find the area of the figure given below.

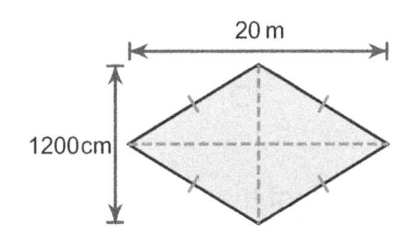

20 m
1200 cm

A) 60 m²   B) 111.1 m²
C) 120 m²   D) 240 m²

27) Find the missing measurement from the figure given below. Round the answer to the nearest tenth.

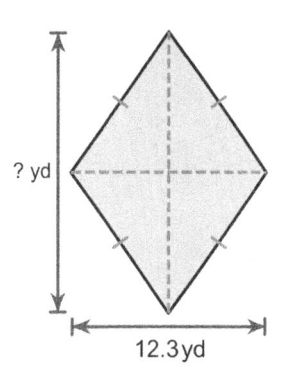

? yd
12.3 yd
Area = 104.55 yd²

A) 20.5   B) 17
C) 11.1   D) 12.1

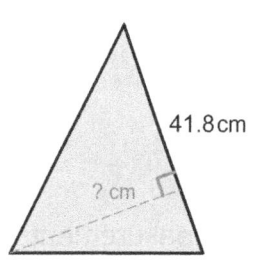

28) Find the missing measurement from the figure given below. Round the answer to the nearest tenth.

41.8 cm
? cm
Area = 679.3 cm²

A) 40.8 cm   B) 33.6 cm
C) 38.6 cm   D) 32.5 cm

29) Find the missing measurement from the figure given below. Round the answer to the nearest tenth.

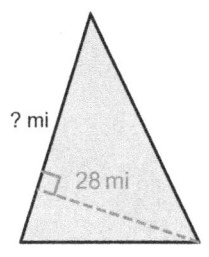

Area =536.2mi²

A) 48 mi B) 38.3 mi

C) 46.3 mi D) 32.6 mi

30) Find the missing measurement from the figure given below. Round the answer to the nearest tenth.

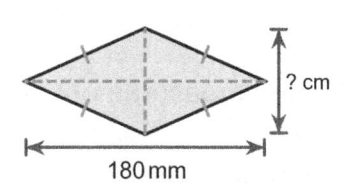

Area =69.3 cm²

A) 6.4 B) 6.6

C) 7.7 D) 0.8

31) Find the missing measurement from the figure given below. Round the answer to the nearest tenth.

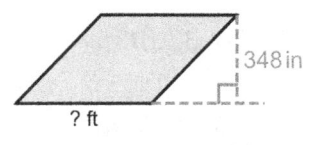

Area =1334 ft²

A) 37.1 ft B) 50.5 ft

C) 3.8 ft D) 46 ft

32) Find the missing measurement from the figure given below. Round the answer to the nearest tenth.

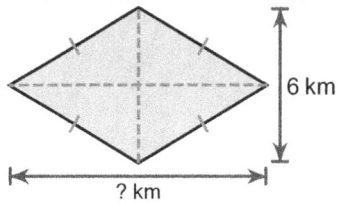

Area =30.6 km²

A) 7.3 B) 13.2

C) 10.2 D) 7.1

33) Find the area of the figure given below.

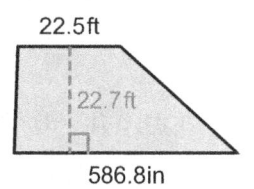

A) 405.2 ft² B) 810.39 ft²

C) 812.69 ft² D) 1620.78 ft²

34) Find the area of the figure given below.

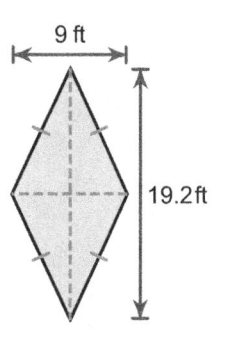

A) 90 ft² B) 86.4 ft²

C) 43.2 ft² D) 172.8 ft²

Grade 7
Volume 1
Week 18

35) Find the area of the figure given below.

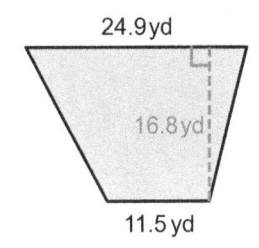

24.9 yd
16.8 yd
11.5 yd

A) 611.52 yd² B) 312.96 yd²

C) 152.9 yd² D) 305.76 yd²

36) Find the missing measurement from the figure given below. Round the answer to the nearest tenth.

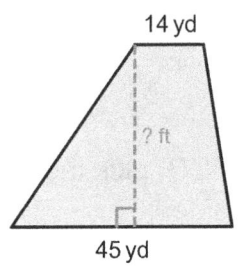

14 yd
? ft
45 yd

Area =1062 yd²

A) 36 ft B) 108 ft

C) 71.3 ft D) 90.9 ft

37) Find the area of the figure given below.

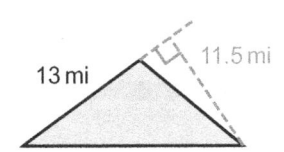

13 mi
11.5 mi

A) 68.05 mi² B) 37.4 mi²

C) 149.5 mi² D) 74.75 mi²

38) Find the area of the figure given below.

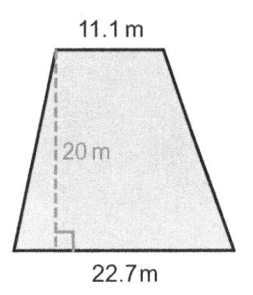

11.1 m
20 m
22.7 m

A) 331.7 m² B) 676 m²

C) 338 m² D) 169 m²

39) Find the missing measurement from the figure given below. Round the answer to the nearest tenth:

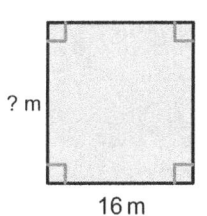

? m
16 m

Area =272 m²

A) 13.7 m B) 18.9 m

C) 22.8 m D) 17 m

40) Find the missing measurement from the figure given below. Round the answer to the nearest tenth.

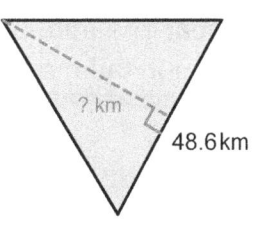

? km
48.6 km

Area =1057.1 km²

A) 43.5 km B) 29 km

C) 58.4 km D) 35.2 km

Grade 7
Volume 1
Week 18

41) The three equal angles of a quadrilateral measure 70° each. Find the measure of fourth angle.

42) The three angles of a quadrilateral are 50°, 70° and 125°. Find the measure of the fourth angle.

43) ∠D of quadrilateral ABCD is equal to 150° and ∠A = ∠B = ∠C. Calculate the three angles of the quadrilateral.

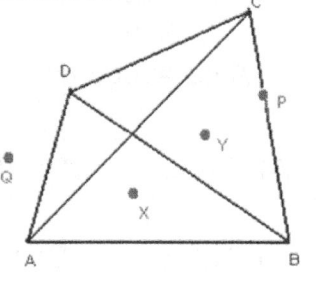

44) The two angles of a quadrilateral measure 65° each. The third angle measures 135°. Find the measure of fourth angle.

45) Is it possible to have a quadrilateral whose angles measure 54°, 80°, 116° and 110°?

46) The four angles of a quadrilateral are in the ratio 6:2:3:4. What is the measure of each angle?

47) In quadrilateral ABCD, pick the pair of adjacent sides.
A) AB & CD
B) AD & CD
C) AC & BD

48) In quadrilateral ABCD, pick the pair of adjacent sides
A) AB & CD
B) AC & BD
C) AB & BC

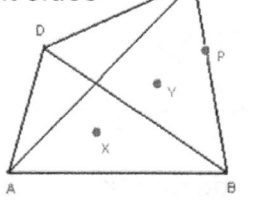

49) In the figure below, Y is a point in the interior of ∠AOB. YZ ⊥ OA and YX ⊥ OB. If ∠XYZ = 115°, what is the measure of ∠AOB?

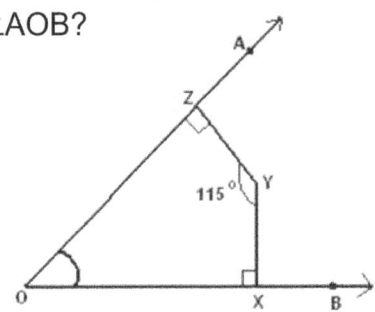

50) The two angles of a quadrilateral measure 70° each. The other two angles are also equal. Find their measures.

51) In the figure below, ABCD is a quadrilateral in which AB ∥ CD. If ∠A = ∠B = 55°, what are the measures of the other two angles?

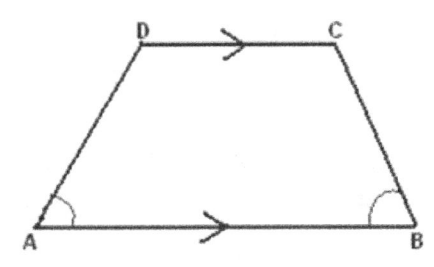

52) A hall is 12 ft long, 5 ft high and 8 ft wide. Find the cost of whitewashing its 4 walls and the ceiling at the rate of $1.20 per ft².

53) In the figure below, Y is a point in the interior of ∠AOB. YZ ⊥ OA and YX ⊥ OB. If ∠AOB = 30°, what is the measure of ∠XYZ?

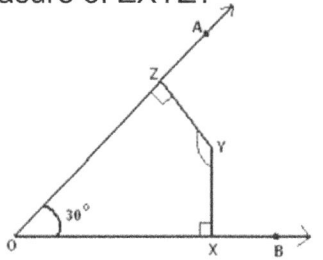

54) How many pairs of adjacent angles are there in a quadrilateral?
A) One
B) Two
C) Three
D) Four

55) Find the area of the shaded region:

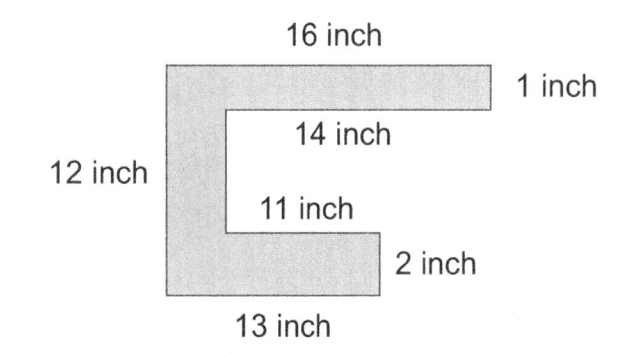

16 inch
1 inch
14 inch
12 inch
11 inch
2 inch
13 inch

56) The perimeter of the floor of a room is 64 ft. Find the height of the room if the area of four walls of the room is 512 ft².

57) The total surface area of the cuboid is 1144 ft². Find the area of 4 sides of the cuboid if the area of its base is 180 ft².

58) Find the perimeter of the base of a carton box whose height is 3 ft and area of four sides of the carton is 66 ft².

59) Find the area of four sides of a cube whose one edge is 5 inches.

60) In the figure below, ABCD is a quadrilateral in which AB || CD. If ∠A = 45° and ∠B = 55°, what are the measures of the other two angles?

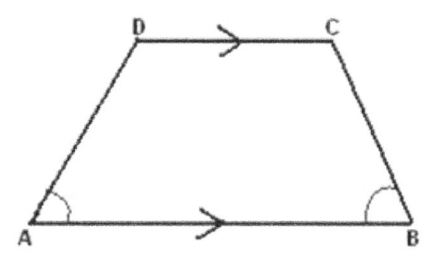

61) In the figure below, ABCD is a quadrilateral in which AB || CD. If ∠A = 50° and ∠B = 45°, what are the measures of the other two angles?

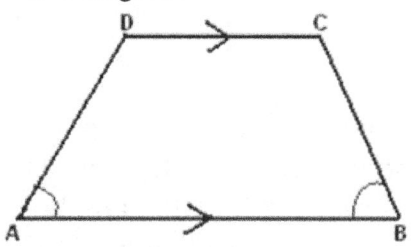

62) In the figure below, ABCD is a quadrilateral in which AB || CD. If ∠A = 60° and ∠B = 70°, what are the measures of the other two angles?

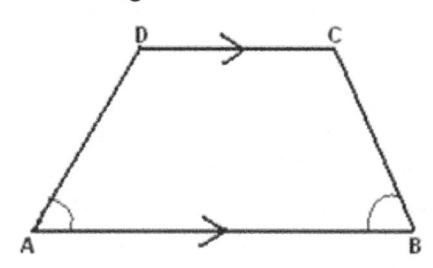

63) A painting is pasted on a square cardboard of length 24 inches, such that there is a margin of 1.5 inches along each of its sides. Find the total area of the margin.

64) The side of a square flower bed is 150 yards. It is enlarged by digging a strip 10 yards wide all around it. Find the area of enlarged flower bed.

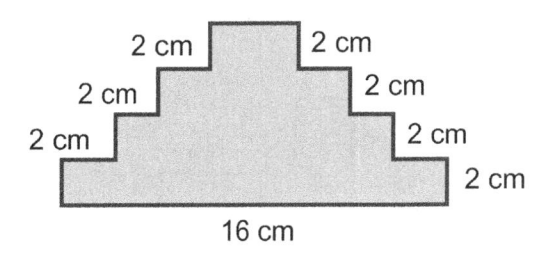

65) A rectangular lawn 35 ft long and 28 ft wide has two cross paths in the center of the park, which are at right angles to each other. The path parallel to the length is 2.5 ft wide and the path parallel to width is 2 ft wide. Find the cost of graveling the paths at $0.70 per ft^2.

66) A path 3 ft wide, is running inside along the boundary of a rectangular field which is 34 yards long and 30 yards wide. How much should be paid to plow the remaining field at the rate of $1.25 per 10 yards^2.

67) A room is 10.4 ft long and 8.4 ft wide. Its height is 12.5 ft. Find the cost of fixing the wall paper on the walls of the room at $3 per square foot.

68) Find the area of the shaded region:

69) A lawn 9.5 ft long and 10 ft wide is surrounded by a 1.75 ft wide path. Calculate the cost of leveling the path at $0.50 per square ft.

70) Each side of a kitchen measures 12 ft. Find the cost of fixing a 2 ft wide granite slab along two adjacent walls of the kitchen at the rate of $2.50 per ft^2.

71) Find the area of the shaded region:

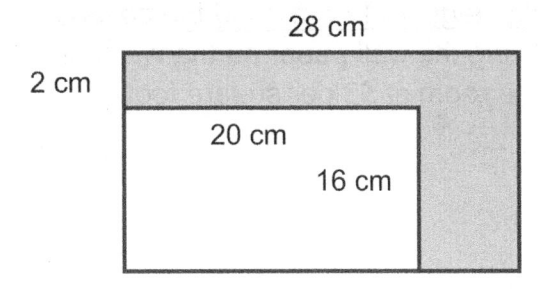

28 cm
2 cm
20 cm
16 cm

72) Find the area of the shaded region:

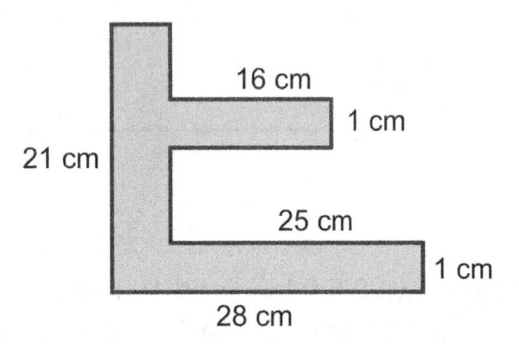

16 cm
1 cm
21 cm
25 cm
1 cm
28 cm

73) In the figure given below, the shaded rectangular path has equal width throughout. Find the area of the path.

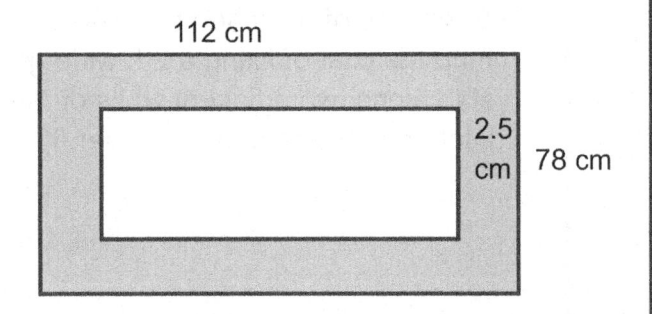

112 cm
2.5 cm
78 cm

Grade 7

Vol 1 Answer Key

Grade 7 Answer Keys

Grade 7

Vol 1 Answer Key

Grade 7

Vol 1 Answer Key

Week 1		Week 1		Week 1		Week 1		Week 1	
1.	B	30.	B	59.	D	88.	B	117.	C
2.	A	31.	A	60.	D	89.	B	118..	C
3.	B	32.	C	61.	B	90.	A	119.	D
4.	B	33.	B	62.	C	91.	B	120.	C
5.	A	34.	C	63.	B	92.	A	121.	C
6.	C	35.	D	64.	D	93.	D	122.	B
7.	D	36.	D	65.	B	94.	A	123.	C
8.	B	37.	D	66.	A	95.	D	124.	B
9.	C	38.	A	67.	D	96.	D	125.	D
10.	D	39.	A	68.	D	97.	D	126.	A
11.	D	40.	A	69.	A	98.	C	127.	D
12.	A	41.	B	70.	C	99.	D	128.	C
13.	B	42.	B	71.	B	100.	D	129.	C
14.	C	43.	B	72.	A	101.	D	130.	B
15.	D	44.	A	73.	A	102.	C	131.	C
16.	C	45.	C	74.	D	103.	B	132.	B
17.	C	46.	C	75.	B	104.	D	133.	A
18.	D	47.	D	76.	B	105.	D	134.	B
19.	B	48.	D	77.	A	106.	D	135.	C
20.	D	49.	A	78.	B	107.	C	136.	D
21.	A	50.	A	79.	C	108.	D	137.	B
22.	B	51.	A	80.	A	109.	D	138.	B
23.	C	52.	D	81.	A	110.	D	139.	C
24.	D	53.	C	82.	B	111.	A	140.	B
25.	D	54.	B	83.	A	112.	A	141.	D
26.	B	55.	B	84.	D	113.	C	142.	A
27.	D	56.	A	85.	B	114.	A	143.	D
28.	C	57.	B	86.	C	115.	C	144.	B
29.	B	58.	C	87.	B	116.	D	145.	A

Grade 7

Vol 1 Answer Key

Week 1		Week 1		Week 1		Week 1		Week 2	
146.	C	175.	C	204.	D	233.	B	1.	C
147.	D	176.	B	205.	D	234.	C	2.	A
148.	B	177.	C	206.	D	235.	A	3.	C
149.	C	178.	C	207.	B	236.	C	4.	B
150.	B	179.	D	208.	B	237.	A	5.	A
151.	D	180.	D	209.	C	238.	D	6.	A
152.	D	181.	D	210.	A	239.	C	7.	B
153.	C	182.	C	211.	D	240.	A	8.	C
154.	D	183.	B	212.	C			9.	B
155.	C	184.	A	213.	C			10.	D
156.	D	185.	C	214.	B			11.	C
157.	B	186.	C	215.	D			12.	C
158.	C	187.	B	216.	A			13.	A
159.	D	188.	B	217.	D			14.	B
160.	C	189.	C	218.	D			15.	A
161.	A	190.	C	219.	D			16.	A
162.	C	191.	B	220.	C			17.	B
163.	A	192.	A	221.	B			18.	A
164.	D	193.	C	222.	B			19.	B
165.	B	194.	A	223.	A			20.	B
166.	D	195.	A	224.	D			21.	B
167.	D	196.	A	225.	C			22.	C
168.	B	197.	A	226.	D			23.	D
169.	C	198.	A	227.	D			24.	C
170.	C	199.	A	228.	A			25.	A
171.	C	200.	D	229.	B			26.	B
172.	C	201.	A	230.	D			27.	D
173.	A	202.	C	231.	C			28.	A
174.	C	203.	A	232.	D			29.	C

Grade 7

Vol 1 Answer Key

Week 2	Week 2	Week 2	Week 2	Week 2
30. C	59. C	88. D	117. B	145. $\dfrac{-27}{125}$
31. D	60. A	89. C	118. A	
32. A	61. A	90. A	119. B	146. $-\left(\dfrac{3}{4}\right)^4$
33. C	62. A	91. A	120. B	
34. D	63. C	92. B	121. C	147. $\left(\dfrac{2}{9}\right)^{14}$
35. A	64. D	93. A	122. C	
36. A	65. D	94. B	123. B	148. $\left(\dfrac{11}{5}\right)$
37. D	66. A	95. B	124. D	
38. D	67. B	96. B	125. C	149. $\dfrac{1}{15625}$
39. D	68. A	97. C	126. B	
40. A	69. D	98. C	127. A	
41. A	70. D	99. A	128. B	150. $\left(\dfrac{1}{6}\right)^3$
42. C	71. A	100. C	129. B	
43. C	72. C	101. B	130. A	
44. D	73. A	102. A	131. D	151. $\left(\dfrac{-23}{17}\right)$
45. B	74. D	103. D	132. B	
46. D	75. A	104. A	133. A	152. 2^8
47. A	76. C	105. D	134. A	
48. B	77. B	106. C	135. D	153. $(-17)^7$
49. B	78. B	107. B	136. D	
50. C	79. C	108. D	137. A	154. $\dfrac{2401}{625}$
51. A	80. A	109. B	138. C	
52. B	81. A	110. D	139. A	155. $\dfrac{-1}{2}$
53. D	82. A	111. B	140. C	
54. A	83. C	112. B	141. -8	156. $\left(\dfrac{1}{2}\right)^{20}$
55. D	84. D	113. D	142. 5^6	
56. B	85. A	114. C	143. 256	157. -3
57. B	86. B	115. A	144. $\dfrac{1}{216}$	158. $\dfrac{15}{49}$
58. B	87. D	116. B		

Grade 7

Vol 1 Answer Key

Week 2		Week 2		Week 3		Week 3		Week 3	
159.	3^8	177.	$\left(\frac{2}{7}\right)^2$	1.	A	30.	B	59.	A
160.	$9^{\left(\frac{4}{5}\right)}$	178.	$\left(\frac{5}{7}\right)^5$	2.	B	31.	A	60.	A
				3.	B	32.	A	61.	A
				4.	A	33.	A	62.	A
161.	3^9			5.	A	34.	A	63.	A
162.	3^5	179.	$\frac{125}{3}$	6.	B	35.	C	64.	A
				7.	A	36.	D	65.	D
163.	69	180.	$\frac{5}{64}$	8.	B	37.	B	66.	C
164.	-1			9.	A	38.	A	67.	A
165.	$\frac{2}{3}$	181.	5^6	10.	A	39.	C	68.	D
				11.	B	40.	B	69.	A
166.	2^4			12.	B	41.	C	70.	B
167.	5^{-6}			13.	A	42.	A	71.	D
				14.	A	43.	D	72.	A
168.	-3^{25}			15.	A	44.	D	73.	C
169.	2^{10}			16.	B	45.	D	74.	D
				17.	A	46.	B	75.	D
170.	$\frac{125}{216}$			18.	A	47.	A	76.	C
				19.	A	48.	D	77.	A
171.	$\frac{40}{81}$			20.	A	49.	B	78.	B
				21.	B	50.	D	79.	B
172.	175			22.	A	51.	B	80.	A
173.	$\left(\frac{1}{3}\right)^{10}$			23.	A	52.	C	81.	A
				24.	B	53.	B	82.	A
174.	$\left(\frac{13}{14}\right)^2$			25.	B	54.	D	83.	C
				26.	A	55.	C	84.	C
175.	$\left(\frac{3}{5}\right)^5$			27.	A	56.	D	85.	C
				28.	B	57.	C	86.	B
176.	$\frac{64}{15625}$			29.	B	58.	A	87.	B

Grade 7

Vol 1 Answer Key

Week 3		Week 3		Week 3		Week 3		Week 4	
88.	C	117.	A	146.	D	175.	C	1.	D
89.	A	118.	B	147.	A	176.	D	2.	D
90.	C	119.	B	148.	D	177.	A	3.	D
91.	D	120.	D	149.	D	178.	D	4.	A
92.	C	121.	C	150.	A	179.	D	5.	B
93.	A	122.	D	151.	B	180.	A	6.	D
94.	D	123.	B	152.	D	181.	B	7.	A
95.	A	124.	C	153.	C			8.	B
96.	A	125.	B	154.	D			9.	D
97.	C	126.	B	155.	D			10.	C
98.	B	127.	A	156.	B			11.	A
99.	D	128.	D	157.	B			12.	D
100.	D	129.	B	158.	D			13.	B
101.	D	130.	A	159.	A			14.	B
102.	D	131.	A	160.	B			15.	A
103.	A	132.	D	161.	B			16.	A
104.	A	133.	B	162.	D			17.	B
105.	A	134.	C	163.	A			18.	D
106.	A	135.	C	164.	D			19.	C
107.	A	136.	B	165.	C			20.	B
108.	D	137.	D	166.	A			21.	D
109.	C	138.	B	167.	D			22.	B
110.	A	139.	B	168.	A			23.	C
111.	C	140.	C	169.	D			24.	D
112.	D	141.	B	170.	C			25.	D
113.	C	142.	C	171.	D			26.	D
114.	A	143.	A	172.	B			27.	A
115.	D	144.	C	173.	B			28.	A
116.	B	145.	C	174.	D			29.	C

©All rights reserved-Math-Knots LLC., VA-USA

Grade 7

Vol 1 Answer Key

Week 4		Week 4		Week 4		Week 5		Week 5	
30.	C	59.	A	88.	C	1.	G	30.	A & C
31.	B	60.	A	89.	D	2.	A	31.	C & D
32.	C	61.	C	90.	C	3.	C	32.	D
33.	D	62.	A	91.	A	4.	E	33.	B
34.	C	63.	B	92.	D	5.	C	34.	D
35.	C	64.	A	93.	D	6.	G	35.	A
36.	A	65.	C	94.	C	7.	A	36.	A & C
37.	C	66.	C	95.	A	8.	C	37.	C
38.	D	67.	C	96.	D	9.	G	38.	A & B
39.	B	68.	B	97.	C	10.	C	39.	Yes
40.	C	69.	B	98.	D	11.	f	40.	Yes
41.	A	70.	D	99.	B	12.	D	41.	Yes
42.	D	71.	B	100.	A	13.	D	42.	No
43.	C	72.	B			14.	I	43.	No
44.	C	73.	A			15.	F	44.	Yes
45.	C	74.	C			16.	C	45.	Yes
46.	D	75.	C			17.	A	46.	Yes
47.	D	76.	D			18.	I	47.	Yes
48.	B	77.	B			19.	G	48.	No
49.	B	78.	A			20.	G	49.	No
50.	C	79.	B			21.	I	50.	No
51.	C	80.	C			22.	G	51.	Yes
52.	A	81.	D			23.	G	52.	Yes
53.	A	82.	D			24.	A	53.	No
54.	B	83.	C			25.	C	54.	No
55.	A	84.	A			26.	A	55.	Yes
56.	B	85.	A			27.	A	56.	No
57.	B	86.	A			28.	B	57.	Yes
58.	D	87.	B			29.	C & D	58.	y = 5x

Grade 7

Vol 1 Answer Key

Week 5

59.	y = 9x
60.	y = 25x
61.	y = 1x
62.	y = 16x
63.	y = 14x
64.	y = 18x
65.	A, B, D
66.	A, B, D
67.	A, B
68.	A, B, D
69.	C
70.	A, B, C
71.	B, C
72.	$3.97
73.	$11.01
74.	4710
75.	$101.25
76.	$99.36
77.	$315.75
78.	176
79.	$31.84
80.	1½ miles
81.	1½ bags
82.	1½ hours
83.	3 bottles
84.	1 ½ hours
85.	2½ boxes
86.	1½ bags
87.	6 baskets

Week 5

88.	2½ minutes
89.	8 bags.
90.	7 bags
91.	294 miles
92.	15 lbs.
93.	27 kgs
94.	360 miles
95.	87 gallons.
96.	1 hr
97.	16 km
98.	225 km
99.	17 liters
100.	15 gallons
101.	$66
102.	6 gallons
103.	34 miles
104.	20 seconds
105.	0.01 miles/sec.
106.	54,000 miles/hour
107.	0.285 miles
108.	6 hours 15 minutes.
109.	52 minutes.
110.	Direct proportion.
111.	297.5 miles
112.	10 miles/sec.
113.	25 minutes.
114.	106 miles/hour
115.	43 miles/hour
116.	84 miles/hour

Week 5

117.	60 miles/hour.
118.	48 minutes
119.	9 sec.
120.	90 km/hr.
121.	52 miles/hour.
122.	400 miles/hour.
123.	0.25 miles/sec.
124.	108 km/hr.
125.	0.025 miles/sec.
126.	10 sec.
127.	10 seconds.
128.	1 minute.
129.	72 miles/hour
130.	60 miles/hour
131.	4 days.
132.	48 minutes.
133.	12 days.
134.	A
135.	C
136.	D
137.	C
138.	C
139.	B
140.	C
141.	B
142.	D
143.	B
144.	A
145.	D

Grade 7

Vol 1 Answer Key

Week 5		Week 6		Week 6		Week 6		Week 6	
146.	C	1.	A	30.	B	59.	D	88.	A
147.	C	2.	A	31.	A	60.	D	89.	C
148.	B	3.	A	32.	D	61.	A	90.	A
149.	B	4.	B	33.	B	62.	A	91.	A
150.	A	5.	A	34.	B	63.	B	92.	B
151.	D	6.	B	35.	A	64.	C	93.	D
152.	A	7.	B	36.	D	65.	D	94.	D
153.	147.92	8.	A	37.	B	66.	B	95.	A
154.	39.45	9.	B	38.	B	67.	B	96.	D
155.	$1208.28	10.	A	39.	D	68.	C	97.	A
156.	$8,976.63	11.	B	40.	B	69.	A	98.	A
157.	$3023.28	12.	D	41.	C	70.	D	99.	C
		13.	C	42.	C	71.	B	100.	C
		14.	D	43.	B	72.	D	101.	B
		15.	D	44.	C	73.	C	102.	B
		16.	A	45.	D	74.	D	103.	C
		17.	B	46.	C	75.	A	104.	A
		18.	B	47.	B	76.	B	105.	A
		19.	C	48.	B	77.	D	106.	D
		20.	C	49.	D	78.	D	107.	A
		21.	B	50.	B	79.	C	108.	C
		22.	A	51.	B	80.	C	109.	C
		23.	A	52.	D	81.	C	110.	C
		24.	A	53.	C	82.	C	111.	A
		25.	A	54.	D	83.	B	112.	A
		26.	B	55.	C	84.	D	113.	C
		27.	D	56.	A	85.	D	114.	B
		28.	A	57.	D	86.	D	115.	C
		29.	B	58.	C	87.	A	16.	D

Grade 7

Vol 1 Answer Key

Week 6

#	Answer
117.	C
118.	A
119.	B
120.	C
121.	D
122.	B
123.	B
124.	D
125.	4 days.
126.	15 days.
127.	162 km/hr.
128.	10 m/sec.
129.	D
130.	90 miles/hour.
131.	126 km/hr.
132.	144 km/hr.
133.	12.5 m/sec.
134.	21
135.	$18.27 = p ; $274.05
136.	13.5 m/sec.
137.	13.5 m/sec.
138.	A
139.	20 m/sec.
140.	$6700.59
141.	14400 miles/hour.
142.	2700 miles/hour.
143.	80 miles
144.	10 hours.
145.	18 hours.

Week 6

#	Answer
146.	30 m/sec.
147.	E
148.	C
149.	G
150.	E
151.	C
152.	No
153.	Yes
154.	15 m/sec.
155.	50 km/hr.
156.	$5.65
157.	720 miles
158.	51.2 km/hr.
159.	$207.56
160.	30.2 km/hr.
161.	B
162.	$4.08
163.	$86
164.	4320 miles/hour.
165.	210 miles.
166.	$69.40
167.	A, C
168.	-1,400
169.	$25.13
170.	34
171.	y = 9x
172.	y = 20x
173.	408
174.	$11.07

Week 6

#	Answer
175.	103
176.	144
177.	10.17
178.	27.56
179.	32.76
180.	22.36
181.	49.20
182.	232.50
183.	24.80
184.	84.53
185.	29.90
186.	32.80
187.	B
188.	3 containers
189.	3 bottles
190.	5
191.	47x
192.	10x
193.	12x
194.	4x
195.	19x
196.	30x
197.	18x
198.	2x
199.	42x

Grade 7

Vol 1 Answer Key

Week 7		Week 7		Week 7		Week 7		Week 7	
1.	850 %	30.	B	59.	D	88.	D	117.	A
2.	95.5 %	31.	A	60.	C	89.	D	118.	D
3.	0.600 %	32.	C	61.	D	90.	B	119.	B
4.	69.$\overline{69}$ %	33.	B	62.	B	91.	A	120.	A
5.	161 %	34.	B	63.	D	92.	B	121.	A
6.	30.$\overline{30}$ %	35.	C	64.	A	93.	B	122.	B
7.	3.$\overline{03}$ %	36.	A	65.	C	94.	C	123.	A
8.	71 %	37.	C	66.	D	95.	B	124.	D
9.	933.$\overline{3}$ %	38.	B	67.	A	96.	A	125.	C
10.	90 %	39.	D	68.	D	97.	D	126.	A
11.	10 %	40.	B	69.	A	98.	C	127.	C
12.	66.$\overline{6}$ %	41.	B	70.	A	99.	D	128.	D
13.	80 %	42.	B	71.	D	100.	C	129.	A
14.	50 %	43.	D	72.	A	101.	D	130.	D
15.	0.7 %	44.	B	73.	B	102.	D	131.	D
16.	25 %	45.	D	74.	D	103.	D	132.	D
17.	72.$\overline{72}$ %	46.	B	75.	C	104.	D	133.	C
18.	87.5 %	47.	C	76.	D	105.	D	134.	A
19.	B	48.	C	77.	A	106.	A	135.	B
20.	C	49.	C	78.	D	107.	A	136.	A
21.	29 %	50.	C	79.	A	108.	B	137.	C
22.	68.75 %	51.	A	80.	D	109.	A	138.	A
23.	C	52.	A	81.	B	110.	D	139.	A
24.	B	53.	C	82.	C	111.	A	140.	A
25.	D	54.	A	83.	D	112.	D	141.	B
26.	D	55.	A	84.	A	113.	D	142.	C
27.	B	56.	A	85.	B	114.	C	143.	D
28.	A	57.	A	86.	D	115.	B	144.	C
29.	D	58.	C	87.	A	116.	D	145.	A

Grade 7

Vol 1 Answer Key

Week 7		Week 7		Week 8		Week 8	
146.	D	175.	D	1.	A		
147.	A	176.	B	2.	D	28.	$28\frac{1}{3}$ %
148.	C	177.	B	3.	320 oz.	29.	$850.
149.	A	178.	B	4.	$4000.	30.	10%.
150.	D	179.	D	5.	400 students.	31.	$50,000.
151.	A	180.	A	6.	600.	32.	54 cents.
152.	A	181.	C	7.	$33\frac{1}{3}$ %	33.	$710.
153.	A	182.	D			34.	160.
154.	B	183.	A	8.	400 fruits.	35.	160.
155.	D	184.	D	9.	4%.	36.	10%
156.	D	185.	C	10.	72%	37.	40%.
157.	A	186.	B	11.	25% of 500 > 20% of 600.	38.	60%.
158.	D	187.	D	12.	20,152.	39.	40%.
159.	B	188.	A	13.	$1,225.	40.	60%.
160.	C	189.	C	14.	3,867.	41.	34%.
161.	D	190.	B	15.	340.	42.	66%.
162.	D	191.	C	16.	$2,205.20.		
163.	A	192.	C	17.	95%.	43.	$33\frac{1}{3}$ %
164.	A	193.	D	18.	70%.		
165.	B	194.	A	19.	20%	44.	$26\frac{2}{3}$ %
166.	C	195.	C	20.	Mathematics.	45.	40%
167.	D	196.	A	21.	26	46.	62.5%.
168.	C	197.	C	22.	25%.	47.	25%.
169.	C	198.	B	23.	81.88 gallons.	48.	12.5%
170.	D			24.	2075	49.	16
171.	C			25.	50	50.	18
172.	A			26.	15.07%.	51.	22
173.	D			27.	$66\frac{2}{3}$ %	52.	36
174.	C					53.	$4,176.00

Grade 7

Vol 1 Answer Key

Week 8

54. $624
55. **$630**
56. $1050
57. $735
58. $2835
59. $632.80
60. $406.80
61. $1220.40
62. $11,700
63. 1767 and 2883
64. 155 points
65. 31%
66. $154
67. $160
68. $301
69. 140
70. 405 kg
71. 101.25 miles
72. 118 inches
73. 237.5 lb
74. $8000
75. $120
76. $76
77. $54
78. 150 oranges
79. 517
80. 42
81. 23
82. 18

Week 8

83. 60
84. $180
85. $\dfrac{17}{20}$
86. $26\dfrac{2}{3}$ %
87. 25%
88. $10,000
89. 20%
90. $53
91. $5,880
92. $43.75
93. $925
94. $1875
95. $544.50 each
96. 5%
97. 20%
98. 33.33%
99. 20%
100. $750
101. 25%
102. $1,400
103. 4%
104. 60 cents
105. $19.65
106. $200
107. $367.50
108. $400
109. $150

Week 8

110. $228
111. 20%
112. $124.80
113. 18 %
114. $560
115. $600
116. Lost $2\dfrac{2}{9}$ %
117. 5%
118. $90
119. $396
120. $1500
121. Loss 12.5%
122. Loss 20%
123. Loss 6.25%
124. 5 at $1.50
125. $205
126. $6359.50
127. 28%
128. 5 gallons at $21
129. 50 for $19
130. 10 for $50.60
131. $902
132. $4.20
133. Gain 1.65%

Grade 7

Vol 1 Answer Key

Week 9
1. A
2. B
3. D
4. A
5. C
6. D
7. B
8. C
9. B
10. A
11. B
12. C
13. C
14. B
15. B
16. A
17. D
18. A
19. C
20. A
21. C
22. B
23. B
24. C
25. A
26. D
27. C
28. A
29. A

Week 9
30. C
31. D
32. C
33. D
34. D
35. B
36. D
37. C
38. C
39. C
40. A
41. C
42. B
43. B
44. D
45. A
46. A
47. C
48. B
49. B
50. B
51. D
52. C
53. D
54. A
55. D
56. 9.3×10^7 miles.
57. 40,000,000 meters.
58. city b, city c, city d, city a.

Week 9
59. 0.0000112.
60. 0.00032256.
61. 8×10^{-1}
62. 1,120,000 km.
63. 0.0055874.
64. 0.08.
65. -0.0044
66. 7.2×10^{-7}
67. 5.6×10^{12}
68. 3×10^8 m/sec.
69. 1.49×10^8 km
70. $(-5.2) \times 10^{-9}$
71. 3.1×10^5, 2.5×10^{-1}, 2.4×10^{-3}
72. 0.001364.
73. 2.385×10^{11} miles.
74. 3.72×10^{-8}, 7.45×10^{-5}, 6.38×10^{-4}, 5.12×10^{-3}
75. 2.3×10^{10}
76. No.
77. 0.00000000013 m.
78. 4.32×10^5 miles.
79. 5.899×10^{12} miles.
80. 1.5×10^7 K.
81. 0.00001276 m.
82. 2,460,000
83. 340,000,000.
84. 0.000145.
85. 4.072×10^7
86. 6.7×10^5

Grade 7

Vol 1 Answer Key

Week 9

87. 3.45×10^{-5}
88. Yes.
89. No.
90. $1.78 \times 10^8 > 4.18 \times 10^6 > 2.34 \times 10^5 > 6.29 \times 10^4$
91. 10%
92. $12000 & $6000
93. 35%
94. $1120
95. 2 years 6 months
96. $1020
97. 8%
98. 16 months
99. $3600
100. 6%
101. 2 years
102. $318.60
103. 7% per annum
104. $180
105. 10%
106. 2.5 years
107. $66.50
108. $127.75
109. 6 years
110. $787.50 in 6 years at 12.5% per annum
111. $1728 and the rate of interest is 25%
112. $450
113. $4000 & $2000

Week 9

114. $560 invested amount to $595
115. Sum of $870 invested amount to $1457.25
116. $750 invested amount to $915
117. Sum is $3410.00

Grade 7

Vol 1 Answer Key

Week 10

1. >
2. $\left(\dfrac{-6}{7}\right), \left(\dfrac{-3}{4}\right), \left(\dfrac{-5}{11}\right), \left(\dfrac{-1}{3}\right)$
3. False
4. >
5. <
6. >
7. >
8. <
9. $\left(\dfrac{-11}{4}\right), \left(\dfrac{5}{-6}\right), \left(\dfrac{-1}{-8}\right), \left(\dfrac{1}{4}\right)$
10. $(-1), \left(\dfrac{17}{-20}\right), \left(\dfrac{-4}{5}\right), \left(\dfrac{-3}{4}\right)$
11. $\left(\dfrac{4}{7}\right), 0, \left(\dfrac{1}{-2}\right), \left(\dfrac{-4}{5}\right)$
12. $\left(\dfrac{-5}{8}\right), \left(\dfrac{7}{-9}\right), \left(\dfrac{-8}{7}\right), \left(\dfrac{4}{-3}\right)$
13. True
14. False
15. 5.72
16. $\dfrac{1}{9}$
17. $\left(\dfrac{-13}{6}\right), \left(\dfrac{-8}{12}\right), \left(\dfrac{-3}{5}\right), \left(\dfrac{-6}{20}\right)$
18. <

Week 10

19. B
20. False
21. A
22. $\dfrac{12821}{1000}$
23. =
24. False
25. <
26. =
27. True
28. True
29. False
30. >
31. <
32. =
33. >
34. =
35. $\left(\dfrac{5}{7}\right), \left(\dfrac{3}{7}\right), \left(\dfrac{2}{5}\right), \left(\dfrac{-7}{8}\right)$
36. >
37. $\left(\dfrac{3}{-4}\right), \left(\dfrac{4}{-9}\right), \left(\dfrac{-2}{-3}\right), \left(\dfrac{-5}{-7}\right)$
38. 0.2121.....
39. $\dfrac{5}{18}$

Grade 7

Vol 1 Answer Key

Week 10

40. B
41. <
42. <
43. <
44. >
45. False
46. True
47. $\dfrac{-13}{16}$
48. $\dfrac{11}{12}$
49. True
50. A
51. True
52. False
53. False
54. False
55. $\dfrac{17}{10}$
56. $\dfrac{113}{36}$
57. $\left(\dfrac{-21}{10}\right)$
58. $\left(\dfrac{4}{11}\right)$
59. $\left(\dfrac{-53}{37}\right)$

Week 10

60. $\left(\dfrac{-277}{156}\right)$
61. $\dfrac{-7}{18}$
62. $\dfrac{-23}{30}$
63. $\dfrac{73}{36}$ ft.
64. $\left(\dfrac{-17}{24}\right)$
65. $\left(\dfrac{-19}{35}\right)$
66. $\dfrac{131}{60}$ miles
67. $\dfrac{-3}{10}$
68. $\left(\dfrac{-13}{19}\right)$
69. $\dfrac{1}{2}$
70. $\dfrac{-17}{18}$
71. $\dfrac{73}{30}$ pounds
72. $\dfrac{62}{21}$
73. $\dfrac{7}{20}$

Week 10

74. $\dfrac{61}{55}$
75. $\left(\dfrac{29}{40}\right)$
76. $\left(\dfrac{-1}{15}\right)$
77. $\left(\dfrac{-19}{24}\right)$
78. True
79. $\dfrac{13}{7}$
80. $\dfrac{19}{8}$
81. $\dfrac{-43}{30}$
82. $\dfrac{-19}{20}$
83. $\dfrac{-38}{35}$
84. $\dfrac{1}{4}$
85. $\dfrac{-19}{60}$
86. $\dfrac{1}{18}$
87. $\dfrac{3}{5}$

Grade 7

Vol 1 Answer Key

Week 10

88. $\dfrac{-1}{5}$

89. $\dfrac{8}{35}$

90. $\dfrac{-3}{2}$

91. $\dfrac{50}{27}$

92. $\left(\dfrac{-163}{88}\right)$

93. $\dfrac{5}{12}$

94. $\dfrac{193}{14}$

95. $\dfrac{117}{250}$

96. $\dfrac{29}{9}$

97. False

98. $\left(\dfrac{-3}{5}\right)$

99. $\dfrac{8}{15}$

100. $\dfrac{39}{8}$ inches

101. $\left(\dfrac{-1}{4}\right)$

Week 10

102. $\left(\dfrac{15}{11}\right)$

103. $\left(\dfrac{15}{14}\right)$

104. $\dfrac{37}{5}$

105. $\left(\dfrac{4}{15}\right)$

106. $\dfrac{33}{25}$

107. $\dfrac{40}{63}$

108. $\dfrac{35}{8}$

109. $\dfrac{11}{10}$

110. $\left(\dfrac{13}{6}\right)$

111. $\dfrac{23}{72}$

112. $\left(\dfrac{-11}{35}\right)$

113. $\left(\dfrac{17}{9}\right)$

114. True

115. $\dfrac{7}{12}$

116. $\left(\dfrac{-17}{14}\right)$

Week 10

117. $\left(\dfrac{11}{13}\right)$

118. $\left(\dfrac{12}{17}\right)$

119. $\left(-\dfrac{4}{9}\right)$

120. $\left(-\dfrac{13}{15}\right)$

121. $\left(\dfrac{-47}{45}\right)$

122. $\dfrac{3}{35}$

123. $\dfrac{31}{20}$

124. $\dfrac{2}{5}$

125. $\left(\dfrac{-37}{11}\right)$

126. $\left(\dfrac{-3}{2}\right)$

127. $\left(\dfrac{1}{10}\right)$

128. $\left(\dfrac{9}{7}\right)$

129. $\left(\dfrac{-16}{3}\right)$

Week 10

130. $\dfrac{1}{6}$

131. $\dfrac{14}{5}$

132. 39

133. $\left(\dfrac{71}{90}\right)$

134. 707 miles

135. False

136. True

137. True

138. True

139. $\dfrac{25}{42}$

140. -4

141. 60

142. $\left(\dfrac{-40}{3}\right)$

143. $\left(\dfrac{2}{21}\right)$

144. $\left(-\dfrac{8}{9}\right)$

145. -2

146. 1

147. $\left(\dfrac{-40}{3}\right)$

148. $\left(\dfrac{-109}{3}\right)$

Grade 7

Vol 1 Answer Key

Week 10 | Week 10 | Week 10 | Week 10

149. $\left(\dfrac{-11}{12}\right)$

150. $\dfrac{46}{13}$

151. 4 ft

152. $\dfrac{18}{55}$

153. $\left(-\dfrac{5}{42}\right)$

154. $\left(\dfrac{1}{24}\right)$

155. $\dfrac{15}{4}$

156. $\dfrac{17}{12}$

157. $\dfrac{27}{56}$

158. 1

159. $\left(\dfrac{16}{35}\right)$

160. $\dfrac{140}{9}$

161. True

162. $\left(\dfrac{-17}{20}\right)$

163. $\left(\dfrac{-295}{504}\right)$

164. $\left(\dfrac{-63}{22}\right)$

165. $\left(\dfrac{5}{126}\right)$

166. $\left(\dfrac{-16}{3}\right)$

167. $\left(\dfrac{5}{14}\right)$

168. $\left(\dfrac{1}{2}\right)$

169. $\left(\dfrac{-14}{3}\right)$

170. $\left(\dfrac{10}{3}\right)$

171. $\left(\dfrac{-1}{3}\right)$

172. $\left(\dfrac{-3}{16}\right)$

173. $\left(\dfrac{3}{4}\right)$

174. $\left(\dfrac{-27}{28}\right)$

175. $\left(\dfrac{-78}{49}\right)$

176. $\left(\dfrac{-20}{27}\right)$

177. $\left(-\dfrac{1}{3}\right)$

178. $\left(\dfrac{-83}{35}\right)$

179. $\left(\dfrac{22}{15}\right)$

180. -15

181. False

182. $\dfrac{7}{3}$

183. $\dfrac{13}{5}$

184. False

185. $\dfrac{80}{81}$

186. -13

187. $\left(\dfrac{-3}{5}\right)$

188. $\left(\dfrac{-2}{5}\right)$

189. $\left(\dfrac{-11}{7}\right)$

190. $\left(\dfrac{16}{5}\right)$

191. $\dfrac{3}{7}$

192. $\dfrac{9}{14}$

193. $\dfrac{63}{5}$

194. $\left(\dfrac{-24}{77}\right)$

195. C

196. B

197. -2

198. $\dfrac{89}{61}$

199. $\dfrac{29}{7}$

200. -9

201. $\left(\dfrac{-7}{6}\right)$

202. $\left(\dfrac{-11}{9}\right)$

203. $\left(\dfrac{-1}{7}\right)$

204. $\left(\dfrac{12}{7}\right)$

205. $\dfrac{10}{3}$

206. $\dfrac{5}{2}$

207. False

208. True

Grade 7

Vol 1 Answer Key

Week 10

209. $\left(\dfrac{48}{-52}\right)$
210. True
211. False
212. True
213. False
214. True
215. True
216. False
217. True
218. Yes
219. A
220. C
221. D
222. $\left(\dfrac{3}{8}\right)$
223. $\left(\dfrac{-3}{4}\right)$
224. $\left(\dfrac{30}{50}\right)$
225. $\left(\dfrac{-28}{-49}\right)$
226. $\left(\dfrac{-30}{48}\right)$
227. $\left(\dfrac{36}{-135}\right)$

Week 10

228. 13.5555....
229. $\dfrac{16}{99}$
230. 0.4375
231. $\dfrac{7}{12}$
232. 3
233. $\dfrac{433}{999}$
234. 0.27777....
235. 0.06060....
236. Not equal
237. Not equal
238. True
239. $\dfrac{641}{250}$
240. $\dfrac{81}{20}$
241. $\dfrac{349}{990}$
242. $\dfrac{7}{999}$
243. True
244. True
245. 0.9
246. $\dfrac{7}{6}$

Week 10

247. $\dfrac{8}{33}$
248. 4.3333.....
249. B
250. $\dfrac{1}{45}$
251. No
252. 4.75
253. Yes
254. 5.91666....
255. 0.2727....
256. Yes
257. No
258. 2.2727....
259. Not equal
260. Yes
261. D
262. $x = -2$
263. $a = -40$
264. D
265. D
266. C
267. $a = -8$

Week 10

268. -4
269. $\dfrac{18}{25}$
270. No
271. B
272. $\dfrac{2}{5}$
273. -72
274. yes
275. $x = 51$
276. $\left(\dfrac{4}{7}\right)$
277. $\left(\dfrac{-5}{7}\right)$
278. D
279. $\left(\dfrac{-4}{7}\right)$
280. -40
281. $\left(\dfrac{4}{7}\right)$
282. No
283. A
284. $\left(\dfrac{13}{7}\right)$

Grade 7

Vol 1 Answer Key

Week 10

285. $\left(\dfrac{-1}{7}\right)$

286. C

287. $\dfrac{8}{7}$

288. Yes

289. C

290. 35

291. True

292. C

293. -30

294. $\left(\dfrac{-22}{48}\right)$

295. B

296. C

297. D

298. A,B,C,D

299. True

Assessment 2 Week 11

1. C
2. C
3. C
4. C
5. A
6. B
7. D
8. C
9. C
10. A
11. C
12. B
13. B
14. D
15. C
16. B
17. D
18. C
19. A
20. B
21. C
22. B
23. A
24. D
25. A
26. C
27. D
28. D
29. A
30. B
31. C
32. A
33. B
34. A
35. A
36. D
37. D
38. C
39. A
40. B
41. B
42. D
43. D
44. C
45. C
46. C
47. A
48. D
49. D
50. C
51. B
52. B
53. A
54. C
55. D
56. C
57. D
58. C
59. B
60. True
61. True
62. <
63. False
64. $\left(\dfrac{-8}{75}\right), \left(\dfrac{11}{-25}\right), \left(\dfrac{-7}{15}\right), \left(\dfrac{9}{-10}\right)$
65. $\left(\dfrac{11}{21}\right)$
66. $\dfrac{37}{144}$
67. >
68. <
69. $\dfrac{1}{3}$
70. $\dfrac{583}{168}$
71. C
72. $\left(\dfrac{-79}{126}\right)$
73. $\dfrac{2}{5}$
74. $\dfrac{-14}{15}$
75. $\dfrac{-23}{24}$

Grade 7

Vol 1 Answer Key

Assessment 2 Week 11

76. $\left(\dfrac{13}{8}\right)$

77. $\left(\dfrac{-13}{45}\right)$

78. $\dfrac{243}{20}$ ft

79. $\dfrac{17}{15}$ pounds.

80. $\left(\dfrac{22}{5}\right)$

81. $\left(\dfrac{16}{21}\right)$

82. $\dfrac{5}{28}$

83. $\dfrac{17}{15}$

84. $\dfrac{28}{15}$

85. -7.

86. $\left(\dfrac{-128}{15}\right)$

87. 0.25

88. $\dfrac{7}{12}$

89. 1

90. False

91. -48

Assessment 2 Week 11

92. $\dfrac{5}{4}$

93. False

94. $\dfrac{7}{2}$

95. A

96. B

Week 12

1. A
2. D
3. B
4. C
5. B
6. D
7. B
8. D
9. B
10. C
11. B
12. B
13. A
14. C
15. D
16. C
17. D
18. A
19. B
20. A
21. C
22. B
23. A
24. A
25. A
26. B
27. A
28. C
29. A

Week 12

30. D
31. A
32. A
33. A
34. C
35. C
36. A
37. B
38. A
39. D
40. A
41. C
42. D
43. A
44. C
45. D
46. D
47. D
48. D
49. C
50. C
51. C
52. A
53. C
54. B
55. A
56. C
57. C
58. C

Grade 7

Vol 1 Answer Key

Week 12		Week 12		Week 12		Week 13		Week 13	
59.	A	88.	D	117.	C	1.	A	30.	B
60.	D	89.	B	118.	A	2.	D	31.	C
61.	D	90.	C	119.	B	3.	C	32.	A
62.	C	91.	A	120.	B	4.	A	33.	C
63.	B	92.	A			5.	C	34.	C
64.	A	93.	C			6.	D	35.	A
65.	C	94.	C			7.	D	36.	D
66.	A	95.	B			8.	A	37.	D
67.	A	96.	C			9.	B	38.	C
68.	D	97.	B			10.	B	39.	B
69.	C	98.	A			11.	A	40.	D
70.	D	99.	C			12.	B	41.	A
71.	B	100.	A			13.	D	42.	A
72.	A	101.	C			14.	A	43.	C
73.	A	102.	C			15.	B	44.	C
74.	B	103.	C			16.	A	45.	C
75.	D	104.	A			17.	C	46.	C
76.	B	105.	A			18.	D	47.	A
77.	C	106.	B			19.	B	48.	A
78.	B	107.	B			20.	B	49.	C
79.	D	108.	A			21.	B	50.	A
80.	A	109.	C			22.	D	51.	A
81.	B	110.	C			23.	A	52.	A
82.	A	111.	A			24.	D	53.	D
83.	C	112.	B			25.	C	54.	A
84.	C	113.	C			26.	A	55.	C
85.	A	114.	B			27.	C	56.	A
86.	B	115.	D			28.	A	57.	C
87.	D	116.	A			29.	B	58.	D

Grade 7

Vol 1 Answer Key

Week 13

59. B
60. D
61. C
62. B
63. C
64. A
65. A
66. D
67. A
68. C
69. B
70. B
71. B
72. A
73. C
74. C
75. B
76. D
77. B
78. D
79. A
80. C

Week 14

1. $(3y - 6x)$.
2. $(a + 3b + 4c)$
3. $-2x - 8y$
4. $(-2x - 1)$
5. $\left(\dfrac{-15}{4}\right)x^2 + \left(\dfrac{8}{15}\right)x + \left(\dfrac{317}{42}\right)$
6. $\left(-11y^2 - \dfrac{13}{3}y + 18\right)$
7. $(-x^2 - 4x + 4)$
8. $(-5x^2 + x + 7)$
9. $(20x + 6y)$
10. $(32 - 15b^2 - 3b^3)$
11. $(-16x^3 + 4x^2 + x - 4)$.
12. $15ab$.
13. $-(x + 10)$
14. $10z$.
15. $5a + 2b - 2c$
16. $(2x^2)$
17. $(5a - 2b + 2)$.
18. $(-2a - 7b + 6c)$.
19. $(5x^2 - 5x + 8)$.
20. $(p^2 - 11p + 9)$
21. $2x^2 + 17x + 9$
22. $(-8m^2 + 20m - 11)$

Week 14

23. $(2x^2 - 11x - 3)$
24. $-5a^2 - 11a + 7$.
25. $(4a + 4b - 9c)$
26. $(10x^2 + 3x + 7)$
27. $3x^2 + \left[\dfrac{4}{5}\right]x + \left[\dfrac{7}{2}\right]$
28. $11x^2y + 2xy^2$
29. $(6x^2 - 2y^2)$
30. $(3a - 7b + 7c)$
31. $(x^2 + 11x + 10)$
32. $(4x + 5y - 11z)$
33. $4ab$.
34. $-5y$.
35. $-5x - 4y$
36. $9x - 12y$
37. $[-10b(b + a)]$.
38. $-3a^3 + 3a^2 - 2a + 9$
39. $10b - 2a - 14c$
40. $5m + 2n - 14$
41. $(8a^2 - 15a - 6)$
42. $2a + 4b - 13c$
43. $2p^2 - pq - q^2$
44. $8x^3 - y^3$
45. $11a - b - 9c$

Grade 7

Vol 1 Answer Key

Week 14

46. $7 - 10x + 7x^2 - 10x^3$
47. $2(2x + 1)$.
48. $4x^2 - 3x + 4$
49. $7a^2 + 4ab - 4b^2$
50. $4xy + 2yz + 10xz$
51. $-8a^2 + 9a - 7$
52. $2p^2 - 13pq + pr$
53. 45
54. 39 yards and 34 yards.
55. 12 years.
56. A
57. C
58. B
59. $\dfrac{-11}{20}$
60. $104.
61. 114° and 66°
62. 30, 32 and 34.
63. $\left(\dfrac{4}{9}\right)$
64. 10
65. 6
66. 40
67. 60
68. $640
69. 40 yards and 20 yards.
70. 17 and 23.

Week 14

71. 4
72. 7
73. 42 and 44.
74. 12 and 20.
75. 19 and 21.
76. $\left(\dfrac{2}{5}\right)$
77. 21, 23 and 25.
78. 45 and 47.
79. 12 inches, 20 inches and 16 inches.
80. 40
81. 30 and 70.
82. 17
83. 32 and 34.
84. 3
85. 62
86. 14
87. 24
88. 33
89. 35
90. 70°
91. 16
92. 23 years.
93. 40 pounds.
94. 150 pounds.
95. 200
96. 120
97. 43 and 44.
98. 100

Grade 7

Vol 1 Answer Key

Week 14

99. 30
100. 14
101. 44 yards.
102. $\left(\dfrac{1}{5}\right)$
103. 51° and 39°
104. 108° and 72°
105. 57, 59 and 61.
106. 165 pounds.
107. 461.
108. 12
109. $1584.
110. 14
111. 36 years.
112. 40 years and 10 years.
113. 68 yards and 34 yards.
114. 36
115. 270 yards and 90 yards.
116. 18 years and 54 years
117. 38 and 51
118. 36, 38 and 40
119. 11, 12 and 13
120. 15
121. 35
122. 8
123. 102
124. 33
125. 17 years and 10 years.
126. $35.

Week 14

127. 100 mL
128. Chair $165 and table $125.
129. 35
130. $\left(\dfrac{3}{7}\right)$
131. 40 years and 10 years.
132. 26
133. 13
134. $\left(\dfrac{3}{5}\right)$
135. 48
136. 33
137. 15
138. 46
139. 60
140. 50 and 20.
141. 9
142. 200
143. 31 and 32.
144. 37, 39.
145. 28, 30 and 32
146. 42 boys and 7 girls
147. 49° and 41°
148. 112° and 68°
149. 72 and 112
150. 50 yards and 25 yards.
151. 21 inches, 21 inches and 13 inches.
152. 500
153. 35 years and 25 years.

Grade 7

Vol 1 Answer Key

Week 14

154. 33 years old and 3 years old.
155. 33 years old and 52 years old.
156. 12
157. 33 and 36.
158. 36
159. $680.
160. $450.
161. 30 kgs
162. True
163. $\left(\dfrac{14}{5}\right)$
164. $\dfrac{104}{21}$
165. True
166. False.
167. True
168. True
169. $\dfrac{21}{10}$
170. $\dfrac{19}{3}$
171. $\dfrac{18}{11}$
172. $\dfrac{47}{24}$
173. $\dfrac{549}{154}$

Week 14

174. $\dfrac{1}{2}$
175. $\dfrac{102}{5}$
176. $\left(\dfrac{3}{16}\right)$
177. $\left(\dfrac{23}{24}\right)$
178. $\left(\dfrac{2}{21}\right)$
179. $\left(\dfrac{65}{24}\right)$
180. 7
181. 6
182. $\dfrac{7}{24}$
183. 2
184. 2
185. 1
186. $\dfrac{8}{15}$
187. $\dfrac{7}{20}$
188. $\dfrac{21}{44}$
189. $\dfrac{5}{2}$
190. $\dfrac{19}{7}$

Week 14

191. $\left(\dfrac{7}{12}\right)$
192. True
193. 1
194. True
195. $\left(\dfrac{15}{8}\right)$
196. $\left(\dfrac{39}{14}\right)$
197. $\left(\dfrac{1}{12}\right)$
198. >
199. <
200. >
201. >
202. $\dfrac{11}{8}$
203. $\left(\dfrac{2}{33}\right)$
204. False

Grade 7

Vol 1 Answer Key

Week 15		Week 15		Week 15		Week 15		Week 15	
1.	D	30.	C	59.	D	88.	C	117.	B
2.	B	31.	D	60.	A	89.	A	118.	D
3.	D	32.	A	61.	D	90.	A	119.	A
4.	B	33.	D	62.	A	91.	D	120.	A
5.	C	34.	A	63.	C	92.	C	121.	C
6.	C	35.	B	64.	D	93.	D	122.	B
7.	A	36.	D	65.	C	94.	D	123.	B
8.	D	37.	C	66.	A	95.	C	124.	C
9.	D	38.	C	67.	D	96.	D	125.	A
10.	C	39.	D	68.	C	97.	D	126.	D
11.	A	40.	D	69.	C	98.	D	127.	B
12.	B	41.	D	70.	D	99.	A	128.	B
13.	B	42.	C	71.	D	100.	D	129.	C
14.	C	43.	D	72.	A	101.	A	130.	C
15.	D	44.	D	73.	D	102.	A	131.	A
16.	A	45.	C	74.	C	103.	A	132.	A
17.	C	46.	A	75.	B	104.	A	133.	C
18.	C	47.	C	76.	D	105.	C	134.	D
19.	D	48.	C	77.	A	106.	B	135.	C
20.	A	49.	D	78.	C	107.	B	136.	B
21.	A	50.	C	79.	C	108.	C	137.	C
22.	B	51.	A	80.	C	109.	D	138.	A
23.	B	52.	B	81.	C	110.	C	139.	D
24.	D	53.	B	82.	A	111.	D	140.	B
25.	B	54.	D	83.	C	112.	D	141.	D
26.	C	55.	A	84.	D	113.	C	142.	D
27.	A	56.	B	85.	D	114.	A	143.	D
28.	B	57.	A	86.	A	115.	C	144.	B
29.	A	58.	C	87.	B	116.	B	145.	C

©All rights reserved-Math-Knots LLC., VA-USA

Grade 7

Vol 1 Answer Key

Week 15		Week 15		Week 16		Week 16	
146.	C	175.	D	1.	17.9 ft	28.	184 inches^2
147.	B	176.	B	2.	300 m	29.	22 inches
148.	D	177.	D	3.	180 ft^2	30.	15 inches and 10 inches
149.	D	178.	D	4.	15 inches.	31.	16 inches
150.	C	179.	D	5.	14 inches	32.	16 ft
151.	C	180.	B	6.	30 inches.	33.	156 ft^2
152.	D	181.	D	7.	242 inches^2	34.	15 inches
153.	A	182.	C	8.	120 inches^2	35.	157.5 inches^2
154.	A	183.	D	9.	30 inches	36.	360 inches^2
155.	D	184.	B	10.	$195.50	37.	4 inches
156.	B	185.	B	11.	$540.	38.	22.5 cm
157.	C			12.	48 inches^2	39.	12 cm
158.	A			13.	40 inches	40.	11.25 cm
159.	C			14.	6.928 inches^2	41.	5 cm
160.	A			15.	19 ft	42.	30 cm
161.	A			16.	$391	43.	20 dm
162.	B			17.	10 ft	44.	4,500 cm
163.	A			18.	8.5 inches	45.	216 inches^2
164.	A			19.	8.5 inches	46.	100 inches^2
165.	A			20.	6 inches	47.	22 inches
166.	C			21.	135 inches	48.	25 inches
167.	B			22.	216 inches	49.	12.5 inches
168.	B			23.	24 inches	50.	9 ft and 18 ft
169.	A			24.	5 inches	51.	28 inches and 42 inches
170.	C			25.	8 inches	52.	392 inches^2
171.	A			26.	40 inches	53.	405 inches^2
172.	B			27.	20 inches		
173.	D						
174.	D						

Grade 7

Vol 1 Answer Key

Week 16

54. 15 ft and 25 ft
55. 132 inches^2
56. 18 ft and 24 ft.
57. 307.5 inches^2
58. 18 ft
59. 20 inches and 10 inches
60. 1,040 inches^2
61. 72 inches^2
62. 65 inches^2
63. 217.5 inches^2
64. 196 inches^2
65. 162 inches^2
66. 44 inches^2
67. 136 inches^2
68. 228 inches^2
69. 152 inches^2
70. 24.4 inches^2
71. 192 inches^2
72. 8,800 cm^2
73. 15,000 cm^2
74. 5,760 ft^2

Week 16

75. 11.25 m^2
76. 1,560 inches^2
77. 13 inches
78. 11 inches
79. 17 inches
80. 5 cm
81. 22 inches
82. 11 inches
83. 4 inches
84. 12 inches
85. 24 inches
86. 7 inches
87. 360°
88. 40°, 60°, 100° and 160°
89. 4
90. 2
91. 4
92. 4
93. 4
94. 360°
95. 80° each
96. 120°
97. 36°, 72°, 108° and 144°
98. True
99. 90°
100. 105°
101. 81°
102. 86° and 98°.

©All rights reserved-Math-Knots LLC., VA-USA

Grade 7

Vol 1 Answer Key

Week 16

103. 97°
104. 72°, 72°, 96° and 120°
105. 10 inches, 15 inches, 25 inches and 35 inches.
106. 26 inches and 32 inches
107. 12 inches
108. 120°
109. 85°
110. 16 inches, 32 inches, 40 inches and 56 inches
111. 40°, 80°, 100° and 140°
112. ∠R is 130°
113. 85 ft
114. 110°
115. 90°
116. A
117. B
118.. B
119. A
120. A
121. A
122. Convex
123. True
124. False
125. True
126. True
127. False
128. True
129. True
130. True

Week 16

131. 130°
132. 100°
133. x = 100° & y = 60°
134. x = 70° & y = 90°
135. 60°, 80°, 100° and 120°
136. A
137. B
138. B
139. A
140. A
141. C
142. A
143. 45°, 75°, 105° and 135°
144. No
145. 45°, 90°, 135° and 90°.
146. 110°.
147. 65°
148. 135°
149. ∠C = ∠D = 120°
150. ∠C = ∠D = 105°
151. 75°
152. Two
153. Four
154. Two
155. 105°
156. 84°
157. 100°
158. 115°
159. ∠C = ∠D = 140°

Grade 7

Vol 1 Answer Key

Week 16
- 160. 140°
- 161. 75°
- 162. ∠C = 125° and ∠D = 135°
- 163. 180°
- 164. 160°

Week 17
1. 105 ft².
2. $13,120
3. 4,640 inches²
4. $12,500
5. 259.25 yards²
6. 7 ft
7. 5.5 ft
8. $2,592
9. 160 ft²
10. $640
11. 38 ft
12. 3 ft
13. $3,920
14. 360 ft²
15. 7 ft
16. $918.40
17. $1,170.40
18. $1,188
19. $782
20. 9 ft
21. 10 ft
22. $126.00
23. 924 ft²
24. 12 ft
25. 4,240 ft²
26. 372 ft²
27. 20 yards and 30 yards.
28. 12 ft
29. $385.00

Grade 7

Vol 1 Answer Key

Week 17

30.	$1,056
31.	104 m^2
32.	12 inches
33.	6 ft
34.	136 ft^2
35.	2.6 ft
36.	$1,230
37.	$5800
38.	$78400
39.	361 ft^2
40.	$1037.50
41.	$12,804
42.	$25,308
43.	$1,380
44.	3.4 ft
45.	$826
46.	3,360
47.	$810
48.	8 ft
49.	606.3 ft^2
50.	$13200.00
51.	3,850 yards^2
52.	176 yards^2
53.	$3380.
54.	416 yards^2
55.	2,784 yards^2
56.	576 yards^2
57.	$525
58.	3036 ft^2

Week 17

59.	6,400 yards^2
60.	304 ft^2
61.	$845.
62.	$234.00
63.	$840.00
64.	88 inches^2
65.	524 ft^2
66.	26 ft^2
67.	372 ft^2
68.	52 yards^2
69.	441 inches^2
70.	225 inches^2
71.	$169.20.
72.	161 m^2
73.	$64.40.
74.	96 inches^2
75.	$645.
76.	216 yards^2
77.	$88,000.
78.	24.2 ft.
79.	$262.80.
80.	169 inches^2
81.	72 ft^2
82.	$315.
83.	$3,549.
84.	1071 ft^2
85.	$202.50.
86.	$14.
87.	40 inches^2

Week 17

88.	2761.5 yards^2.
89.	126 cm^2
90.	60 cm^2.
91.	756 m^2.
92.	4 cm^2.
93.	1300 m^2
94.	2,200 cm^2
95.	500 m^2
96.	168 m^2
97.	300 m^2
98.	52 cm^2.
99.	88 cm^2.
100.	22 cm^2.
101.	84 cm^2.
102.	5 m.
103.	3 m.
104.	3.5 m.
105.	5.5 m
106.	24 m
107.	14 m
108.	192 m^2
109.	18 m^2.
110.	$1,188.
111.	$3
112.	360 ft^2
113.	$540
114.	$720

Grade 7

Vol 1 Answer Key

	Week 17		Week 18		Week 18
115.	$290.	1.	B	30.	C
116.	$2340.	2.	D	31.	D
117.	11.56 m^2	3.	B	32.	C
118.	3.72 m^2.	4.	D	33.	B
119.	$11570.	5.	B	34.	B
120.	$1,392.	6.	B	35.	D
121.	40 m and 16 m.	7.	B	36.	B
122.	$1560.	8.	D	37.	D
123.	11700 m^2.	9.	A	38.	C
124.	9,450 yards^2	10.	D	39.	D
125.	82 cm^2.	11.	A	40.	A
126.	136 inch^2.	12.	C	41.	150°
127.	94 cm^2.	13.	A	42.	115°
		14.	A	43.	70° each
		15.	C	44.	95°
		16.	B	45.	Yes
		17.	C	46.	144°, 48°, 72° and 96°
		18.	C	47.	B
		19.	C	48.	C
		20.	C	49.	65°
		21.	A	50.	110° each
		22.	B	51.	∟C = ∟D = 125°
		23.	C	52.	$355.20
		24.	B	53.	150°
		25.	D	54.	D
		26.	C	55.	60 inch^2
		27.	B	56.	8 ft
		28.	D	57.	784 ft^2
		29.	B	58.	22 ft

©All rights reserved-Math-Knots LLC., VA-USA   www.math-knots.com | www.a4ace.com

Grade 7

Week 18

59. 100 inches^2

60. ∠C = 125° and ∠D = 135°

61. ∠C = 135° and ∠D = 130°

62. ∠C = 110° and ∠D = 120°

63. 135 inches^2

64. 28,900 yards^2

65. $96.95

66. $112.00

67. $1,410

68. 80 cm^2

69. $40.25

70. $110

71. 184 cm^2

72. 104 cm^2

73. 925 cm^2

Made in the USA
Monee, IL
26 February 2025